SHORT COURSE
in
SURVEYING

BOOKS BY THE SAME AUTHORS

Davis and Foote

SURVEYING: THEORY AND PRACTICE

A one-volume text giving a full presentation of theory and practice. Third edition.

Davis, Foote, and Rayner

ELEMENTS OF SURVEYING

An abridgment for short courses.

Davis

MANUAL OF SURVEYING FOR FIELD AND OFFICE

A manual presenting the problems of field and office encountered in surveying practice.

ELEMENTARY PLANE SURVEYING: TEXT AND MANUAL

A combined text and field manual for non-civil engineering students. Second edition.

Davis and Kelly

SHORT COURSE IN SURVEYING

Essential principles and practice for rapid training during the national emergency.

SHORT COURSE
in
SURVEYING

BY

RAYMOND E. DAVIS, M.S., C.E., D.Eng.
Professor of Civil Engineering, University of California

AND

JOE WALLACE KELLY, B.S.
Assistant Professor of Civil Engineering,
University of California

McGRAW-HILL BOOK COMPANY, Inc.
NEW YORK AND LONDON
1942

SHORT COURSE IN SURVEYING

PRINTED IN THE UNITED STATES OF AMERICA

PREFACE

Extremely condensed, this book presents the essential principles and practice of elementary plane surveying in useful form for rapid training of rodmen, chainmen, instrumentmen, and draftsmen in short courses necessitated by the national emergency. It is a condensation of more advanced textbooks long and widely used in engineering colleges; those subjects have been selected which will lead to a reasonable proficiency in the various phases of surveying in the shortest possible time. For courses designed to train men for a specific job, selective assignments can be made and certain sections of the book omitted.

To supplement the classwork, a variety of field problems are stated. Even if not all these problems are assigned for field work, it is recommended that they be read in connection with the text, for they describe many useful details of procedure.

Material for some of the illustrations and tables was originally furnished by the U. S. Coast and Geodetic Survey, the U. S. Geological Survey, the U. S. Naval Observatory, and the General Land Office. Photographs of various surveying instruments have been furnished by the following manufacturers: C. L. Berger and Sons, Boston; W. and L. E. Gurley, Troy, N.Y.; Keuffel and Esser Company, New York; and The A. Lietz Company, San Francisco.

R. E. D
J. W. K.

BERKELEY, CALIF.,
September, 1942.

CONTENTS

TABLES

FIELD AND OFFICE PROBLEMS

SHORT COURSE IN SURVEYING

CHAPTER I

FIELD AND OFFICE WORK; ERRORS

1. Surveying.—Surveying is the art of determining horizontal distances, differences in elevation, directions, angles, positions, areas, and volumes at or near the surface of the earth. It comprises the *field work* of measuring and the *office work* of computing and drawing. Surveys are made for the purpose (1) of establishing the boundaries of land, (2) of providing information necessary to the construction of engineering works, or (3) of portraying land and water forms for purposes of navigation, mining, construction, and general use.

In surveying that extends over a relatively small part of the earth's surface, generally the spheroidal shape of the earth is neglected, and the surface of the earth is considered as a plane. This is called *plane surveying*. This book deals with the methods of plane surveying.

2. Kinds of Plane Surveying.—*Land surveying* consists in subdividing land into parcels, reestablishing old or obliterated land lines, calculating areas, constructing land maps, and writing descriptions for deeds. *Topographic surveying* consists in securing field data for plotting a map that shows the configuration of the earth's surface and the location of natural and artificial objects. *Route surveying* consists in determining the ground configuration and location of objects along a proposed line such as that of a highway, railway, canal, or pipe line; establishing the line on the ground; and calculating volumes of earthwork.

Photogrammetric surveying is the application to surveying of the science of measurement by means of photographs taken with specially designed cameras either from airplanes or from ground stations. The principles of perspective are employed in the projection of details from the photographs onto maps drawn to scale. In conjunction with limited ground surveys to locate control points visible from the air, *aerial photogrammetry* is employed on many topographic surveys because of its outstanding advantages of wealth of detail, speed, accuracy, and low cost. Aerial photogrammetry is rapidly being developed and simplified, and it may soon replace the more laborious ground methods on most surveys.

With regard to complete surveys, this book deals only with land surveying, topographic surveying, and route surveying.

FIELD WORK

3. Typical Surveys.—In a typical case where points are to be established or existing objects are to be located, the horizontal distances are measured with a tape by two chainmen and the horizontal angles are measured with an engineer's transit by a transitman, who may also act as chief of the survey party. Sometimes a surveyor's compass is used instead of the transit. If elevations are to be determined or established, the survey party includes a levelman and a rodman, who measure the vertical distances with an engineer's level and a graduated leveling rod.

In another method of surveying, a transitman and one or more rodmen are employed. Horizontal angles are measured with the transit, and at the same time the horizontal distances and differences in elevation are measured by the stadia method (Chap. VIII).

In a third method, which is widely employed for map making, a plane table (Chap. XIII) is used instead of the transit. The map is drawn in the field by the plane-table man. Directions are laid off graphically; and distances and elevations are determined by stadia.

4. Field Work.—Field work consists in (1) adjusting instruments and caring for field equipment, (2) measuring horizontal and vertical distances and angles, and (3) recording the field measurements, usually in the field notebook. The field problems given herein might in practice become parts of extended surveys.

In the two succeeding articles are given some general suggestions regarding the care, handling, and adjustment of surveying instruments. Other details of field practice are given in connection with the individual field problems.

5. Care and Handling of Instruments.—The following suggestions apply to such instruments as the transit, level, surveyor's compass, and plane table.

1. Handle the instrument with care, especially when removing it from the box.
2. Before climbing over a fence or similar obstacle, place the instrument on the other side, with the tripod legs well spread.
3. Never leave the instrument while it is set up in the street, on the sidewalk, near construction work, in fields where there are livestock, or in any other place where there is a possibility of accident.
4. Just before setting up the instrument, adjust the wing nuts controlling the friction between tripod legs and head, so that each leg when placed horizontally will barely fall from its own weight.

5. Do not set the tripod legs too close together, and see that they are firmly planted. Push *along* the leg, not vertically downward. So far as possible, select solid ground for instrument stations. On soft or yielding ground, do not step near the feet of the tripod.

6. While an observation is being made, do not touch the instrument except as necessary to make a setting; and do not move about.

7. In tightening the various clamp-screws, adjusting screws, and leveling screws, bring them only to a firm bearing.

8. Before observations are begun, focus the eyepiece on the cross-hairs, and (by moving the eye slightly from side to side) see that no parallax is present (Art. 41).

9. *When the magnetic needle is not in use, see that it is raised off the pivot.* While the needle is resting on the pivot, impact is likely to blunt the point of the pivot or to chip the jewel, thus causing the needle to be sluggish.

10. Remove grit from exposed movable parts such as the threads of tangent-screws by wiping them with an oiled rag. If the threads or slides work hard, clean them in gasoline and oil lightly. Use no abrasives.

11. Use only the best quality of clock or watch oil. Oil sparingly, and wipe off the excess oil.

12. Never rub the objective or the eyepiece of a telescope with the fingers or with a rough cloth. Use a camel's-hair brush to remove dust, or use clean chamois or lint-free soft cloth if the dust is caked or damp.

Chaining Equipment.—Keep the tape straight when in use; any tape will break when kinked and subjected to a strong pull. Steel tapes rust readily and for this reason should be wiped dry after being used. The method of doing up the tape is given in Art. 23.

Do not use the flagpole as a bar to loosen stakes or stones; such use bends the steel point and soon renders the point unfit for lining purposes.

To avoid losing pins, tie a piece of colored cloth (preferably bright red) through the ring of each.

Leveling Rod.—Do not allow the metal shoe on the foot of the rod to strike against hard objects, for this, if continued, will round off the foot of the rod and thus introduce a possible error in leveling. Keep the foot of the rod free from dirt. When not in use, long rods should be either placed upright or supported for their entire length; otherwise they are likely to warp. When not in use, jointed rods should have all clamps loosened to allow for possible expansion of the wood.

6. Adjustment of Instruments.—By "adjustment" of a surveying instrument is meant the bringing of the various fixed parts into proper relation with one another, as distinguished from the ordinary operations of leveling the instrument, alining the telescope, etc.

The frequency with which adjustments are required depends upon the instrument, its care, and the precision with which measurements are to be taken. On ordinary surveys, it is good practice to test the critical adjustments once each day, especially on long surveys where frequent checks on the accuracy of the field data are impossible.

The adjustments are made by tightening or loosening certain screws. Usually these screws have capstan heads which may be turned by a pin called an adjusting pin. Following are general suggestions:

1. Before adjusting the instrument, see that no parts (including the objective) are loose. When an adjustment is completed, always check it before using the instrument.

2. When several interrelated adjustments are necessary, time will be saved by first making an approximate or rough series of adjustments and by then repeating the series to make finer adjustments. In this way, the several disarranged parts are gradually brought to their correct position. This practice does not refer to those adjustments which are in no way influenced by others.

7. Precision of Measurements.—The degree of precision of a given measurement varies with the conditions surrounding the survey. Although it is desirable that all measurements be made with high precision, the fact must be considered that increasing the precision usually increases the time and labor in more than direct proportion. It therefore becomes the duty of the surveyor to maintain a degree of precision as high as justified by the purpose of the particular survey but not higher. Before undertaking a series of measurements he should consider the sources of error, the instruments and methods employed to keep these errors within allowable limits, the methods of checking, and the most efficient organization of the work.

It is common practice before beginning a survey to fix the permissible error of linear measurement, this being expressed as a ratio, for example, 1/1,000; sometimes it is desired to locate a given point within a specified distance of its true position. If measurements are to be consistent, the precision of angles should correspond to the precision of related distances—in other words, the error in location of a point on account of error in angle should not be greatly different from its error in location on account of error in distance. With the ratio of precision of linear measurements and the average length of sighting to determine angles known, the permissible limit of error in angles can be determined by simple trigonometry. As an illustration, the tangent (or sine) of 01′ is 0.000291; hence the linear error at a distance of 1,000 ft. for an angular error of 01′ is 0.291 ft., and the corresponding ratio of precision in linear measurement is 1/3,440. For other distances from the vertex of the angle the linear error is in direct proportion.

In some surveys in which distances are determined only roughly, it is possible to measure the angles with greater than the required precision without increased effort or loss of time. For example, in rough chaining the ratio of precision might be 1/1,000, corresponding to an angular error of 03′; but with the ordinary transit the angles could be determined to the nearest 01′ as quickly as to the nearest 03′.

8. Field Notes.—All field notes should be recorded *in the field book* at the time the work is being done. Nothing should be left to memory or copied from temporary notes. The field notes consist of numerical data, explanatory notes, and sketches. The record of every survey should include the date, the weather conditions, the names and duties of the members of the party, and a title indicating the location of the survey and its nature or purpose.

Convenient forms of notes in common use are shown in this book, but in many cases it will be necessary for the surveyor to devise his own form of record.

A 4H pencil, well pointed, should be used. The figures used should be plain; one figure should never be written over another. In general, numerical data should not be erased; if a number is in error, a line should be drawn through it, and the corrected value written above. Portions of sketches and explanatory notes may be erased if there is a good reason for erasing them. The number should always show with what degree of precision the measurement was taken; thus a rod reading taken to the nearest 0.01 ft. should be recorded not as 7.4 ft. but as 7.40 ft.

Sketches are made roughly to scale. They are made freehand and of liberal size. Many features may be most readily shown by conventional signs (Figs. 124*a–d*).

Explanatory notes are employed to make clear what the numerical data and sketches fail to do. Usually they are placed on the right-hand page in the same line with the numerical data that they explain.

If a page of notes is abandoned, either because it is illegible or because it contains erroneous or useless data, it should be retained, and the word "void" written in large letters diagonally across the page. The page number of the continuation of the notes should be indicated.

In the field, the recorder should always record at once the uncorrected readings of the instrument and apply any necessary computed corrections afterward.

OFFICE WORK

9. General.—Office work consists in (1) correcting the field measurements to meet known conditions or reducing them to a form suitable for computation; (2) plotting maps, profiles, and cross-sections; and (3) calculating areas of land and volumes of earthwork.

10. Computations.—All computations should be preserved in a notebook for that purpose, preferably about 8½ by 11 in. in size, having cross-ruled pages which facilitate the making of tabulations and sketches. Usually enough of the field notes should be transcribed to make computations possible without further reference to the field notebook.

In practice, *no confidence is placed in results that have not been checked,* and important results are preferably checked by more than one method. Approximate checks to discover large arithmetical mistakes can be obtained in many cases by the use of such mechanical devices as the slide rule, the planimeter, and the protractor. Often graphical methods can be used as an approximate check; they generally take less time than arithmetical or logarithmic solutions, and possible incorrect assumptions in the precise solution may be detected.

11. Significant Figures.—The term *significant figures* is used to refer to those digits in a number which have meaning, that is, those digits the values of which are known. Measured quantities are not exact, and the number of digits that have meaning is limited strictly by the precision with which the measurement has been made. If we measure roughly a given distance with a steel tape, we may find it to be 732 ft.; but if we measure it more carefully, we may find it to be 732.4 ft.; or by still greater refinements we may determine it to be 732.38 ft. But we have not yet reached an exact number, nor can we, for there will always be an error of indeterminable amount. The number of significant figures in these three results is 3, 4, and 5, respectively.

In a decimal the number of significant figures is not necessarily the number of decimal places, as the following examples illustrate:

0.0000065 contains two significant figures.
0.00000650 contains three significant figures.
10.00000650 contains 10 significant figures.

In lengthy computations it is advisable to carry out the intermediate results to one significant figure more than that desired in the final result. In ordinary work, however, it is useless and time-consuming to employ more significant figures in computation than are to be retained in the final result.

The precision of a result is independent of the unit in which it is expressed. Thus, an area of 190,600 sq. ft. is properly expressed in acres as 4.376 acres. There are four significant figures in each value.

12. Logarithms.—For multiplying, dividing, squaring, cubing, or taking the roots of numbers, arithmetic cannot be used advantageously beyond three or perhaps four significant figures. For four or more significant figures there is a decided advantage in the use of *logarithms,* both to save time and to lessen the likelihood of mistakes. Tables VII and VIII give logarithms of numbers and of the functions of angles.

13. Graphical and Mechanical Methods.—Frequently, combined graphical and mechanical methods may be employed in conjunction with algebraic processes. For example, earthwork cross-sections may be plotted to scale (graphical), the area of the cross-sections may be measured with the planimeter (mechanical), and the volume of earthwork may be determined by arithmetic.

The ordinary 10-in. *slide rule* greatly facilitates computations involving no more than three significant figures and is in every way the equivalent of a three-place table of logarithmic functions.

For results of more than three significant figures the *computing machine* is coming largely to replace other methods of computing. The details of operation vary with the type of machine, but they are simple and can be mastered in a few minutes of instruction.

The *polar planimeter* is of great value in finding the area of figures plotted to scale and is the most efficient means of determining the area of figures with irregular or curved boundaries. In general, results may be determined to three significant figures.

14. Drafting.—It is assumed that the student is familiar with the use of the ordinary drafting instruments and with the elements of mechanical drawing. The drawings of surveying consist of maps, profiles, cross-sections, and, to some extent, graphical computations. The usefulness of all these drawings is dependent largely upon the accuracy with which points and lines are projected upon paper; few dimensions are shown. Much of the drafting calls for a degree of skill and precision unnecessary on dimensional plans, and the student should exercise great care in plotting.

The details of drafting that pertain to the drawings of surveying are described in Chap. IX.

ERRORS

15. Sources of Error.—In the measurements of surveying, *instrumental errors* arise from imperfections or faulty adjustment of the devices with which measurements are taken; *personal errors* occur through the observer's inability to read the instruments exactly; and *natural errors* occur from variations in the phenomena of nature such as temperature, humidity, wind, gravity, refraction, and magnetic declination.

Mistakes are unintentional faults of conduct arising from poor judgment or from confusion in the mind of the observer. They are detected and eliminated by checking.

16. Kinds of Error.—The *resultant error* is the difference between a measurement and the true value. It is made up of individual errors from a variety of sources. If the measurement is too large the error is said to be positive; if too small, the error is said to be negative.

A *discrepancy* is the difference between two measurements of the same quantity.

A *systematic error* is one that, so long as conditions remain unchanged, always has the same magnitude and the same algebraic sign (which may be either positive or negative). If conditions do not change during a series of measurements, the error is termed a *constant* systematic error; for example, a line may be measured with a tape that is too short. If conditions change, resulting in corresponding changes in the magnitude of the error, it is termed a *variable* systematic error; for example, a line may be chained during a period in which the temperature varies. A systematic error always follows some definite mathematical or physical law, and a correction can be determined and applied. The error may be instrumental, personal, or natural.

An *accidental error* is an error due to a combination of causes beyond the ability of the observer to control and for which it is impossible to make correction; for each observation the magnitude and algebraic sign of the accidental error are matters of chance and hence cannot be computed as can the magnitude and algebraic sign of a systematic error. However, accidental errors taken collectively obey the law of probability (see Art. 17). Since each accidental error is as likely to be positive as negative, a certain compensative effect exists; and accidental errors are sometimes called "compensating" errors. As an example of the occurrence of an accidental error, in chaining it is impossible to set the chaining pin exactly at the proper graduation on the tape. Accidental errors remain after mistakes have been eliminated by checking and systematic errors have been eliminated by correction.

The total systematic error in any given number of measurements is the algebraic sum of the individual errors of the individual measurements. According to the mathematical theory of probability, accidental errors tend to increase in proportion to the *square root* of the number of opportunities for error.

If the discrepancy between a check measurement and the original measurement is small, it is an indication that no mistakes have been made and that the accidental errors are small, but it is not an indication that the systematic errors are small.

17. Theory of Probability.—For surveys of higher precision special effort is made to eliminate systematic errors, and the accuracy of a measured quantity is governed by the accidental error that it contains. To form a judgment of the probable value or the probable precision of a quantity, *from which systematic errors have been eliminated,* it is necessary to rely upon the theory of probability, which deals with accidental errors of a series of like or related measurements. It is assumed that the true value of a quantity is the mean of an infinite number of like observations. In practice it is impossible either to eliminate systematic errors entirely or to take an infinite number of observations; hence the value of a quantity is never known exactly. However, in the discussions to follow, it is assumed that systematic errors are so far eliminated as to be a negligible factor.

18. Probable Value.—The *most probable value* of a quantity is a mathematical term used to designate that adjusted value which, according to the principles of least squares, has more chances of being correct than has any other.

Same Quantity.—For a series of measurements of the same quantity made under identical conditions, the most probable value is the mean of the measurements.

Related Quantities.—For related measurements (taken under identical conditions) the sum of which should equal a mathematically exact quantity, the most probable values are the observed values corrected by an equal part of the total error. (This situation can arise only in the case of angles about a point or angles in a closed figure.) The correction is in proportion to the *number* of related measurements and not to the *magnitude* of the individual measurements.

Example 1: The angles about a point have the following observed values: 130°15′20″, 142°37′30″, and 87°07′40″. The sum of the measurements is 360°00′30″; therefore the total error is 30″. Since there are three angles, the error is assumed to be 10″ for each measurement. The most probable values are

$$
\begin{array}{rcrcr}
130°15'20'' & - & 10'' & = & 130°15'10'' \\
142°37'30'' & - & 10'' & = & 142°37'20'' \\
87°07'40'' & - & 10'' & = & 87°07'30'' \\
\hline
360°00'30'' & - & 30'' & = & 360°00'00''
\end{array}
$$

For related angular measurements (taken under identical conditions) the sum of which should equal a single measurement taken under the same conditions, the most probable values are obtained by dividing the discrepancy equally among all the measurements, including the sum. If the correction is added to each of the related measurements, it is subtracted from the measurement representing their sum; and vice versa.

Example 2: Measurements of three angles about a point O are AOB = 12°31′50″, BOC = 37°29′20″, and COD = 47°36′30″. The measurement of the single angle AOD is 97°37′00″. The discrepancy between the sum of the three measured angles and the measurement of the angle representing their sum is 40″. Since the size of the errors is independent of the size of the angle, the discrepancy is divided into equal parts: $40/4 = 10''$. This correction is to be subtracted from measurements of each of the angles *AOB*, *BOC*, and *COD* and is to be added to the measurement of the angle *AOD*. The most probable values are

$$\begin{array}{lllll} AOB & = 12°31'50'' & - 10'' & = 12°31'40'' & \\ BOC & = 37°29'20'' & - 10'' & = 37°29'10'' & \\ COD & = 47°36'30'' & - 10'' & = 47°36'20'' & \\ \hline \text{Sum} & \;\; 97°37'40'' & & \;\; 97°37'10'' & \multirow{2}{*}{\Big\} \textit{Check}} \\ AOD & = 97°37'00'' & + 10'' & = 97°37'10'' & \end{array}$$

19. Probable Error.—*Probable error* is a plus or minus quantity within which limits the actual accidental error is as likely as not to fall. Thus, if 6.23 represents the mean of several measurements and 0.11 represents the probable error of the mean value, the chances are even that the true value lies between the limits $6.23 - 0.11 = 6.12$ and $6.23 + 0.11 = 6.34$. The quantity would be written 6.23 ± 0.11.

In the adjustment of observations, the probable error of the most probable value of each quantity can be estimated from a series of measurements; and the probable errors can then be used in computing weights and/or the corrections to be applied to related quantities (Ref. 7, p. 205). Consideration of probable error is also useful in choosing methods of surveying to produce desired degrees of precision.

Same Quantity.—It has been stated that the mean of a series of observations of a single quantity is the most probable value. For the purpose of determining the probable error this mean value is mathematically regarded as being the best value, and the difference between each of the individual measurements and the mean value is determined. These differences are termed *residuals* or *deviations*. For the ordinary measurements of surveying, the relations stated below may be applied to series of perhaps 10 or more measurements.

The probable error of a *single observation* is calculated by the equation

$$E = 0.6745 \sqrt{\frac{\Sigma v^2}{n - 1}} \tag{1}$$

in which Σv^2 is the sum of the squares of the residuals, and n is the number of observations.

The probable error of a single observation may be calculated approximately by the expression

$$E = 0.845\, \bar{v} \text{ (approx.)} \tag{2}$$

in which $\bar{v}$ is the mean value of the residuals without regard to signs.

The probable error of the *mean* of a number of observations is inversely proportional to the square root of the number of observations, or

$$E_m = \frac{E}{\sqrt{n}} \tag{3}$$

Related Quantities.—The probable error of the sum of observations each having the same probable error is equal to the probable error of a single observation multiplied by the square root of the number of observations (opportunities for error), or

$$E_s - E\sqrt{n} \tag{4}$$

20. Numerical Problems.

1. The following values were observed in a series of rod readings under identical conditions. What is the most probable value? What is the probable error of a single observation? Of the mean?

Rod Reading, Ft.

3.187	3.181	3.186	3.181
3.182	3.184	3.183	3.188
3.179	3.176	3.178	3 179

2. Adjust the following angles measured at station O:

Angle	Observed value
AOB	46°14′45″
BOC	71°32′29″
COD	85°54′38″
AOD	206°41′28″

CHAPTER II

MEASUREMENT OF DISTANCE

21. Methods.—In surveying, the distance between two points is understood to mean the *horizontal* distance. Though frequently slope distances are measured, they are reduced to their equivalent on the horizontal projection before being used.

The most common method of determining distance is by direct measurement with the tape (Arts. 22 to 29); the operation is called *taping* or *chaining.*

The *stadia* method, described in detail in Chap. VIII, offers a rapid indirect means of determining distance and is particularly useful in topographic surveying. The telescope of the transit, level, or plane-table alidade is equipped with two horizontal hairs, one above and the other an equal distance below the horizontal cross-hair; the distance from the instrument to a given point is indicated by the intercept between these stadia hairs as shown on a graduated rod held vertically at the point.

The method of *pacing* furnishes a rapid means of approximately checking more accurate measurements of distance. It is used on reconnaissance surveys, for locating details in small-scale mapping, and in traversing with the plane table. Either the natural pace (about 2¾ ft.), a 2½-ft. pace, or a 3-ft. pace may be employed according to the preference of the surveyor, who standardizes his pace before beginning a survey. 2 paces = 1 stride.

Distance may be measured by observing the *number of revolutions of a wheel* of a vehicle. The mileage recorder attached to the ordinary automobile speedometer registers distance to 0.1 mile. The odometer is a simple device which is attached to any vehicle and which registers directly the number of revolutions of the wheel. Although the distance indicated by a wheel is somewhat greater than the true horizontal distance, this method of measurement is sufficiently precise for many purposes, especially for small-scale maps. In hilly country, rough corrections for average slope may be applied.

By *graphical* or *algebraic* methods, unknown distances may be determined through their relation to known distances. These methods are used in triangulation, as described in Art. 76.

22. Equipment for Chaining.—The *steel tape* usually employed is 100 ft. long and is graduated in feet, with the first and last foot graduated to tenths or hundredths of feet. Other lengths and various types of graduation are available. Some tapes have an extra graduated foot at one or both ends; some have ends graduated so that corrections for slope can be applied directly; and some have ends graduated so that temperature corrections can be applied directly. The ordinary widths of tape are ¼ and 5⁄16 in. A steel tape often used in land surveying is 66 ft. long and is graduated into 100 "links" each 7.92 in. long, corresponding to the old Gunter's chain. One common type of metric tape is graduated to half meters, with the first and last meter graduated to decimeters.

The *metallic tape* is a ribbon of waterproofed fabric into which are woven small brass or bronze wires to prevent its stretching. It is usually 50 or 100 ft. long and is graduated to feet, tenths, and half-tenths. It is used principally in earthwork cross-sectioning and in similar work where a light, flexible tape is desirable and where small errors in length are not of consequence.

For very precise measurements the *invar tape* has come into general use. Invar is a composition of steel and nickel which is affected but little by temperature changes; the thermal expansion of commercial invar tapes is about one-tenth that of steel tapes. It is a soft metal, and the tape must be handled very carefully to avoid bends and kinks.

Steel *chaining pins*, also called surveyor's arrows, are employed to mark the ends of the tape during the process of chaining between two points more than a tape length apart. A set consists of 11 pins. (For more precise chaining and for future reference, small nails may be set into the earth.) Steel or wood *range poles*, also called flags, flagpoles, or lining rods, are used as signals to indicate the location of points or the direction of lines.

23. Handling the Steel Tape.—The steel tape will break if kinked and subjected to a strong pull. Some tapes are wound on a reel, but usually the tape is done up in 5-ft. lengths into a figure 8 and then "thrown" into the form of a circle with diameter about 10 in., as follows: Stand beside the zero end of the tape, take the end of the tape in the left hand, and—allowing the tape to slide loosely through the right hand—extend the arms. As the 5-ft. mark is reached, grasp it with the right hand. Bring the hands together, and lay the 5-ft. mark of the tape in the fingers of the left hand without permitting the tape to turn over. Then grasp this loop with the left hand, and again extend the arms for another 5-ft. length; and so on. When the last mark is reached, tie the loop tightly where the ends of the tape come together, by means of the rawhide thongs. Grasp the loop with the right hand, and at the opposite point with the left hand. Twist the loop in such a manner that it will be thrown into circular form, with diameter half that of the loop. To undo the tape, reverse the operation of throwing; untie the thongs;

remove the first loop in such a way as not to twist the tape; and walk in the direction of measurement, removing one loop at a time and watching for kinks.

24. Chaining on Smooth Level Ground.—The following represents the usual method of chaining between two fixed points over smooth level ground when the measurements are of ordinary precision, say 1/5,000. Errors and mistakes in chaining are discussed in Arts. 28 and 29.

The rear chainman *with one pin* stations himself at the point of beginning. The head chainman, with the zero (graduated) end of the tape and 10 pins, advances toward the distant point. When the head chainman has gone nearly 100 ft., the rear chainman calls "Chain," a signal for the head chainman to halt. The rear chainman holds the 100-ft. mark at the point of beginning and, by hand signals or by speaking, lines in a chaining pin (held by the head chainman) with the range pole or other signal marking the distant point. The head chainman sets the pin vertically on line and a short distance to the rear of the zero mark. With his left hand he then pulls the tape taut and, making sure that it is straight, brings it in contact with the pin. The rear chainman, when he observes that the 100-ft. mark is at the point of beginning, calls "Stick" or "All right." The head chainman pulls the pin and sticks it at the zero mark of the tape, with the pin sloping away from the line. As a check, he again pulls the tape taut and notes that the zero point coincides with the pin at its intersection with the ground. He then calls "Stuck" or "All right"; the rear chainman releases the tape; the head chainman moves forward as before; and so the process is repeated.

As the rear chainman leaves each intermediate point he pulls the pin. Thus there is always one pin in the ground, and the number of pins held by the rear chainman at any time indicates the number of hundreds of feet, or *stations,* from the point of beginning to the pin in the ground. At the end of each 10 full stations, the head chainman signals for pins, the rear chainman comes forward and gives the 10 pins to the head chainman, both check the tally, and the head chainman records it. The procedure is then repeated.

When the end of the course is reached, the head chainman halts, and the rear chainman comes forward to the last pin set. The head chainman holds the zero mark at the terminal point. The rear chainman pulls the tape taut and observes the number of *whole feet* between the last pin and the end of the line. For a tape having the end foot graduated, he then holds the next *larger* foot mark at the pin; the head chainman pulls the tape taut and reads the decimal by means of the finer graduations of the end foot. The decimal is counted from the 1-ft. mark; thus the distance between the last pin and the end of the line is 1 ft. *less* than indicated at the pin, plus the decimal read by the head chainman.

For tapes having an extra graduated foot beyond the zero point, it is not necessary to subtract 1 ft. as just described. The rear chainman holds the next *smaller* foot mark at the pin, and the head chainman reads the decimal.

Where the transit is set up on line, the transitman usually directs the head chainman in placing the pins on line.

In some surveys it is required that stakes be set on line at short intervals, usually 100 ft. Such stakes may be driven by the rear chainman after he has pulled the pin. On surveys of low precision no pins are used, and the head chainman sets the stakes in the manner previously described for pins, measuring the distance between centers of stakes at their junction with the ground. On more accurate surveys, measurements are carried forward by setting a tack in the head of each stake.

25. Horizontal Measurements over Uneven or Sloping Ground.—The measurements are carried forward much as just described for level ground. The tape is held horizontal, and a plumb line is used by either, or at times by both, chainmen for projecting from tape to pin or from pin to tape.

Where the slope is less than 5 or 6 ft. in 100, the head chainman advances a full tape length at a time. If the course is downhill, he holds the plumb line at the zero point of the tape, with the tape horizontal and the plumb bob a few inches from the ground, then pulls the tape taut and is directed to line by the rear chainman. When the plumb bob comes to rest, he lowers it carefully to the ground; he then sets a pin in its place. As a check, the measurement is repeated. If the course is uphill, the rear chainman holds a plumb line suspended from the 100-ft. mark and signals the head chainman to give or take until the rear chainman's plumb bob comes to rest over the pin; the head chainman then sets a pin.

Where the course is steeper and is downhill, the head chainman advances a full tape length and then returns to an intermediate point from which he can hold the tape horizontal. He suspends the plumb line at a foot mark, is lined in by the rear chainman, and sets a pin at the indicated point. The rear chainman comes forward, *gives the head chainman a pin*, and at the pin in the ground holds the tape at the foot mark from which the plumb line was previously suspended. The head chainman proceeds to another point from which he can hold the tape horizontal, and so the process is repeated until the zero mark on the tape is reached. At each *intermediate* point of a tape length the rear chainman gives the head chainman a pin, but not at the point marking the full tape length. In this manner the tape is always advanced a full length at a time, and the rear chainman's count of pins is not confused. The operation described in this paragraph is termed "breaking chain."

Usually the tape is estimated to be horizontal by eye. This commonly results in the downhill end's being too low, sometimes causing a large error in horizontal measurement. The safe procedure in rough country is to use a hand level (Art. 44).

In horizontal measurements over uneven or sloping ground, the tape sags between supports and becomes effectively shorter. The effect of sag can be eliminated by standardizing the tape, by applying a computed correction, or by increasing the pull sufficiently to offset the effect of sag. In breaking chain, or when the tape is supported for part of its length, the difference in effect of sag as between a full tape length and the unsupported length can be taken into account roughly by varying the pull.

26. Measurements on Slope.—Where the ground is fairly smooth, measurements on the slope may sometimes be made more accurately and quickly than horizontal measurements; and slope measurements are generally preferred. Some means of determining either the slope or the difference in elevation between successive 100-ft. points or breaks in slope is required. For surveys of ordinary precision either the clinometer for measuring slope (Art. 45) or the hand level for measuring difference in elevation may be used to advantage. If only the distance between the ends of the line is desired, the procedure of chaining is the same as on level ground, but a record is kept either of the slope or of the difference in elevation of each 100-ft. length (or less at breaks in slope). The horizontal distances are then computed from the distances measured on the slope.

Where stakes are to be placed every 100 ft., corrections to the slope distances may be applied as the chaining progresses, either by mental computation or by use of the slide rule. Corrections are much more readily calculated than are the horizontal distances themselves.

27. Corrections for Slope.—Where the slope is measured with sufficient precision so to warrant, the corresponding horizontal distances may be calculated by exact trigonometric relations. For measurements of ordinary precision where the slope is not greater than about 20 in 100, it is simpler and sufficiently precise to subtract an approximate correction from the slope distance.

In Fig. 27 let s represent the slope distance between two points, h the difference in elevation, and d the horizontal distance. The correction is $C_h = s - d$. It can be shown that

$$C_h = \frac{h^2}{2s} \text{ (approx.)} \tag{1}$$

For the usual case, where s = 100 ft., this formula can be solved mentally, and it is worth remembering.

Where the *angle* of slope is determined,

$$C_h = 0.015S\ (\theta^\circ)^2 \text{ (approx.)} \tag{2}$$

in which θ° is the slope angle (Fig. 27), in degrees; and S is the slope distance, *in 100-ft. stations.*

For the case in which it is desired to set stakes at a given horizontal distance (as 100 ft.) apart, the corresponding slope distance is determined approximately by *adding* the same correction C_h to the required horizontal distance.

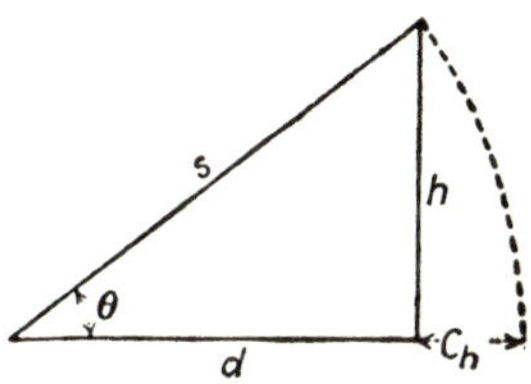

FIG. 27.

28. Errors in Chaining; Corrections. Errors in chaining may be caused by (1) incorrect length of tape, (2) imperfect alinement of tape, (3) imperfections of observing, (4) variations in temperature, (5) variations in tension, or (6) sag in tape. The precision of chaining for transit-tape surveys is discussed in Art. 105.

In applying corrections to the length of a line measured with a tape that is too long, the correction is *added*. In laying out a required distance with a tape that is too long, the correction is *subtracted* from the required distance to determine the distance to be laid out. For a tape that is too short, the corrections are, of course, opposite in direction to those just discussed.

1. *Length of Tape.*—The 100-ft. tape as received from the manufacturer is usually close to the standard length when subjected to the standard pull of 10 lb. at 68°F., the tape being horizontal and supported throughout its length. However, from time to time it is well to refer the tape to a standard of length. Many cities have such standards.

If the tape is not of standard length under the specified conditions a systematic error is produced in measurements To eliminate the error, a correction (per tape length) determined by standardizing the tape is applied to the observed lengths.

2. *Imperfect Alinement of Tape.*—The head chainman is likely to set the pin to one or the other side of the line, producing a positive variable systematic error. The error can be reduced to a negligible

quantity by care in lining. Generally it is the least important of the errors in chaining, and extreme care in lining is not justified.

In chaining through grass and brush or when the wind is blowing it is impossible to have all parts of the tape in perfect alinement with its ends, and a positive variable systematic error is produced. If the head chainman is careful to stretch the tape taut and to observe that it is straight, the error is not of consequence.

If the tape is not horizontal for supposedly horizontal measurement, a positive variable systematic error is produced similar to that due to imperfect alinement. Often slopes are deceptive, even to experienced men; the tendency is to hold the downhill end of the tape too low. In ordinary chaining this is one of the largest contributing errors. It can be reduced to a negligible amount by leveling the tape by means of either a hand level or a clinometer.

3. *Imperfections of Observing.*—Errors in plumbing, reading the tape, and setting the pins are accidental errors; hence they tend to vary as the square root of the number of tape lengths. Only the error due to plumbing is of real importance; it can be avoided by chaining on the slope. Although the errors in reading the tape and setting the pins cannot be eliminated, their effect upon the resultant error is usually not large. All the accidental errors can be kept reasonably low by care in chaining.

4. *Variations in Temperature.*—The tape expands as the temperature rises and contracts as the temperature falls. Steel has a coefficient of thermal expansion about 0.0000065 per 1°F.; hence a 100-ft. steel tape will change in length about 0.01 ft. for a change in temperature of 15°F. Even for measurements of ordinary precision, the error due to thermal expansion becomes of consequence in extremely cold or extremely hot weather.

If the tape is standard at a temperature of T_0 degrees and measurements are taken at a temperature of T degrees, the correction C_x for change in length is given by the formula

$$C_x = 0.0000065L(T - T_0) \tag{3}$$

in which L is the measured length.

Errors due to variations in temperature are greatly reduced by using an invar tape.

5. *Variations in Tension.*—The steel tape, being elastic, stretches when a pull is applied. If the tension is greater or less than that for which the tape is standardized, the tape is correspondingly too long or too short. For the heavy 100-ft. tape a change in tension of 3 lb. changes the length of the tape about 0.001 ft. The error is systematic. It is negligible except in precise work.

The correction for variation in tension in a steel tape is given by the formula

$$C_p = \frac{(P - P_0)L}{AE} \tag{4}$$

in which C_p is the correction per distance L, in feet; P is the applied tension, in pounds; P_0 is the tension for which the tape is standardized, in pounds; L is the length, in feet; A is the cross-sectional area, in square inches; and E is the elastic modulus of the steel, in pounds per square inch. The elastic modulus of steel can be taken as 30,000,000 lb. per square inch without error of consequence.

6. *Sag in Tape.*—If the tape is supported at intervals rather than throughout its full length, it sags between the points of support. If a heavy 100-ft. tape weighing 3 lb., standardized while supported throughout its length, is supported at the ends under a tension of 10 lb., the positive systematic error due to sag alone is about 0.37 ft. per tape length.

The correction for sag is given by the formula

$$C_s = \frac{w^2L^3}{24P^2} = \frac{W^2L}{24P^2} \tag{5}$$

in which C_s is the correction between points of support, in feet; w is the weight of tape, in pounds per foot; W is the total weight of tape *between supports*, in pounds; L is the distance between supports, in feet; and P is the applied tension, in pounds.

28a. In ordinary chaining the systematic errors are likely to be much larger than the accidental errors. Hence the resultant error may be expected to vary as the number of tape lengths or as the length measured.

When every possible method is employed to eliminate or to correct for the systematic errors, the resultant error is almost entirely accidental; hence the resultant error may be expected to vary as the square root of the number of tape lengths or as the square root of the distance.

29. Mistakes in Chaining.—Some of the mistakes commonly made by inexperienced chainmen are:

1. Adding or dropping a full tape length, through failure to handle or count the pins properly.

2. Adding a foot, usually in measuring the fractional part of a tape length at the end of the line.

3. Other points incorrectly taken as 0 or 100-ft. marks on tape.

4. Reading numbers incorrectly, as "6" for "9." As a check, it is good practice to read the numbers on each side of the one indicating the measurement.

5. Calling numbers incorrectly or not clearly. Decimal points and zeros should be called, and the recorder should repeat a number as it is recorded.

29a. Often large mistakes will be prevented or discovered if the chainmen form the habit of pacing distances or estimating them by eye. If a transit is being used to give line, distances can be checked by reading the stadia intervals on a range pole.

30. Erecting Perpendicular to Line with Tape.—To locate the altitude of a triangle or to lay out a right angle as for a building corner, the 3:4:5 method is often used. To erect a perpendicular to the line *AB* (Fig. 30) that will include point *C*, a point *a* on line *AB* is assumed to be on the perpendicular, and a pin is set at *a*. With sides 3, 4, and 5 ft. or a multiple of these numbers, such as 24, 32, and 40 ft., a right triangle *abc* is constructed as follows: A pin is set on line *AB* at *b*, 32 ft. from *a*. The zero end of the tape is fixed with a pin at *a*, and the 100-ft. end at *b*. The head chainman moves to *c* and holds the 24-ft. and the 60-ft. marks of the tape in one hand, with the tape between these marks laid out so as to avoid kinking. He then sets a pin at *c*. The rear chainman moves from *a* to *b* as necessary to check the position of the tape at these points as *c* is established. He then sights along *ac* to *C′* by *C*. If *C* does not lie on this prolongation, the perpendicular distance from the point *C* to the line *aC′* is measured, and the foot *a* of the perpendicular *aC′* is moved along the line *AB* by an equal amount, to the point *a′*. If the trial perpendicular *aC′* fails to include the point *C* by several feet, the process is repeated for *a′*, the new point; otherwise the location of *a′* may be assumed as correct.

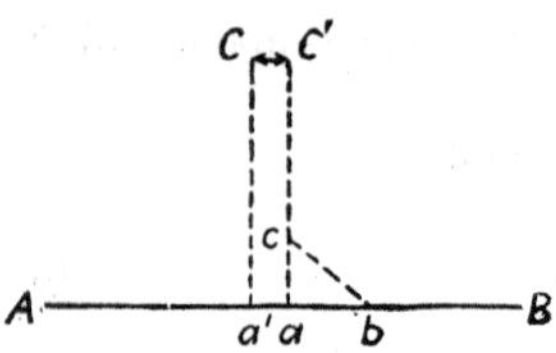

FIG. 30.—Perpendicular by 3:4:5 method.

31. Numerical Problems.

1. The measured length of a line is 1,831.8 ft., when a tape 100.04 ft. long is used. Compute the length of the line.

2. The slope measurement of a line is 600.00 ft. The differences in elevation between successive 100-ft. points, as measured with a hand level, are 1.5, 2.7, 5.3, 6.0, 2.2, and 1.0 ft. Determine the horizontal distance.

3. The slope measurement of a line is 957.2 ft. Slope angles measured with a clinometer are shown below. Determine the horizontal distance.

Chainage, feet........	0		300		700		957.2
Slope angle, degrees...		½		1½		4	

4. Two points at a slope distance of 100 ft. apart have a difference in elevation of 10 ft. What slope distance should be laid off to establish a

horizontal distance of 100.0 ft.? Compute by an exact and an approximate method.

5. Compute the effect of sag *per tape length* for a 100-ft. tape weighing 1½ lb., under a pull of 25 lb., (*a*) supported at the ends only, (*b*) supported at ends and center.

6. A 100-ft. tape weighing 1½ lb. is of standard length under a tension of 10 lb., supported for full length. A line on smooth level ground is measured with the tape under a tension of 25 lb. and found to be 3,218.4 ft. long. $E = 28{,}000{,}000$ lb. per square inch; 3.56 cu. in. of steel weighs 1 lb. Make the correction for increase in tension.

7. A horizontal line 613.24 ft. long is to be laid out along a street having a grade of 3 per cent. The temperature is 86°F. The steel tape is of standard length at 68°F. What slope measurement should be made?

8. A hedge along the line *AB* makes direct measurement impossible. A point *C* is established at an offset distance of 25 ft. from the line *AB* and roughly equidistant from *A* and *B*. The distances *AC* and *CB* are then measured; $AC = 1{,}213.7$ ft.; $CB = 1{,}184.5$ ft. By an approximate method compute the length of the line *AB*.

32. Field Problems.

Problem 1. Chaining over Level Ground with Tape

Object.—To chain with the 100-ft. steel tape over an approximately level course about 1,000 ft. long and to check the distance by chaining in the opposite direction.

Procedure.—Set a flag pole at each end of the line. Follow the procedure indicated in Art. 24, reading the distance to tenths of feet and estimating to hundredths. Record the two lengths, and compute the ratio of discrepancy to length. Measurements should check within 1/5,000. As a check, standardize the pace, and measure the distance by pacing.

Problem 2. Standardization of Tape and Chaining over Uneven Ground

Object.—To standardize the 100 ft. steel tape, and to find the horizontal length of an assigned course about 800 ft. long over uneven ground by two methods.

Procedure.—(1) Standardize the tape before and after the field work by comparing it with the official standard of length. Maintain the required pull by means of a spring balance. Determine the error with a finely divided scale. (2) Chain the course by horizontal measurement after the manner described in Art. 25. (3) Correct for the error in length of tape. (4) Chain the course by measurement on the slope, recording the difference (to the nearest foot) in elevation of each 100-ft. tape length as determined by a hand level. If sharp breaks in slope occur between 100-ft. stations, treat each distance between breaks in the same manner as a full station. Correct the measurements for error in length of tape. Correct the measurements for slope, and determine the horizontal distance. (5) Compare the results obtained by the two methods of measurement.

CHAPTER III

MEASUREMENT OF DIFFERENCE IN ELEVATION

33. Definitions.—The *elevation* of a point near the surface of the earth is its vertical distance above or below an arbitrarily assumed *level surface*, or curved surface every element of which is normal to the plumb line. The level surface (real or imaginary) used for reference is called the *datum*. Most commonly the datum is taken at mean sea level. A *level line* is a line in a level surface.

The *difference in elevation* between two points is the vertical distance between the two level surfaces in which the points lie. *Leveling* is the operation of measuring vertical distances, either directly or indirectly, in order to determine differences in elevation.

A *horizontal line* is a straight line tangent to a level surface.

A *vertical angle* is an angle between two intersecting lines in a vertical plane. In surveying it is commonly understood that one of these lines is horizontal.

34. Curvature of the Earth; Refraction.—In leveling, it is necessary to consider (1) the effect of the curvature of the earth and (2) the

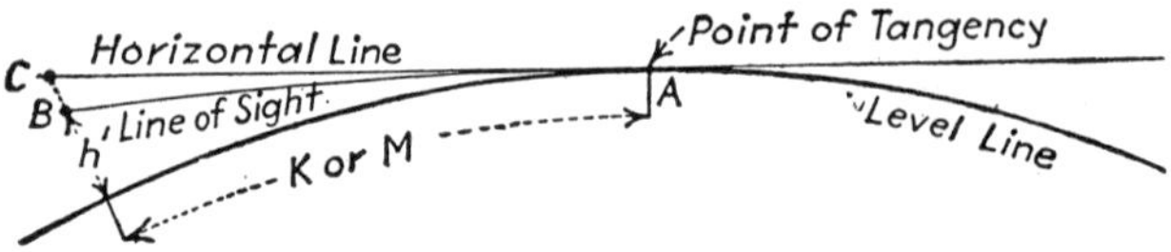

Fig. 34.—Earth's curvature and atmospheric refraction.

effect of atmospheric refraction, which affects the line of sight. Usually these effects are considered together.

In Fig. 34 is shown a horizontal line tangent to a level surface near the surface of the earth. The vertical distance between the horizontal line and the level surface is a measure of the earth's curvature. It varies approximately as the square of the distance from the point of tangency.

Owing to the phenomenon of atmospheric refraction, rays of light transmitted along the surface of the earth are bent downward slightly. Thus, as viewed from point A (Fig. 34), an object actually at B would appear to be at C; the actual line of sight is along a curve AB. Let K be the distance, in miles, from the point of tangency (station of the observer); M the same distance, in thousands of feet; and h' the combined effect of the earth's curvature and of atmospheric refraction, in feet. Then

$$h' = 0.57K^2 = 0.021M^2 \text{ (approx.)} \tag{1}$$

The effect of the earth's curvature alone is about $0.66K^2$. The effect of refraction alone is about $0.09K^2$ in the opposite direction.

35. Methods.—Difference in elevation may be measured by the following methods:

1. *Direct* or *spirit leveling*, by measuring vertical distances directly (Art. 37). Direct leveling is the most accurate method of determining elevations and is the one most commonly used.

2. *Indirect* or *trigonometric leveling*, by measuring vertical angles and horizontal distances (Art. 36).

3. *Barometric leveling*, by measuring the variation in the pressure of the earth's atmosphere with variation in elevation, using a barometer. The method is limited to rough observations, as the pressure may vary even at a given elevation. Barometric leveling is employed principally on exploratory surveys and where differences in elevation are large.

Differential leveling is the operation of determining relative elevations of points some distance apart (Chap. IV). Usually differential leveling is accomplished by direct leveling.

Profile leveling is the operation—usually by direct leveling—of determining elevations of points at short measured intervals along a definitely located line, such as the center line for a highway or a sewer (Art. 62).

36. Indirect Leveling.—In indirect or trigonometric leveling, the difference in elevation between the instrument at A and a given point B is determined as follows: The vertical angle α to B is measured at A, and the *horizontal* distance AB' from A to B is determined by some method of measurement. Then

$$\text{Difference in elevation} = AB' \tan \alpha \tag{2}$$

If the sight is long and if the precision of the work so warrants, a correction for curvature and refraction (Art. 34) must be applied. When the vertical angle is upward or positive, the curvature and refraction correction is added; and when downward or negative, the correction is subtracted.

DIRECT LEVELING

37. General.—In Fig. 37, A represents a point of known elevation, and B a point the elevation of which is desired. In the method of direct or spirit leveling, the level is set up at some intermediate point as L, and the vertical distances AC and BD are observed by holding a leveling rod first at A and then at B, the line of sight of the instrument being horizontal.[1]

If the difference in elevation between the points A and E is designated as H_a, and between E and B as H_b; if the vertical distances read at A and B respectively are designated as h_a and h_b; and if the

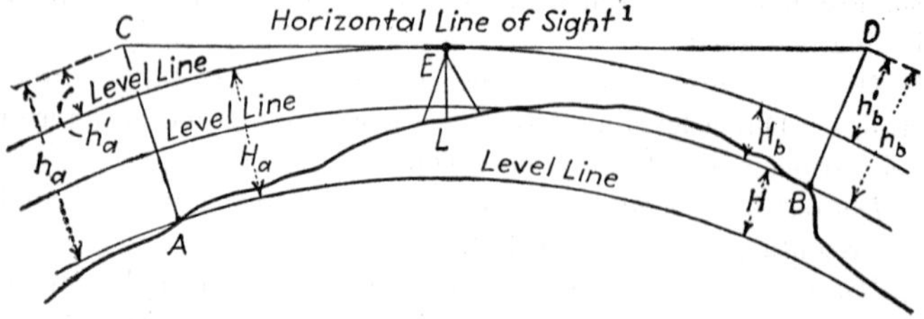

Fig. 37.—Direct leveling.

effects of curvature and refraction for the horizontal distances LA and LB respectively are designated h'_a and h'_b, then

$$H_a = h_a - h'_a \qquad \text{and} \qquad H_b = h_b - h'_b \tag{3}$$

The difference in elevation H between A and B is then

$$H = H_a - H_b = h_a - h_b - h'_a + h'_b \tag{4}$$

If the backsight distance LA is equal to the foresight distance LB, then $h'_a = h'_b$ and

$$H = h_a - h_b \tag{5}$$

Thus if backsight and foresight distances are balanced, the difference in elevation between two points is equal to the difference between the rod readings taken to the two points, and no correction for curvature and refraction is necessary. In direct leveling, usually the work is so conducted that the effect of curvature and refraction is reduced to a negligible amount.

The procedure of direct leveling is described in Arts. 48 and 49 and in Chap. IV.

38. Instruments.—Any of the instruments commonly used for direct leveling has as its essential features (1) a *line of sight* and (2) a *level tube* mounted with its axis parallel to the line of sight and used as a guide in making the line of sight horizontal.

The instrument used principally is the *engineer's level*, either the *dumpy level* for which the telescope tube is permanently fastened to

[1] Owing to refraction, the line of sight is slightly curved as explained in Art. 34.

the supporting bar or the *wye level* for which the telescope is removable and rests in Y-shaped supports. The *hand level* is a simple and useful device for rough work. The *transit* and the telescopic alidade of the *plane table* are not designed primarily for direct leveling but are frequently used for this purpose.

39. Engineer's Level.—The principal parts of the engineer's level are shown in Figs 42 and 43. The details of the instrument are constructed quite differently by the different manufacturers. Essentially the level consists of a *telescope* mounted on a *level bar* which is rigidly fastened to a vertical *center spindle*. Attached to the telescope or the level bar and parallel with the telescope is a *level tube*. The spindle fits into the cone-shaped bearing of a *leveling head*, so that the level is free to revolve about the spindle as a vertical axis. At the lower end of the spindle is a ball-and-socket joint which makes a flexible connection between the movable leveling head and a stationary *foot plate*. Any desired relation between leveling head and foot plate is maintained by means of vertical *leveling screws*, usually four. The foot plate is screwed to a metal *tripod head* which is supported by wooden tripod legs. In the tube of the telescope are *cross-hairs* supported on a metal ring; the intersection of these hairs defines a point on the line of sight.

40. Level Tube.—The level tube is a glass vial with the inside ground barrel-shaped, so that a longitudinal line on its inner surface is the arc of a circle. The tube is nearly filled with sulphuric ether or with alcohol, the remaining space being occupied by a bubble of air. A longitudinal line tangent to the inside surface of the tube at its upper mid-point is called the *axis of the level tube;* when the bubble is centered, the axis of the level tube is horizontal. The metal housing of the tube is attached to the instrument by means of screws that permit adjustment, as shown in Figs. 42 and 43. The level tube is graduated so that the bubble can be centered accurately.

The longer the radius of the circle to which the level tube is ground the more sensitive the tube. However, the more sensitive the tube the longer the time required to center the bubble.

41. Telescope.—Figure 41*a* shows the principal parts of the telescope as it is commonly constructed. Rays of light emanating from an object within the field of view are caught by the objective lens *A* and are brought to a focus and form an image in the plane of the cross-hairs *B*. The lenses of the eyepiece *C* form a microscope which is focused on the image at the cross-hairs. The objective lens is screwed in the outer end of the objective slide *D* which fits in the telescope tube *E*. The objective lens is focused by the screw *F* at the inner end of which is a pinion that engages the teeth of a rack fixed to the objective slide. The eyepiece slide G is held in position transversely by rings H and J, through which it may be moved in a longitudinal direction for focusing. By means of screws the ring J may be moved transversely so that the intersection of the cross-hairs will appear in the center of the field of view.

The line of sight is defined by the intersection of the cross-hairs and the optical center of the objective lens.

Another type of telescope, called the *internal-focusing* telescope, has recently increased greatly in use. It is shown in section in Fig. 41*b*. Its arrangement and operation are similar to that just described, except that the objective lens A is fixed in the end of the telescope tube and that the slide carries a focusing lens L.

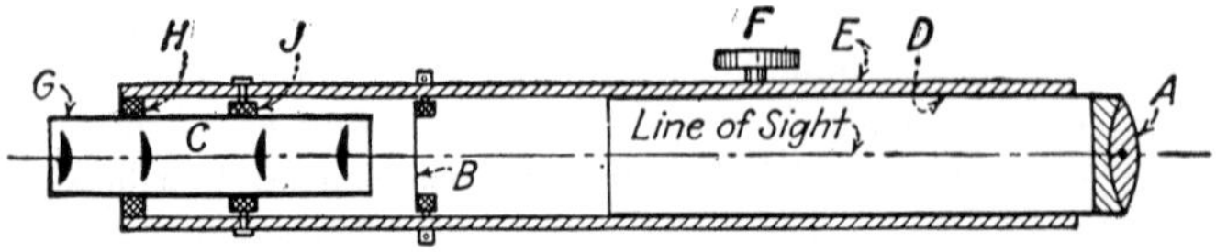

FIG. 41*a*.—Longitudinal section of external-focusing telescope.

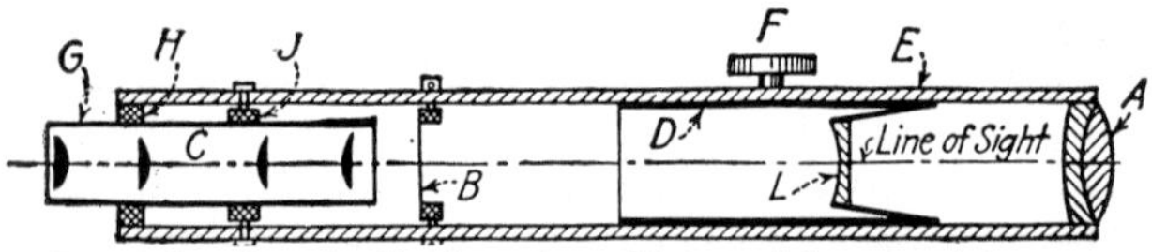

FIG. 41*b*.—Longitudinal section of internal-focusing telescope.

The discussions hereinafter refer to the external-focusing type of telescope, except as specifically stated. With modifications in detail, they apply also to the internal-focusing type.

Focusing.—When the telescope is to be used, the eyepiece is first moved in or out until the cross-hairs appear distinct. This adjustment of the eyepiece should be tested frequently, as the observer's eye becomes tired.

When an object is sighted, the objective slide is moved in or out until the image appears clear, when the image should be in the plane of the cross-hairs. If a slight movement of the eye from side to side produces an apparent movement of the cross-hairs over the image, the plane of the image and the plane of the cross-hairs do not coincide, and *parallax* is said to exist. Since parallax is a source of error in observations, it should be eliminated by refocusing the objective, the eyepiece, or both. The objective lens must be focused for each distance sighted.

Cross-hairs.—Threads from the cocoon of the brown spider are usually used as cross-hairs. In some instruments the cross-hairs are made of very fine platinum wire. One vertical and one horizontal

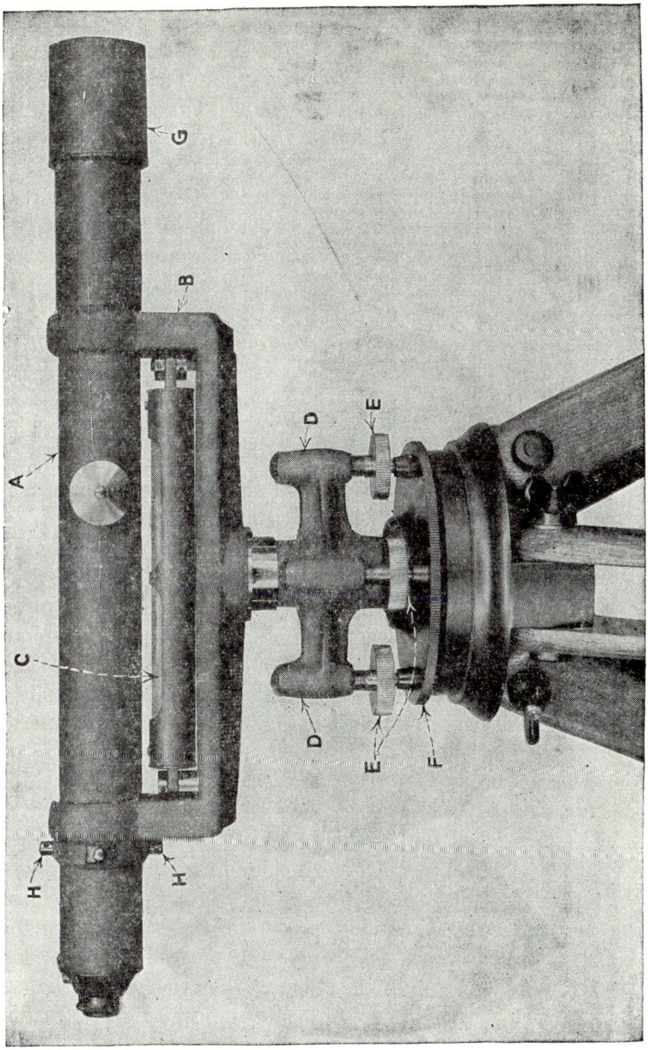

FIG. 42.—Engineer's dumpy level.

hair are fastened to a metal ring called the *cross-hair ring*, or *reticule*, as shown at 46 in Fig. 43. The ring is smaller than the telescope tube. It is held in position by four capstan-headed screws (47, Fig. 43) by means of which it may be moved either horizontally or vertically or may be rotated through a small angle about the axis of the telescope.

To rotate the cross-hair ring about the axis of the telescope without disturbing its centering, loosen any two adjacent screws, tap one lightly, then tighten the same two screws. To move the cross-hair vertically or horizontally, loosen one screw and then tighten the opposite screw.

Eyepiece.—The image formed by the objective is inverted. The eyepiece commonly used, called the *erecting* or *terrestrial* eyepiece, reinverts the image so that the object appears to the eye in its normal position. Usually it consists of four plano-convex lenses placed in the eyepiece slide (*G*, Fig. 41).

The *inverting* or *astronomical* eyepiece simply magnifies the image without reinverting it. It consists of two plano-convex lenses.

42. Dumpy Level.—Figure 42 shows the details of a dumpy level with erecting eyepiece. The telescope *A* is rigidly attached to the level bar *B*, and the instrument is so constructed that the optical axis of the telescope is perpendicular to the axis of the center spindle. The level tube *C* is permanently placed so that its axis lies in the same vertical plane as the optical axis, but it is adjustable in altitude by means of a capstan-headed screw at one end. The spindle revolves in the socket of the leveling head *D* which is controlled in position by the four leveling screws *E*. At the lower end of the spindle is a ball-and-socket joint which makes a flexible connection between the instrument proper and the foot plate *F*. When the leveling screws are turned, the level is moved about this joint as a center (for details, see Fig. 43). The sunshade *G* protects the objective from the direct rays of the sun. The adjusting screws *H* for the cross-hair ring are near the eyepiece end of the telescope.

The advantages of the dumpy level over the wye level are that it is simpler in construction, has fewer parts subject to wear, requires fewer adjustments, and stays in adjustment better. It is the type most commonly employed.

43. Wye Level.—Figure 43 shows the details of a wye level with erecting eyepiece. The telescope rests in Y-shaped bearings called the *wyes*. By means of capstan-headed nuts 30 and 31, the leg of one wye may be raised or lowered. When the wye clips or stirrups 32 are raised, the telescope may be revolved in the wyes, or it may be lifted from the wyes and turned end for end. The enlarged cylin-

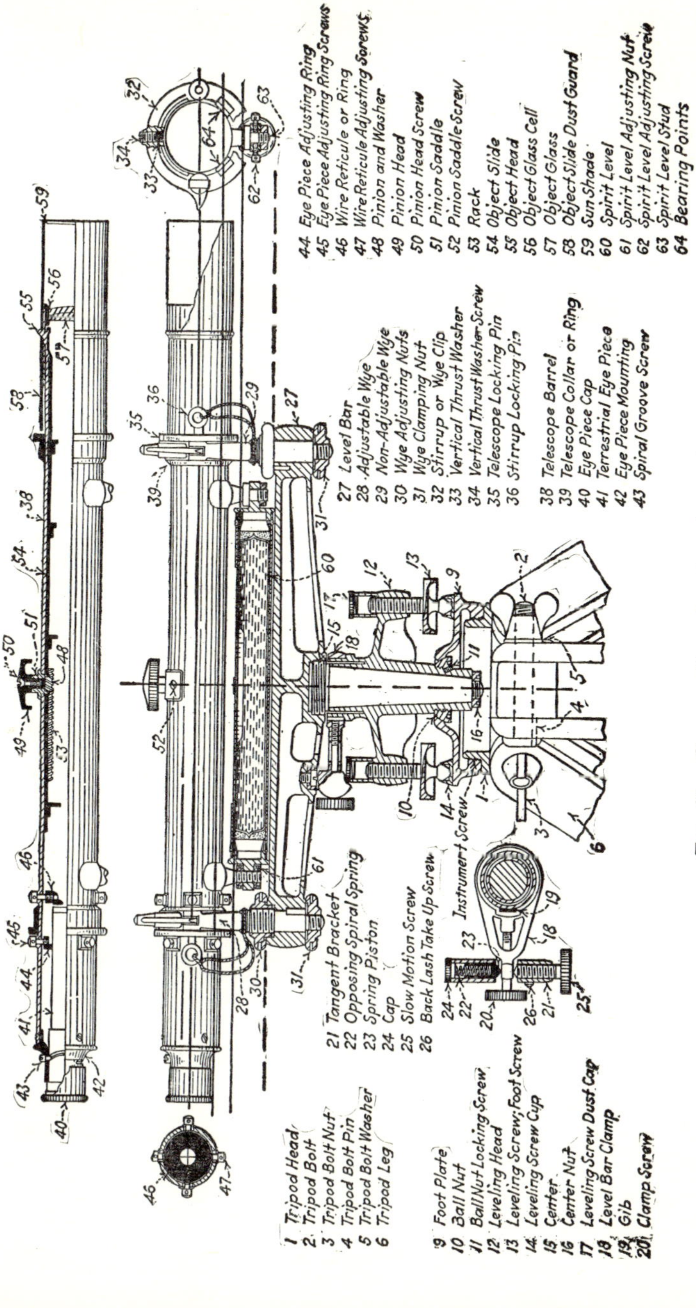

FIG. 43.—Cross-section of engineer's wye level.

drical portions 39 of the telescope barrel which rest in the wyes are called the *rings* or *collars*. The line joining the centers of the rings is defined as the *axis of the collars*. The *axis of the wyes* is a general term used to denote a reference line—sometimes actually the axis of the collars and sometimes actually the axis of the supports—that represents the alinement of the telescope tube in the wye supports.

Fig. 44.—Locke hand level.

The level tube is attached to the telescope and is adjustable in both a horizontal and a vertical plane. Other details are much the same as for the dumpy level just described.

44. Locke Hand Level.—The Locke hand level is widely used for rough leveling. It consists of a metal sighting tube about 6 in. long, on which is mounted a level vial (Fig. 44). In the sighting tube, beneath the vial, is a prism which reflects the image of the bubble to the eye end of the level. Just beneath the level vial is a cross-wire. Mounted on the right half of the eyepiece slide is a lens which magnifies the image of the bubble and cross-wire as reflected by the prism.

Fig. 45.—Abney hand level and clinometer.

In using the level the object is viewed directly through the left half of the sighting tube, without magnification, while the position of the bubble is observed in the right half of the field of view. The level is held with the level vial uppermost and is tipped up or down until the cross-wire bisects the bubble, when the line of sight is horizontal.

45. Abney Hand Level and Clinometer.—This instrument is suitable both for direct leveling and for measuring the angles of slopes. The instrument shown in Fig. 45 is graduated both in degrees and in percentage of slope, or grade. When it is used as a level, the index of the vernier is set at zero, and it is then used in the same way as the Locke hand level. When it is used as a clinometer, the object is sighted, and the level tube is caused to rotate about the axis of the vertical arc until the cross-wire bisects the bubble as viewed through the eyepiece. Either the slope angle or the slope percentage is then read on the vertical arc.

46. Leveling Rods.—These are graduated wooden rods of rectangular cross-section by means of which difference in elevation is measured. The lower end is shod with metal.

The rod is held vertical, and hence the reading of the rod as indicated by the horizontal cross-hair of the level is a measure of the vertical distance between the point on which the rod is held and the line of sight.

Rods are obtainable in a variety of patterns and graduations. Common lengths are 12 and 13 ft. In the United States, ordinarily rods are graduated in hundredths of a foot.

The two general classes of leveling rods are (1) *self-reading* rods, which may be read directly by the leveler, and (2) *target* rods, for which a target sliding on the rod is set by the rodman as directed by the leveler. Under ordinary conditions, observations with the self-reading rod may be made with nearly the same precision and much more rapidly. For ordinary leveling, the self-reading rod is the one commonly employed.

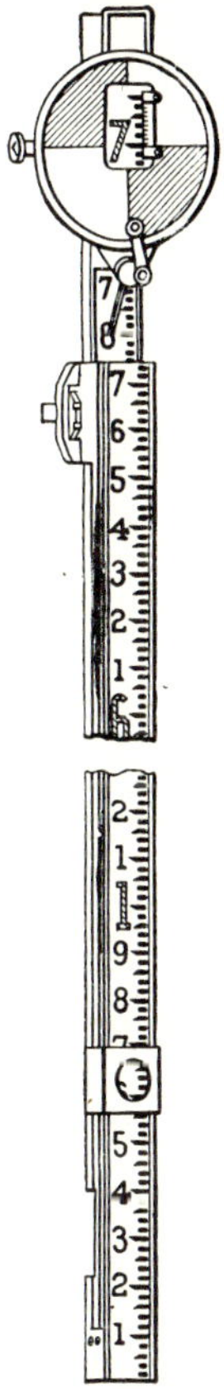

FIG. 46*a*.—Philadelphia rod.

Of the many types of rods, the *Philadelphia rod* (Fig. 46*a*) is the most widely used. It can be used as either a self-reading or a target rod. It is usually in two sections or strips, the strips being held in contact by two brass sleeves. By means of a screw attached to the upper sleeve the two strips may be clamped together in any relative position desired. For readings of 7 ft. or less (on a 13-ft. rod), the back strip is clamped in its normal position. For greater readings, the rod is extended its full length; the graduations on the front face of the back strip are then a continuation of those on the front strip. In this position the rod is called a "long" rod. The background is white with graduations 0.01 ft. wide painted in black as shown in Fig. 46*a*. The numbers indicating feet are in red, and those indicating tenths of feet are in black. If the target is not used, observations closer than the smallest division on the rod are made by estimation.

The usual target (Fig. 46*b*, left) is a circular or elliptical disk about 5 in. in diameter with horizontal and vertical lines formed

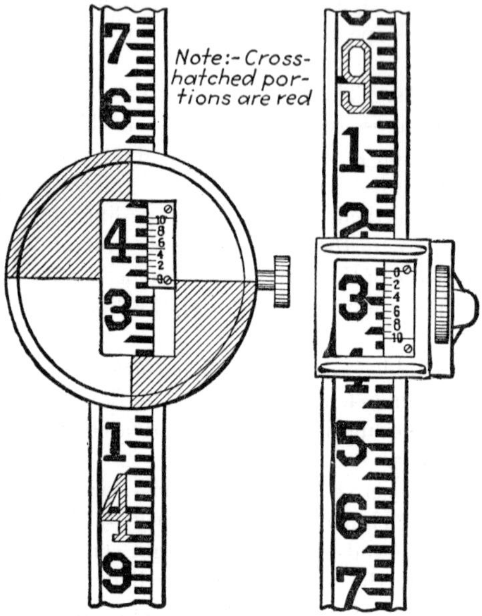

Fig. 46*b*.—Direct vernier settings.

by the junction of alternate quadrants of white and red (shown crosshatched in the figure). A rectangular opening in the front of the target exposes a portion of the rod to view so that readings can be taken. The attached *vernier* (Art. 47) fits closely to the rod, its zero point or index being at the horizontal line of the target. The leveler signals the rodman to slide the target up or down until it is bisected by the line of sight. The target is then clamped, and the rodman, leveler, or both observe the indicated reading. For readings greater than can be taken with the "short" rod, the target is clamped at the same graduation on the face of the rod as the reading of the vernier on the back of the upper sleeve of the rod (Fig. 46*b*, right). The rod is then extended until the target is bisected by the line of sight. The vertical distance from foot of rod to target is then indicated by the reading of the vernier on the back of the rod.

The *New York rod*, used to some extent in building construction, is similar to the Philadelphia rod but has graduations consisting of short, fine lines at intervals of 0.01 ft. and longer, numbered lines at intervals of 0.1 ft. It is not a self-reading rod.

The *architect's rod* sometimes used in building construction is similar to the New York rod but is graduated in ⅛ in. and is equipped with verniers reading to 1/64 in.

47. Verniers.—A vernier is a short auxiliary scale placed parallel and in contact with the main scale, by means of which fractional parts of the least division of the main scale can be measured accurately. The accuracy of the vernier depends on the fact that the eye can determine more closely when two lines coincide than it can estimate the distance between two parallel lines. The scale may be either straight (as on a leveling rod) or curved (as on the horizontal and vertical circles of a transit). The zero of the vernier scale is the index for the main scale.

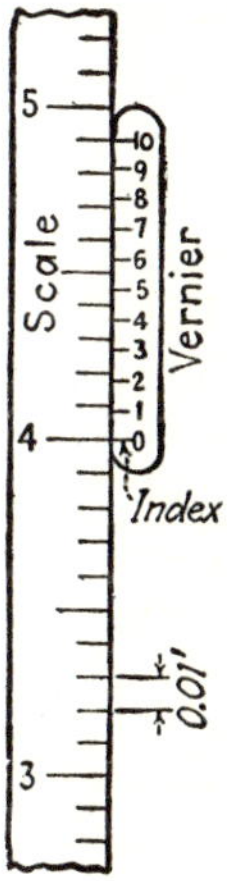

Fig. 47.—Direct vernier.

The *least count*, or fineness of reading, of a vernier is equal to the difference in length between one scale space and one vernier space. Furthermore, the number n of spaces on the vernier is equal to the number of equal parts into which one space s on the main scale can be subdivided by reading the vernier. It follows that the least count is

$$\text{Least count} = \frac{s}{n} \qquad (6)$$

The *direct vernier* has spaces slightly shorter than those of the main scale; n spaces on the vernier are equal in length to $(n - 1)$ spaces on the main scale. Figure 47 represents a scale graduated in hundredths of feet and a direct vernier having each space 0.001 ft. shorter than a 0.01-ft. space on the main scale; thus each vernier space is equal to 0.009 ft., and 10 spaces on the vernier are equal to 9 spaces on the scale. The index, or zero of the vernier, is set at 0.400 ft. on the scale. If the vernier were moved upward 0.001 ft., its graduation numbered 1 would coincide with a graduation on the scale and the index would be at 0.401 ft., and so on. It is thus seen that the position of the index is determined to thousandths of feet without estimation, simply by noting which graduation on the vernier coincides with one on the scale. Note that the coinciding graduation on the main scale does *not* indicate the main-scale reading and that it has no significance.

A *retrograde vernier* has spaces slightly *longer* than those of the main scale; n spaces on the vernier are equal in length to $(n + 1)$ spaces on the main scale. The retrograde vernier extends backward along the main scale and the vernier graduations are also numbered in reverse order; however, the retrograde vernier is read in the same manner as the direct vernier.

Figure 46*b* illustrates settings of direct verniers on the target (at left) and on the back (at right) of a Philadelphia rod. The rod reading indicated by the target (4.347 ft. in figure) is determined by first observing the position of the vernier index on the scale to hundredths of feet (4.34 in figure), next by observing the number of spaces *on the vernier* from the index to the coinciding graduations (7 spaces in figure), and finally by adding the vernier reading (0.007 ft. in figure) to the scale reading (4.34 ft.). On the back of the rod (Fig. 46*b*, right) both the main scale and the direct vernier read *down* the rod. The scale reading is 9.26 ft. and the vernier reading is 0.004 ft.; hence the rod reading is 9.264 ft.

48. Setting Up the Engineer's Level.—The engineer's level is placed in a desired location, with the tripod legs well spread and firmly pressed into the ground and with the tripod head somewhere near level. The telescope is brought over one pair of opposite leveling screws, and the bubble is centered approximately; then the process is repeated with the telescope over the other pair. By repetition of this procedure the leveling screws are manipulated until the bubble remains centered, or nearly so, for any direction in which the telescope is pointed. If the instrument is in adjustment, the line of sight is then horizontal.

Suggestions for the care and handling of the level are given in Art. 5.

49. Reading the Rod.—For accurate observations, the rod is held vertical on some well-defined point. The leveler sights at the rod, focuses the objective, and carefully centers the bubble. If the self-reading rod is used, the leveler records the reading of the cross-hair on the rod. As a check he again observes the bubble and rod. If the target rod is used, the procedure is identical except that the target is set by the rodman as directed by the leveler.

For less accurate leveling, as when rod readings for ground points are determined to the nearest 0.1 ft., the observations usually are not checked, and proportionately less care is exercised in keeping the rod vertical and the bubble centered.

If no rod level is used, in calm air the rodman can plumb the rod accurately by balancing it upon the point. By means of the vertical cross-hair the leveler can determine when the rod is held in a vertical plane passing through the instrument but cannot tell whether it is tipped forward or backward in this plane. If it is in either of these positions, the rod reading will be greater than the true vertical distance, as illustrated by Fig. 49. The larger the rod reading the greater the error. To eliminate the error, the rodman *waves the rod* or tilts it slowly forward and backward, and the leveler takes the least reading.

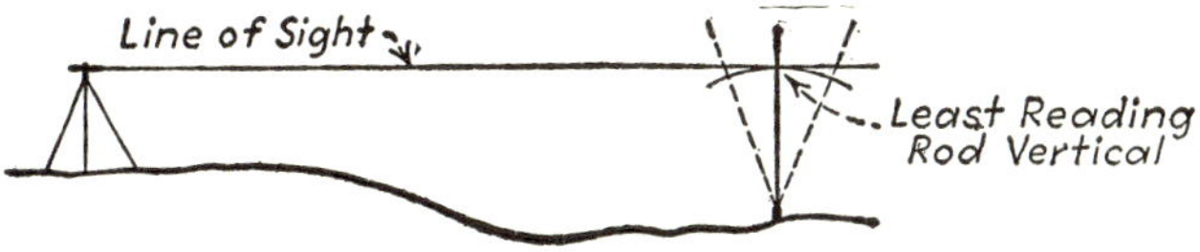

FIG. 49.—Waving the rod.

50. Adjustment of the Dumpy Level.—The axis of the objective slide and the optical axis are permanently fixed perpendicular to the vertical axis by the manufacturer. In addition, for a dumpy level in perfect adjustment the following relations should exist:

a. The axis of the level tube should be perpendicular to the vertical axis.

b. The horizontal cross-hair should lie in a horizontal plane when the instrument is level.

c. The line of sight should be parallel to the axis of the level tube.

The parts capable of and requiring adjustment are the cross-hairs and the level tube. The basis for adjustments is the vertical axis. The adjustments are as follows:

1. *To Make the Axis of the Level Tube Perpendicular to the Vertical Axis.*—Approximately center the bubble over each pair of opposite leveling screws; then carefully center the bubble over one pair. Revolve the level 180° about its vertical axis. If the bubble is displaced, bring it back halfway to the center by means of the capstan nuts at one end of the level tube. Relevel the instrument with the leveling screws, and repeat the process until the adjustment is perfected.

This adjustment involves the principle of *reversion.* If the tube is not in adjustment, the displacement of the bubble indicates double the actual error, as shown by Fig. 50*a*. If $(90° - \alpha)$ represents the angle between the vertical axis and axis of level tube, then when the bubble is centered the vertical axis makes an angle of α with the true vertical. When the level is reversed, the bubble is displaced through the arc whose angle is 2α. Hence the correction is the arc whose angle is α.

2. *To Make the Horizontal Cross-hair Lie in a Plane Perpendicular to the Vertical Axis.*—Sight the horizontal cross-hair on some clearly defined point (as A, Fig. 50b), and rotate the instrument slowly about

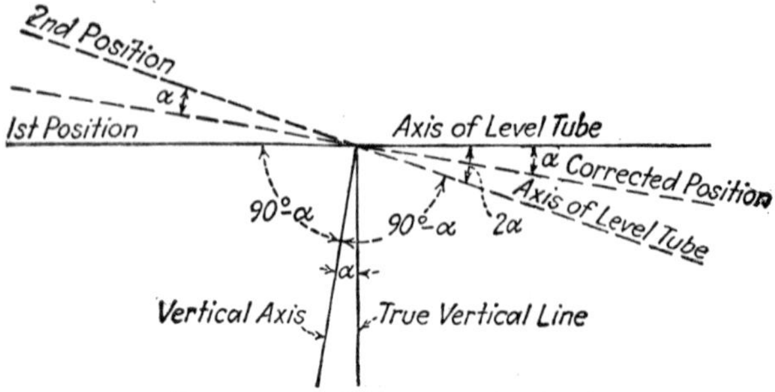

FIG. 50a.—Adjustment of level tube.

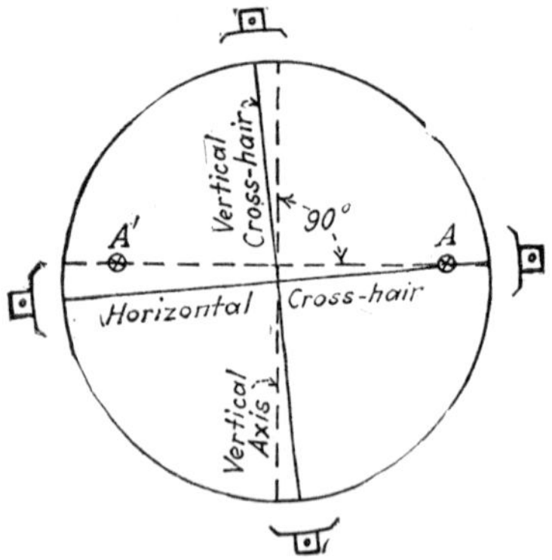

FIG. 50b.—Adjustment of horizontal cross-hair.

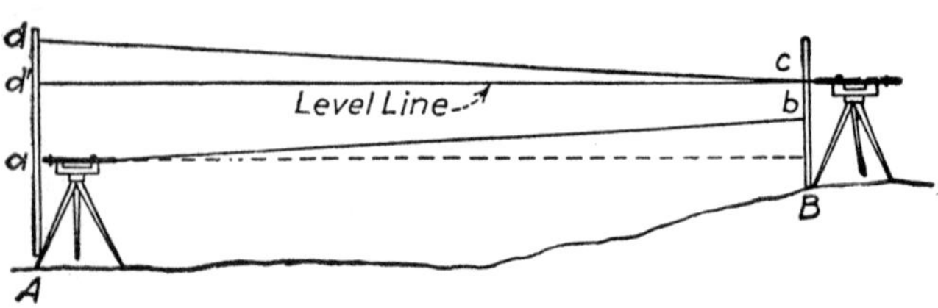

FIG. 50c.—Two-peg test.

its vertical axis. If the point appears to depart from the cross-hair and takes some position as A', loosen two adjacent screws and rotate the cross-hair ring until by further trial the point appears to travel along the cross-hair. Tighten the same two screws. The instrument need not be level when the test is made.

3. *To Make the Line of Sight Parallel to the Axis of the Level Tube (Two-peg Test).*—Set two pegs 200 to 300 ft. apart on approximately level ground. Set up and level the instrument in a location such that the eyepiece is ½ in. or less in front of the rod held on one of the pegs as at A (Fig. 50c). With the rod held at A, take a rod reading a by sighting through the objective end of the telescope (with the eyepiece next to the rod). The center of the field of view may be determined within one or two thousandths of a foot by holding the point of a pencil on the rod.

Move the rod to the other peg B, and take a rod reading b with instrument at A, in the usual manner.

Move the instrument to B, set up as before, and take rod readings c and d.

If the two differences in elevation thus determined are equal, that is, $(a - b) = (d - c)$, the line of sight is in adjustment. If not, the correct rod reading at A for instrument with position unchanged at B is

$$d' = c + \frac{(a - b) + (d - c)}{2} \tag{7}$$

in which

$$\frac{(a - b) + (d - c)}{2}$$

is the true difference in elevation.

Equation (7) must be solved with due regard to signs. Strictly speaking, the effect of curvature and refraction (Art. 34) should be included; but for ordinary leveling the quantities are negligible.

The adjustment is made by moving the cross-hair ring vertically until the line of sight cuts the rod at d'. The preceding steps are then repeated as a check.

Some surveyors prefer to set up midway between the two pegs, in which case the difference between rod readings at A and B is the true difference in elevation. The leveler than moves to A or B and adjusts the line of sight.

51. Adjustment of the Wye Level.—For a wye level in perfect adjustment the following relations should exist:

a. The axis of the level tube should lie in the same plane with, and should be parallel to, the axis of the wyes.

b. The horizontal cross-hair should lie in a horizontal plane when the instrument is level.

c. The line of sight should coincide with the axis of the wyes.

d. For convenience in leveling, the axis of the wyes (and hence the axis of the level tube) should be perpendicular to the vertical axis.

The optical axis and the axis of the objective slide are fixed in the proper relation by the manufacturer, and they are presumed to remain in coincidence with the axis of the wyes (and hence with the line of sight) without further attention. The remaining relations are established by the following adjustments:

1. *To Make the Axis of the Level Tube Lie in the Same Plane with the Axis of the Wyes.*—Raise the wye clips; level the instrument; and rotate the telescope a few degrees in the wyes. If the bubble is displaced, bring it back to the center by means of the lateral adjusting screws at one end of the level tube.

2. *To Make the Axis of the Level Tube Parallel to the Axis of the Wyes.*—Raise the wye clips; level the instrument carefully; lift the telescope from the wyes, and turn it end for end. If the bubble is displaced, bring it back halfway to the center by means of the vertical adjusting nuts at one end of the level tube. Relevel the instrument by means of the leveling screws, and repeat the process until the adjustment is perfected.

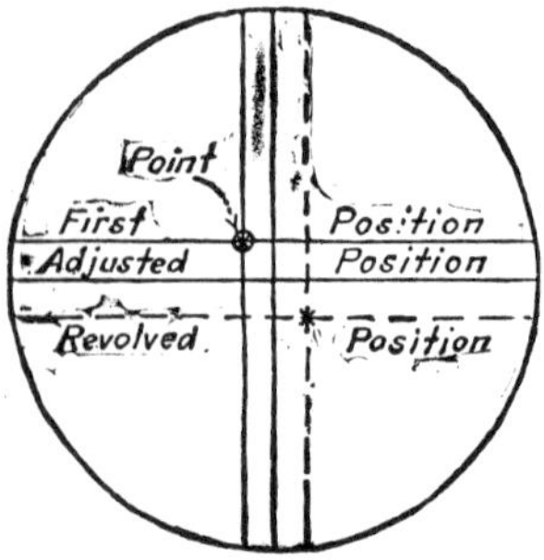

Fig. 51.

3. *To Make the Horizontal Cross-hair Lie in a Plane Perpendicular to the Vertical Axis.*—This adjustment is the same as for the dumpy level. For some instruments the adjustment may be made by rotating the telescope in the wyes instead of rotating the cross-hair ring in the telescope.

4. *To Make the Line of Sight Coincide with the Axis of the Wyes (and Thus Parallel to the Axis of the Level Tube).*—Raise the wye clips, sight the intersection of the cross-hairs on some well-defined point, and clamp the vertical axis. Revolve the telescope 180° (about its own axis) in the wyes. If the line of sight is displaced from the point, adjust the cross-hair ring until the line of sight is midway between its two former positions (Fig. 51). Loosen only one screw at a time, and tighten the opposite screw. Repeat the process until the proper relation is obtained.

5. *To Make the Axis of the Level Tube (and Hence the Axis of the Wyes) Perpendicular to the Vertical Axis.*—This adjustment does not add to the precision of observations but makes it possible to level the instrument so that the bubble will remain centered for any direction in which the telescope may be pointed. Level the instrument, and revolve the telescope 180° about the vertical axis. If the bubble is displaced, bring it back halfway to the center by means

of the capstan nuts controlling the vertical position of one of the wyes.

51a. When either the collars or the bearing points become worn, the adjustments just described are inadequate, and the two-peg test must be made as for the dumpy level. The common procedure is first to perform the adjustments for the cross-hairs in the usual manner, making the line of sight coincide as nearly as may be with the axis of the wyes. The true difference in elevation between two points having been determined by the two-peg test, the line of sight is then set to the proper rod reading for a horizontal line by means of the leveling screws, and the bubble is centered by means of the capstan nuts at one end of the level tube.

52. Adjustment of the Hand Level.—The simplest procedure is to hold the hand level alongside an engineer's level that has been leveled and sighted at some well-defined point. The line of sight of the hand level should strike the same point when the bubble is centered. The Locke hand level is adjusted by means of a screw that moves the cross-wire. The Abney hand level is adjusted by means of a screw that moves one end of the level tube.

The hand level, even if not in adjustment, may be used to establish a horizontal line by employing the principle of the two-peg test (Art. 50).

53. Numerical Problems.

1. What is the combined effect of the earth's curvature and mean atmospheric refraction in a distance of 250 ft.? ½ mile? 10 miles?

2. Two points, A and B, are each 2,000 ft. distant from a third point C. From C, the vertical angle to A is $+3°10'$, and that to B is $+0°18'$. What is the difference in elevation between A and B? Between A and C?

3. Two points, A and B, are 1,200 ft. apart. The elevation of A is 382.19 ft. A level is set up on the line between A and B and at a distance of 300 ft. from A. The rod reading on A is 11.32, and that on B is 3.64. Neglecting curvature and refraction, what is the elevation of B? Considering curvature and refraction, what is the elevation of B?

4. Design a direct vernier reading to $\frac{1}{32}$ in., for an architect's rod graduated in ⅛ in. Draw a neat sketch, full size, of the vernier and a portion of the rod for a reading of $2\frac{19}{32}$ in., with graduations shown and parts labeled.

5. In the two-peg test of a dumpy level the following observations are taken:

	Instrument at A	Instrument at B
Rod reading on A	4.424	2.564
Rod reading on B	6.223	4.239

With the instrument at B, to what rod reading on A should the line of sight be adjusted?

54. Field Problem.

PROBLEM 3. ADJUSTMENT OF THE ENGINEER'S LEVEL

Object.—To make the field adjustments of the engineer's level.

Procedure.—(1) Proceed as outlined in Art. 50 for the dumpy level or Art. 51 for the wye level. (2) Record all field observations in the field book, in columns headed as follows:

Desired relation	Method of test	Amount of error	Screws used in adjustment	Relative amount of adjustment

Include sketches if necessary to make the method clear. Show computations for the two-peg test.

CHAPTER IV

DIFFERENTIAL AND PROFILE LEVELING

55. Definitions.—*Differential leveling* is the operation of determining the elevations of points some distance apart. Usually this is accomplished by direct leveling. Differential leveling requires a series of set-ups of the instrument along the general route and, for each set-up, a rod reading back to a point of known elevation and forward to a point of unknown elevation.

Profile leveling is the operation of determining elevations of points at short measured intervals along a definitely located line.

A *bench mark* (B.M.) is a definite point of more or less permanent character the elevation and location of which are known. Bench marks serve as points of reference for levels in a given locality. Their elevations are established by differential leveling. Permanent bench marks have been established throughout the United States by the U. S. Geological Survey; these consist of bronze plates set in stone or concrete and marked with the elevation above mean sea level. Other agencies have established permanent monuments. For any survey or construction enterprise, local bench marks are established, based on some selected datum and employing natural or artificial objects such as stones, pegs, spikes in trees, and marks painted on street curbs.

A *turning point* (T.P.) is an intervening point between two bench marks, upon which point foresight and backsight readings are taken. The nature of the turning point is generally indicated in the notes, but no record is made of its location. A bench mark may be used as a turning point.

A *backsight* (B.S.) is a rod reading taken on a point of known elevation.

A *foresight* (F.S.) is a rod reading taken on a point whose elevation is to be determined.

The *height of instrument* (H.I.) is the elevation of the line of sight of the telescope when the instrument is leveled.

A *level circuit* is a line of levels that ends at the point of beginning.

The *gradient* or *grade* of a line is its slope, or rate of ascent or descent. The term *grade* is also used to denote an established elevation, as for a point on the surface of an existing or a proposed roadway.

DIFFERENTIAL LEVELING

56. Procedure.—In Fig. 56, B.M.$_1$ represents a point of known elevation (bench mark) and B.M.$_2$ represents a bench mark to be established some distance away. It is desired to determine the elevation of B.M.$_2$. The rod is held at B.M.$_1$, and the level is set up in some convenient location, as L_1, along the general route but not necessarily on the direct line joining B.M.$_1$ and B.M.$_2$. A backsight is taken on B.M.$_1$. The rodman then goes forward and, as directed by the leveler, chooses a turning point T.P.$_1$ at some convenient spot within the range of the telescope along the general route B.M.$_1$ to B.M.$_2$. It is desirable, but not necessary, that each foresight dis-

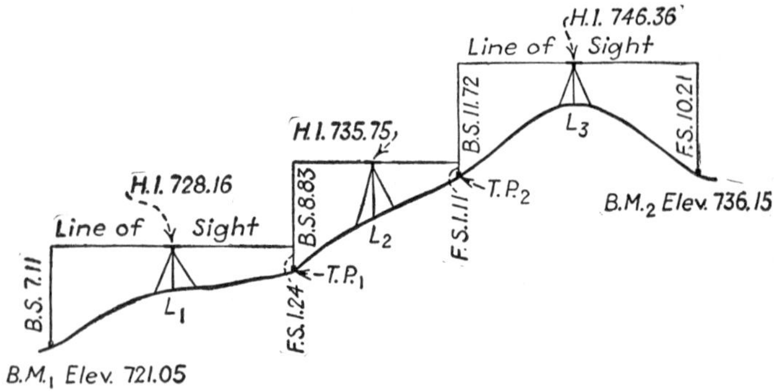

FIG. 56.—Differential leveling.

tance, as L_1–T.P.$_1$, be made approximately equal to its corresponding backsight distance, as B.M.$_1$–L_1 (see Art. 57). The rod is held on the turning point, and a foresight is taken. The leveler then sets up the instrument at some favorable point, as L_2, and takes a backsight to the rod held on the turning point; the rodman goes forward to establish a second turning point T.P.$_2$; and so the process is repeated until finally a foresight is taken on the terminal point B.M.$_2$.

It is seen in Fig. 56 that a *backsight added to the elevation* of a point on which the backsight is taken gives the height of instrument, and that a *foresight subtracted from the height of instrument* determines the elevation of the point on which the foresight is taken. Also, the difference between the backsight taken on a given point and the foresight taken on the following point is equal to the difference in elevation between the two points. It follows that the difference between the sum of all backsights and the sum of all foresights gives the difference in elevation between the bench marks.

57. Balancing Backsight and Foresight Distances.—In Art. 37 it has been shown that if a foresight distance were equal to the corresponding backsight distance, any error in readings due to earth's curvature and to atmospheric refraction (under uniform conditions) would be eliminated.

In leveling, no special attempt is made to balance *each* foresight distance against the preceding backsight distance. Whether or not such distances are approximately balanced between bench marks will depend on the desired precision. The effect of the earth's curvature and atmospheric refraction is slight unless there is an

LEVELS FOR BENCH MARKS ALONG RIDGE ROAD — Dec. 31, 1942, Fair — J.G. Sutter, ⊼; W.R. Knowles, Rod

Sta.	B.S.	H.I.	F.S.	Elev.		Remarks
$B.M._1$	7.11	728.16		721.05		Top of Hydrant Cor. Oak St.
$T.P._1$	8.83	735.75	1.24	726.92		Curb
$T.P._2$	11.72	746.36	1.11	734.64		
$B.M._2$	4.32	740.47	10.21	736.15		Spike in Pole North of Williams House Marked B.M. 736.15
$T.P._3$	3.06	733.57	9.96	730.51		
$T.P._4$	2.74	727.40	8.91	724.66		Stone
$T.P._5$	0.81	716.59	11.62	715.78		
$B.M._3$			12.42	704.17		Concrete Monument No. of Road; County Line
$\Sigma B.S.=$	38.59	$\Sigma F.S.=$	55.47	721.05		
			38.59			
		Diff-	16.88 =	16.88	ck.	

Fig. 58.—Differential level notes.

abnormal difference between the backsight and foresight distances. The effect of instrumental errors is likely to be of considerably greater consequence. The effect of any lack of parallelism between the line of sight and the axis of the level tube is eliminated if, between bench marks, the sum of the foresight distances is made equal to the sum of the backsight distances.

In leveling of ordinary precision, *with an instrument in good adjustment*, the backsight and foresight distances are not measured, and no special attempt is made to equalize them even by estimation. Normally they tend to balance in the long run.

For accurate leveling, it is necessary to equalize backsight and foresight distances between bench marks.

58. Differential-level Notes.—For ordinary differential leveling, usually the record of field work is kept in the form indicated by

Fig. 58. The levels from B.M.$_1$ to B.M.$_2$ are the same as shown by Fig. 56. The heights of instrument and elevations are computed as the work progresses. Usually at the foot of each page of level notes the *computations* are checked by comparing the difference between the sum of the backsights and the sum of the foresights with the difference between the initial and the final elevation, as illustrated at the bottom of Fig. 58. However, this procedure does not check against mistakes in observing or recording.

When backsight and foresight distances are to be balanced, usually they are recorded in the last column of the left-hand page of the field notebook, and the cumulative excess or deficiency of foresight over backsight distances is noted for each turning point and bench mark.

Normally backsights are "plus" sights and foresights are "minus" sights, since the instrument is at a higher elevation than that of the point on which the rod reading is taken. Sometimes, however, in leveling for a building or a tunnel it is necessary to take rod readings on points which are at a higher elevation than the H.I. In such cases, the rod is held inverted, and in the notes each such backsight is indicated with a minus sign and each foresight with a plus sign.

59. Mistakes in Leveling.—Some of the mistakes commonly made in leveling are:

1. Confusion of numbers in reading the rod. To avoid this mistake, the observer should notice the numbers on both sides of the observed reading.
2. Recording backsights in foresight column, and vice versa.
3. Faulty additions and subtractions; adding foresights and subtracting backsights. As a check, the difference between the sum of the backsights and the sum of the foresights should be computed.
4. Rod not held on same point for both foresight and backsight.
5. Not having the Philadelphia rod fully extended when reading the long rod.
6. Wrong reading of vernier on target rod.
7. When the long target rod is used, not having the vernier on the target set exactly the same as the vernier on the back of the rod when the rod is short.

60. Errors in Ordinary Leveling.—In ordinary leveling, errors of consequence may be due to the following causes:

1. *Imperfect Adjustment of the Instrument.*—In so far as results are concerned, the only essential relation is that the line of sight should be parallel to the axis of the level tube. The effect of imperfect adjustment is minimized by balancing backsight and foresight distances.

2. *Parallax.*—This condition produces an accidental error. It can be practically eliminated by careful focusing.

3. *Earth's Curvature and Atmospheric Refraction.*—In ordinary leveling, when backsight and foresight distances roughly balance each other any error due to earth's curvature and atmospheric refraction is accidental in nature and is negligible in amount. If heat waves are apparent, the line of sight should be well above the ground, and the sight distances should be short.

4. *Rod Not Standard Length.*—If the error is distributed along the length of the rod, a systematic error is produced which varies directly as the difference in elevation and bears no relation to the length of the line over which levels are run. The error can be eliminated by comparing the rod with a standard length and applying the necessary corrections. The case is analogous to measurements of distance with a tape that is too long or too short.

If the rod is worn uniformly at the bottom, an erroneous height of instrument is shown at each set-up, but the error in backsight is balanced by that in the following foresight, and no error results in the elevation of the foresight point.

5. *Rod Not Held Plumb.*—This condition produces rod readings that are too large. The error varies directly with the magnitude of rod reading and directly as the square of the inclination. It can be eliminated by waving the rod or by using a rod level.

6. *Faulty Turning Points.*—An accidental error results when turning points are not well defined. If either the level tripod or the turning point settles between readings, as in soft ground, the observed elevations will be too high.

7. *Bubble Not Exactly Centered at Instant of Sighting.*—This produces an accidental error which tends to vary as the distance from instrument to rod. The longer the sight the greater the care that should be observed in leveling the instrument.

8. *Inability of Observer to Read the Rod Exactly or to Set the Target Exactly on the Line of Sight.*—This produces an accidental error, which can be confined within reasonable limits through proper choice of length of sight.

60a. If the proper leveling procedure is observed, the important errors are accidental. Hence the resultant error may be expected to vary as the square root of the number of set-ups or as the square root of the distance. It is customary to express the errors of leveling in terms of the square root of the distance, usually in miles.

61. Precision of Differential Leveling.—The conditions affecting the precision of differential leveling are many and variable, but experience indicates that under average conditions and with a level in good adjustment the *maximum* error may be kept within the limits shown below. Usually the *average* error will be materially less.

1. *Rough leveling*, as practiced on preliminary surveys. Sights up to 1,000 ft. in length. Rod readings to tenths of feet. No particular attention paid to balancing backsight and foresight distances. Maximum error, in feet, $\pm 0.4 \sqrt{\text{distance, in miles}}$.

2. *Ordinary leveling*, as necessary in connection with most engineering works. Sights up to 500 ft. in length. Rod readings to hundredths of feet. Backsight and foresight distances roughly balanced only when running for long distances uphill or downhill. Turning points on solid objects. Maximum error, in feet, $\pm 0.1 \sqrt{\text{distance, in miles}}$.

3. *Accurate leveling* for the principal bench marks on extensive surveys. Sights up to 300 ft. in length. Rod readings to thousandths of feet. Backsight and foresight distances measured by pacing and approximately balanced between bench marks. Rod waved for large readings. Bubble carefully centered before each sight. Turning points on well-defined points of solid objects. Tripod set on firm ground. Maximum error, in feet, $\pm 0.05 \sqrt{\text{distance, in miles}}$.

4. *Precise leveling* for establishing bench marks with great accuracy at widely distributed points. Maximum error, in feet, $\pm 0.02 \sqrt{\text{distance, in miles}}$.

PROFILE LEVELING

62. Profile Leveling.—During the location and construction of highways, railroads, canals, sewers, etc., stakes or other markers are placed at regular intervals along an established line, usually the center line. Ordinarily the interval between stakes is 100 ft., 50 ft., or 25 ft. The 100-ft. points, reckoned from the beginning of the line, are called *full stations*, and all other points are called *plus stations*. Thus a stake set at 1,600 ft. from the point of beginning is numbered "16 + 00," and one set at 1,625 ft. from the point of beginning is numbered "16 + 25." Elevations by means of which the profile may be constructed are obtained by taking level-rod readings on the ground at each stake and at intermediate points where marked changes in slope occur.

Figure 62 illustrates in plan and elevation the steps in leveling for profile In this case stakes are set every 100 ft. The instrument is set up in some convenient location not necessarily on the line (as at L_1), the rod is held on a bench mark (B.M. 28, el. 564.31); a backsight (1.56) is taken, and the height of instrument (565.87) is obtained as in differential leveling. Readings are then taken with the rod held on the ground at successive stations along the line. These rod readings are foresights, frequently designated as *intermediate foresights* to distinguish them from foresights taken on turning points or bench marks. The intermediate foresights (0.7, 2.9 . . . 11.9) subtracted from the H.I. (565.87) give ground elevations of stations. When the rod has been advanced to a point beyond which further readings to ground points cannot be observed, a turning point (T.P.$_1$) is selected, and a foresight (11.63) is taken to establish its elevation. The level is set up in an advanced position (L_2), and a backsight (0.41) is taken on the turning point (T.P.$_1$) just established. Rod readings on ground points are then continued as before. The rodman

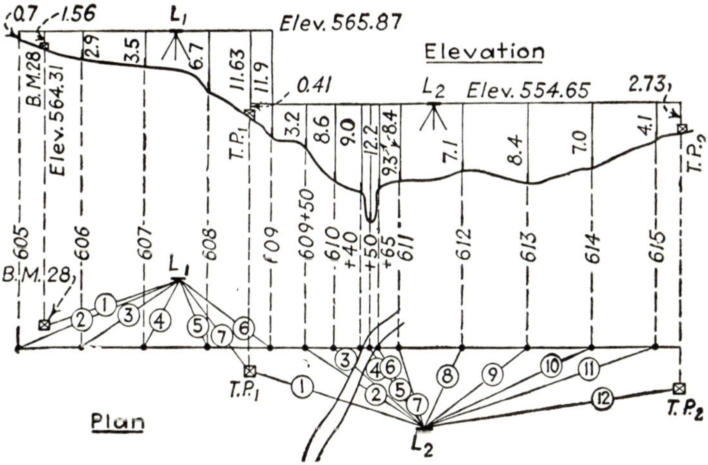

FIG. 62.—Profile leveling.

observes where changes of slope occur (as 609 + 50 · · · 610 + 65), and readings are taken to these intermediate stations. The "plus," or distance from the preceding full station to the intermediate point, is measured by pacing or with a tape or the rod according to the precision required. Usually the backsights and foresights are read to hundredths of feet, and intermediate foresights to tenths of feet only.

63. Profile-level Notes.—The notes for profile leveling may be recorded as shown in Fig. 63. The values shown in the notes are the same as those illustrated in Fig. 62. The notes for turning points are the same as for differential leveling. Elevations of ground points are obtained by subtracting the corresponding intermediate foresights from the preceding height of instrument and are recorded only to the number of decimal places in the intermediate foresights.

64. Vertical Curves.—On highways and railways, adjacent segments of differing gradient are connected by a curve in a vertical plane, called a *vertical curve*. Usually the vertical curve is the arc of a parabola.

The length of the vertical curve cannot be less than the algebraic difference in gradient between the two segments, divided by the maximum allowable change in grade per station (usually established by specifications). Usually it is some convenient whole number of feet in highway work or an even number of stations in railway work.

PROFILE LEVELS FOR						I. N. RY. LOCATION	
	Cox Brook to Big Forks					Buff Dumpy Level	J.C. Brown, ⊼
							F. Graham, Rod
							Sept. 16, 1942
Sta.	B.S.	H.I.	I.F.S.	F.S.	Elev.		Fair
B.M.28	1.56	565.87			564.31	On spruce root 50 ft. lt. Sta. 605.	
605			0.7		565.2		
606			2.9		563.0		
607			3.5		562.4		
608			6.7		559.2		
609			11.9		554.0		
T.P.	0.41	554.65		11.63	554.24	On stone.	
609+50			3.2		551.5		
610			8.6		546.1		
+40			9.0		545.7	Bank Cox Brook.	
+50			12.2		542.5	Ctr. " "	water 1.5 ft. deep
+65			9.3		545.4	Bank " "	
611			8.4		546.3		
612			7.1		547.6		
613			8.4		546.3		
614			7.0		547.7		
615			4.1		550.6		
T.P.	8.02	559.94		2.73	551.92	On plug.	
616			9.7		550.2		
+40			6.3		553.6	Ctr. highway to St. Leonards.	
	9.99		564.31	14.36			
			559.94	9.99			
			4.37 =	4.37	ck.		

FIG. 63.—Profile-level notes.

The station and plus of the vertex, or point of intersection of the two segments, is computed; the elevations of stations along the grade lines are known. The length of vertical curve is then computed or is taken at some convenient value that meets the specification requirements; and the stations and elevations of the beginning and end of curve are calculated. Then the offsets from the uniform gradients to the curve are computed, and the elevations of points along the curve are thus determined. In the field the vertical curve is laid out by setting grade stakes at the computed elevations, just as along a uniform grade.

One method of calculation for a vertical curve is as follows: The elevation of the mid-point of the "long chord" connecting the points of beginning and end of the vertical curve is computed. As the curve is a parabola, the elevation of the mid-point of the vertical curve is the mean of the elevation of the vertex and the elevation of the mid-point of the long chord. The tangent offsets to various points along the curve are then computed, employing the known property of a parabola that the tangent offset varies as the square of the distance from the tangent point. The computations are illustrated by the following example:

Example: On a railroad a +0.8 per cent grade meets a −0.4 per cent grade at sta. 90 + 00 and at el. 100.00 (Fig. 64). The maximum allowable change in grade per station is 0.2. It is desired to establish a vertical curve connecting the two grades.

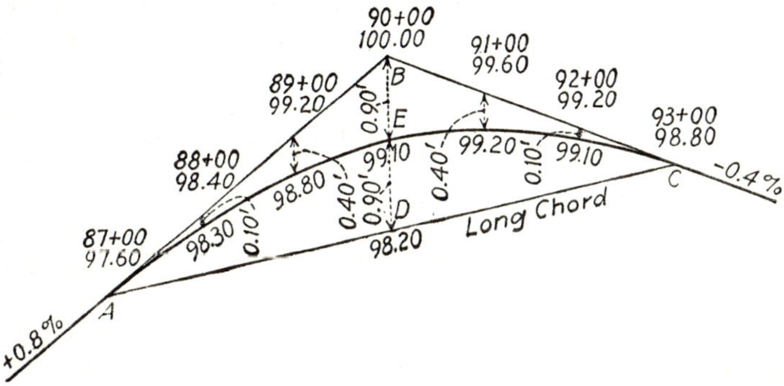

FIG. 64.—Vertical curve.

The algebraic difference in gradient is +0.8 − (−0.4) = 1.2 per cent. The minimum length of curve is then 1.2/0.2 = 6 stations or 600 ft. The length on either side of the vertex (*AB* = *BC*, Fig. 64) is 600/2 = 300 ft. The station at *A* is therefore 90 − 3 = 87, and the station at *C* is 90 + 3 = 93. The elevation of *A* is 100.00 − 3 × 0.80 = 97.60, and the elevation of *C* is 100.00 − 3 × 0.40 = 98.80 ft.

The elevation of the mid-point *D* of the long chord *AC* is the mean of the elevations *A* and *C*:

$$\tfrac{1}{2}(97.60 + 98.80) = 98.20 \text{ ft.}$$

The mid-point *E* of the vertical curve is midway between *D* and the vertex *B*:

$$\tfrac{1}{2}(98.20 + 100.00) = 99.10 \text{ ft.}$$

The offset from vertex to curve is

$$100.00 - 99.10 = 0.90 \text{ ft.}$$

The tangent offsets at stations 89 and 91 are

$$\frac{2^2}{3^2} \times 0.90 = 0.40 \text{ ft.}$$

and the offsets at stations 88 and 92 are

$$\frac{1^2}{3^2} \times 0.90 = 0.10 \text{ ft.}$$

The elevations of points on the curve are shown in Fig. 64.

65. Plotting Profiles.—The profile is plotted from profile-level notes or from elevations taken from a topographic map. Usually it is plotted on regular *profile paper*, which is ruled with vertical lines at intervals of ¼ or ½ in. and with horizontal lines at intervals of $\frac{1}{20}$ or $\frac{1}{10}$ in. Figure 65 illustrates to reduced scale a portion of the profile for a proposed railroad; in the original drawing the vertical scale was 1 in. = 20 ft., and the horizontal scale was 1 in. = 400 ft. Other scales may be used, depending on the purpose of the profile. The vertical scale is exaggerated because the vertical distances on the ground are relatively small as compared with the horizontal distances.

The station numbers increase from left to right. The ground line is drawn freehand, usually as the elevations are plotted. The profile should not be unduly rounded at summits and depressions.

Notes on the profile show the station and plus of important objects, as streams and roads, crossed by the line (Fig. 65). Generally an alinement diagram is drawn near the bottom of the sheet, with points on the diagram directly below corresponding points on the profile. The alinement diagram is not a true plan view. Sometimes general drawings show the profile on the same sheet with the map or plan of the line. The profile is finished in ink.

66. Fixing Grades.—The ground profile furnishes the basis for the study of economic grade location. For road location, the elevation of grade is fixed at certain controlling points as at terminals and at stream and roadway crossings. In addition, maximum permissible rates of grade are fixed by considerations of traffic. Conforming to these limitations, the grade of the proposed road is fitted to the ground until so far as possible the volume of earthwork in cuts will balance that in adjacent fills. For sewers and drains certain minimum permissible rates of grade are fixed by considerations of flow, and these with the profile of the ground determine the grade between controlling points.

Grade lines are established on the plotted profile until by trial a satisfactory solution is obtained. The elevations of points of change are noted on the profile, and the rate of grade is noted just above each segment of the grade line (Fig. 65).

The operation of setting grades in the field is similar to profile leveling. The rod reading to be employed in setting a given stake to grade is computed by subtracting the established grade elevation (taken from the plotted profile) from the H.I.

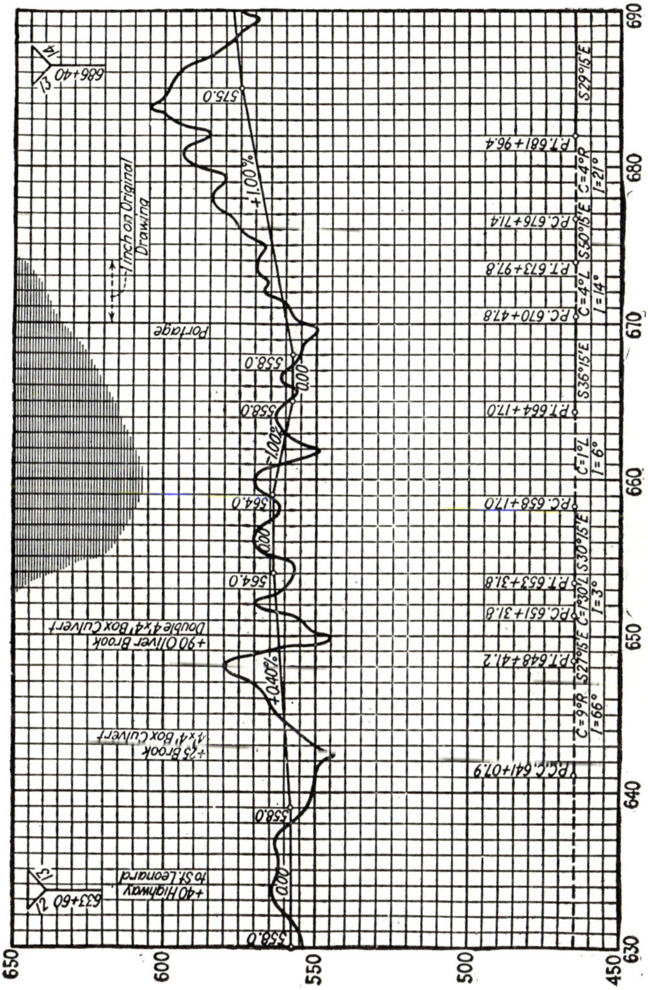

FIG. 65.—Profile and alinement diagram of railroad.

67. Leveling for Earthwork.—Four general situations arise in connection with field measurements to determine volumes of earthwork:

1. *Excavation to Predetermined Surface.*—A given area is to be cut or filled to a predetermined surface, for example, in excavating the basement for a building or in grading a piece of land for an airport. The shape of the existing surface may be obtained by staking out the

area into a system of 100-ft., 50-ft., or 25-ft. squares and then determining the elevations of the corners and of other points where changes in slope occur. Directions of the lines may be obtained with either the tape or the transit, distances may be laid off either with the tape or by stadia, and elevations may be determined with either the engineer's level or the hand level, all depending on the required precision. The elevations are determined as in profile leveling. When the grade of the finished surface has been established, the cut or fill at each station will be known, and the volume of earthwork can be calculated.

2. *Borrow-pit Cross-sections.*—An irregular mass of unknown volume is to be excavated at a given site, as, for example, to furnish material for a highway fill. Sufficient data for the calculation of volume can be obtained by taking cross-sections of the site before and after the material is removed. Usually a base line is staked out near one side, and crosslines are established at regular intervals. Levels are run over the crosslines. When the material has been removed, levels are rerun over the crosslines. The difference between the original cross-section and the final cross-section shows the area cut at each crossline, from which the volume can be determined.

3. *Excavation for Trench.*—A trench is to be excavated, as when a sewer or pipe line is to be laid. Profile levels are run along the proposed line. When the grade of the bottom of the trench has been fixed, the cut at each station can be computed. With the necessary width of trench at top and bottom and also its depth at each station known, the volume of excavation can be calculated.

4. *Road or Canal Cross-sections.*—Earth must be cut or filled to a given grade line along some route, as a highway, railroad, or canal, and, furthermore, must have a prescribed shape of cross-section (see Arts. 197–204).

68. Numerical Problems.

1. A line of differential levels was run between two bench marks 12 miles apart, and the measured difference in elevation was found to be 1,684.7 ft. Later the rod whose nominal length was 13 ft. was found to be 0.004 ft. too short, the error being distributed over its full length. Correct the measured difference in elevation for erroneous length of rod.

2. Suppose that the line of levels of problem 1 were continued to form a circuit closing on the initial bench mark. What error of closure due to erroneous length of rod would be expected?

3. Suppose that the levels of problem 1 had been run by using a rod that was 0.004 ft. too short due to wear on the lower end. What would have been the error?

4. Complete the following differential-level notes. Perform the customary check.

Sta.	B.S.	H.I.	F.S.	El.
$B.M._1$	7.10		....	672.18
$T.P._1$	8.42		6.36	
$T.P._2$	6.87		4.60	
$B.M._2$	12.33		8.73	
$T.P._3$	9.25		2.18	
$T.P._4$	10.82		3.55	
$T.P._5$	7.73		4.29	
$B.M._3$	.....		2.17	

69. Field and Office Problems.

Problem 4. Differential Leveling with Engineer's Level and Self-reading Rod

Object.—Given the elevation of an initial bench mark, to determine the elevations of points in an assigned line of levels.

Procedure.—Follow the procedure outlined in Art. 56. Keep notes in the form of the sample notes shown in Fig. 58, but estimate rod readings to thousandths of feet. Check each rod reading by taking a second observation. Close the circuit, and compute the error of closure.

Hints and Precautions.—(1) Sight in the direction of the rod before centering the bubble exactly. (2) Center the bubble before each reading, and check its position immediately afterward. (3) Test for parallax (see Art. 41). (4) Choose the turning points with an eye to simplicity of field operations, but roughly balance the backsight and foresight distances between bench marks; keep no record of these distances. (5) Keep the foot of the rod free from dirt. (6) Be sure that the rod is held vertical while a sight is being taken. When the "long rod" is used, wave the rod, and before each reading check the position of the back vernier. (7) Describe each bench mark in the field notes. (8) Check the computations.

Problem 5. Profile Leveling for a Road

Object.—To determine the elevations necessary for plotting the profile of a line. It is assumed that the center line has already been laid out with numbered stakes every 100 ft. (see field problem 13, p. 85; also field problem 29, p. 178, and office problem 6, p. 54).

Procedure.—Select a bench mark, with datum below every point of the proposed route. Adapt the procedure indicated in Arts. 62 and 63 to the field conditions encountered. Keep the notes in the form of the sample notes (Fig. 63).

Hints and Precautions.—(1) Read the rod carefully to the nearest 0.01 ft. on bench marks and turning points, and quickly to the nearest 0.1 ft. on ground points. (2) Take rod readings on the ground at all full stations and at such other points on line (plus stations) as are necessary to obtain a sufficiently accurate profile. In general these plus stations will be at points where the slope of the ground changes noticeably and at highways, railroads, and streams. (3) If the line is long, establish bench marks every 1,500 or 2,000 ft., and mark each bench mark. (4) Make the computations as the work progresses, and check each page of notes.

Problem 6. Plotting Profile

Object.—To plot a profile from level notes (field problem 5) and to fix the grade line for a road or pipe line.

Procedure.—(1) Choose a horizontal and vertical scale in keeping with the purpose of the profile. (2) Examine the field notes to determine the range between points of maximum and minimum elevation. Number each of the heaviest horizontal lines of the profile paper with its elevation. Number each of the heaviest vertical lines with its station number. (3) From the profile-level notes plot the profile; through the plotted points draw a freehand curve (see Fig. 65). Show the names of streams and roads crossed. Check the profile, and ink it in black. (4) Fix the grade line, and ink it in red. Show elevations of points of change in grade and rates of grade. (5) Near the bottom of the sheet indicate the horizontal alinement, using a scheme similar to that shown in the figure. (6) Make a title showing the name and location of the work, the horizontal and vertical scale, the date, and the name of the draftsman.

Hints and Precautions.—(1) Avoid the common mistake of reading elevations of turning points and bench marks as ground elevations, by enclosing the elevations of turning points and bench marks in the field notebook with a circle. (2) In checking the profile, read stations and elevations back from the profile. (3) Ink the profile freehand preferably with a ruling pen; do not round off the summits and depressions unduly.

CHAPTER V

MEASUREMENT OF ANGLES AND DIRECTIONS

70. General.—This chapter treats only of angles and directions in the horizontal plane.

The location of a point is fixed if measurements are made of (1) its direction and distance from a known point, (2) its direction from two known points, (3) its distance from two known points, or (4) its direction from one known point and its distance from another.

The direction of any line (as fixed by two points) is defined by the horizontal angle that it makes either with an adjacent line of the survey or with some real or imaginary reference line of fixed direction called a *meridian*. If the meridian is arbitrarily chosen it is called an *assumed meridian*. If it is a true north-and-south line passing through the geographical poles of the earth, it is called a *true meridian*. If it lies parallel with the magnetic lines of force of the earth it is called a *magnetic meridian*.

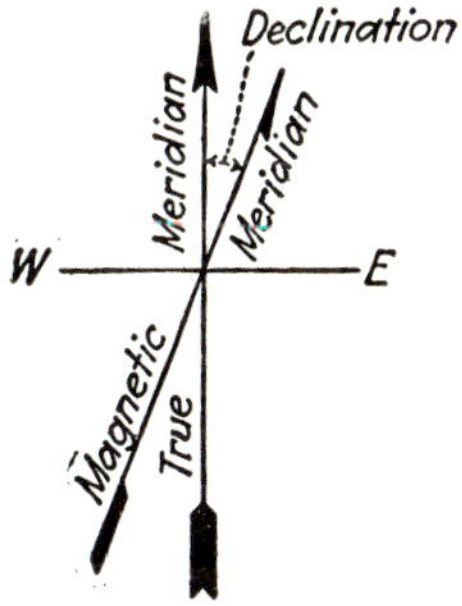

FIG. 72*a*.—Declination east.

Methods of determining the meridian are described in Art. 73 and Chap. XVII.

71. Magnetic Meridian.—The direction of the magnetic meridian is that taken by a freely suspended magnetic needle. The magnetic poles are at some distance from the true poles; hence in general the magnetic meridian is not parallel to the true meridian. The location of the magnetic poles is constantly changing; hence the direction of the magnetic meridian is not constant.

72. Magnetic Declination.—The angle between the true meridian and the magnetic meridian is called the *magnetic declination*. If the north end of the compass needle points to the east of the true meridian, the declination is said to be east (Fig. 72*a*); if it points to the west of the true meridian, the declination is said to be west.

If a true north-and-south line is established, the mean declination of the needle for a given locality can be determined by compass observations extending over a period of time. The declination may be estimated with sufficient accuracy for most purposes from an isogonic chart of the United States, published by the United States Coast and Geodetic Survey (Fig. 72*b*). The chart shows *isogonic lines*, or lines of equal magnetic declination for the date of issue. Also it indicates the rate of change in magnetic declination from year to year.

73. Establishing the Meridian.—The true meridian is established by astronomical observations, as described in Chap. XVII. For surveys of ordinary precision the transit is employed.

On compass surveys, in order to determine the magnetic declination, sometimes the true meridian is established by ranging two plumb lines with Polaris (the north star), usually when the star is at elongation (farthest east or farthest west). The time of elongation may be taken from published astronomical tables, such as Table IV. The true meridian is established by laying off from the line established at elongation an angle equal to the horizontal angle between the star and the meridian, as given in published tables such as Table V. If care is exercised in setting the ground points and if the plumb lines are at least 15 ft. apart, the error need not exceed 05′.

A magnetic meridian can be established by setting up the compass over any convenient point and then sighting to set a point on a stake or other object that marks another point on the meridian.

At many of the triangulation stations established throughout the United States by the United States Coast and Geodetic Survey, reference lines of known true direction have been established for use by local surveyors.

74. Angles and Directions.—Angles and directions may be defined by means of *bearings*, *azimuths*, *deflection angles*, or *interior angles*, as described in the following paragraphs. These quantities are said to be *observed* when obtained directly in the field and *calculated* when obtained indirectly by computation. Conversion from one means of expressing angles and directions to another is a simple matter if a sketch is drawn to show the existing relations.

74a. Bearings.—The direction of any line with respect to a given meridian may be defined by the *bearing*. Bearings are called *true bearings*, *magnetic bearings*, or *assumed bearings* according as the meridian is true, magnetic, or assumed. The bearing of a line is indicated by the quadrant in which the line falls and the acute angle that the line makes with the meridian in that quadrant. Thus in Fig. 74*a* the bearing of the line OA is read North 37° East and is written N37°E. The bearings of OB, OC, and OD are respectively S62°E, S50°W, and N20°W.

74b. Azimuths.—The *azimuth* of a line is its direction as given by the angle between the meridian and the line, measured in a clockwise direction usually from the south point of the meridian. On any given survey the direction of zero azimuth is either always south or always north. Azimuths are called *true azimuths*, *magnetic azimuths*, or *assumed azimuths* according as the meridian is true, magnetic, or assumed. Azimuths may have values between 0° and 360°. In Fig. 74*b*, the directions of the lines are the same as in Fig. 74*a*, but the azimuths from north are indicated. Az. $OA = 37°$, Az. $OB = 118°$, Az. $OC = 230°$, Az. $OD = 340°$.

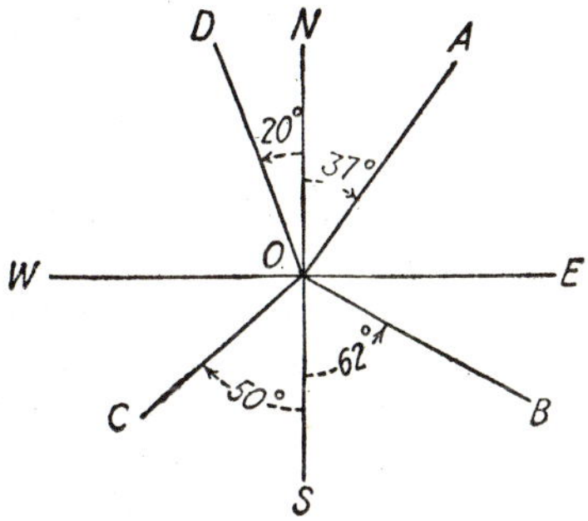

FIG. 74*a*.—Bearings.

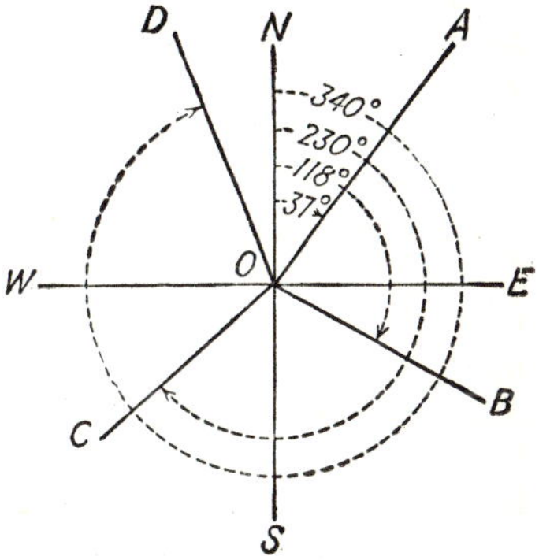

FIG. 74*b*.—Azimuths from north.

74c. Deflection Angles.—The angle between a line and the prolongation of the preceding line is called a *deflection angle*. Deflection angles are recorded as *right* or *left* according as the line to which measurement is taken lies to the right (clockwise) or left (counter-clockwise) of the prolongation of the preceding line. Thus in Fig. 74*c* the deflection angle at *B* is 22°R, and at *C* is 33°L. Deflection angles may have values between 0° and 180°, but usually they are not employed for angles greater than 90°. In any closed polygon the

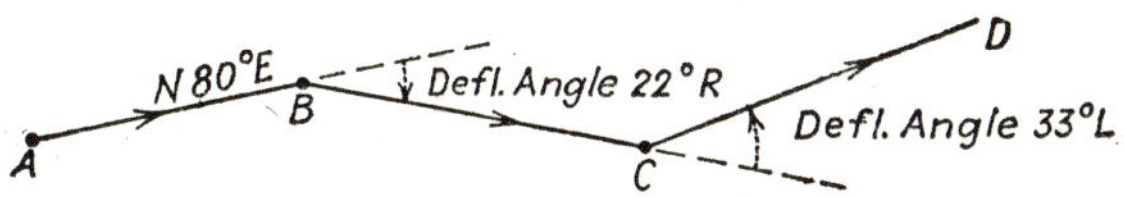

FIG. 74*c*.—Deflection angles.

algebraic sum of the deflection angles (considering right deflections as of sign opposite to left deflections) is 360°.

74d. Interior Angles.—In a closed polygon the angles inside the figure between adjacent lines are called *interior angles*. If n is the number of sides in a closed polygon, the sum of the interior angles is $(n - 2)\ 180°$.

75. Traverses.—The succession of straight lines connecting a succession of established points along the route of a survey is called a *traverse* or *traverse line.* Distances along the line between successive points are determined either by direct measurement or by stadia. At each point where the traverse changes direction an angular measurement is taken. If the traverse forms a closed figure, as. for example, the boundary of a parcel of land, it is called a *closed traverse;* if it does not form a closed figure, as, for example, the line for a highway, it is called an *open traverse* or *continuous traverse.* Traverses are also designated according to the purpose of the survey (as preliminary), the field instrument or method employed (as transit-tape), or the kind of angular measurements observed (as azimuth).

For details of the method of traversing, see Chap. VII. Traversing is the method of surveying in most common use where favorable routes are available.

76. Triangulation.—Where the lines of a survey form triangular figures whose angles are measured and whose distances are determined by trigonometric computations, the operation of making the necessary field observations is called *triangulation.* The simplest case is that of a single triangle, one of whose sides is of known length; if any two angles of the triangle are measured, sufficient data are obtained for computing the length of the other two sides. Furthermore, if the third angle is measured, the angular measurements may be checked.

A *triangulation system* is made up of a series of triangles so connected that, having measured the angles of the triangles and the length of one line, the length of other lines may be computed. The line of known length, upon which all computed distances are based, is called a *base line.*

The advantage of triangulation over traversing lies in the small number of linear measurements that are necessary; the disadvantage lies in the greater amount of computing required. Triangulation is superior to traversing where the terrain offers many obstacles (such as hills, vegetation, or marsh) to traverse work.

77. Instruments for Measuring Angles and Directions; the Compass.—Angles and directions may be determined by means of the *transit* (Art. 85), the *plane table* (Chap. XIII), the *sextant,* and the magnetic *compass.* The compass is useful alone in making rough surveys and is useful on the transit as a means of approximately checking horizontal angles measured by more precise methods.

The essential features of the compass used by the surveyor are (1) a compass box with circle graduated from 0° to 90° in both directions from the N and S points and usually having the E and W points interchanged as shown in Fig. 77, (2) a line of sight in the direction of the SN points of the compass box, and (3) a magnetic needle.

When the line of sight is pointed in a given direction, the compass needle (when pivoted and brought to rest) gives the magnetic bearing. Thus in the figure the bearing of AB is N60°E.

The varieties of compasses are (1) various *pocket compasses* which are generally held in the hand; (2) the *surveyor's compass* (Art. 78) which is mounted usually on a light tripod or sometimes on a Jacob's staff (a pointed stick about 5 ft. long); and (3) the *transit compass* (Art. 82), a compass box similar to that of the surveyor's compass, mounted on the upper vernier plate of the engineer's transit.

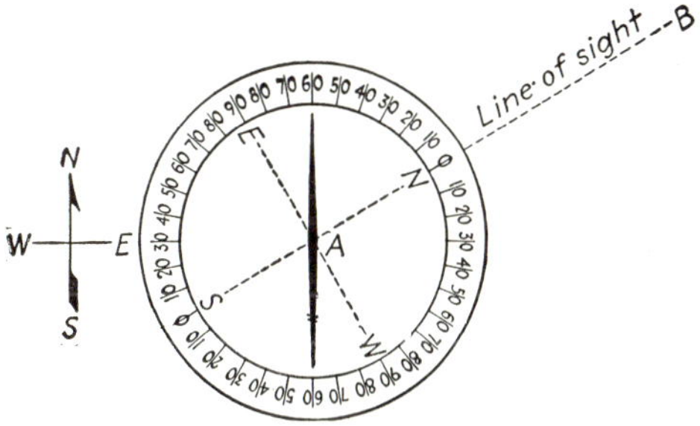

FIG. 77.

78. Surveyor's Compass.—This consists of a compass box to opposite sides of which are fastened vertical sight vanes (Fig. 78). The box is rigidly connected to a vertical spindle which is free to revolve in a conical socket. Below the spindle is a leveling head consisting of a ball-and-socket joint, by means of which the compass may be leveled; the upper portion of the socket is a thumb nut which is tightened until the ball is held securely by friction. The leveling head may be screwed on to a wooden tripod or a Jacob's staff. Within the compass box is a circular, or "bull's eye," level. The compass is provided with a screw for lifting and clamping the needle and with a screw for clamping the vertical spindle. To counteract the effect of magnetic dip so that the needle will be horizontal while in use, an adjustable counterweight of fine brass wire is attached to the south end of the needle. The compass circle is graduated usually in half degrees, and bearings may be read by estimation to 05′ or 10′.

In order that *true* bearings may be read directly, some compasses, as the one shown in the illustration, are so designed that the compass circle may be rotated with respect to the box in which it is mounted. If the magnetic declination is set off by means of the circle, the observed bearings will be true.

When the direction of a line is to be determined, the compass is set up on line and is leveled. The needle is released, and the compass is rotated about its vertical axis until a range pole or other object on line is viewed through the slits in the two sight vanes. When the needle comes to rest, the bearing is read. Ordinarily the sight vane at the end of the compass box marked "S" is held next to the eye; in this case the bearing is given by the north end of the needle.

Surveying with the compass is usually by traversing. Only alternate stations need be occupied, but a check is secured and local attraction is detected if both a backsight and a foresight are taken from each station. Unlike a transit traverse, in which an error

FIG. 78.—Surveyor's compass.

in any angle affects the observed or calculated directions of all following lines, an error in the observed bearing of one line in a compass traverse has no effect upon the observed *directions* of any of the other lines. One form of field notes for a closed compass traverse has column headings as follows:

Sta.	Object (adjacent stations)	Distance	Observed bearing	Interior angle (computed)	Corrected interior angle	Corrected bearing

79. Local Attraction.—Objects of iron or steel, some kinds of iron ore, and electric wires alter the direction of the lines of magnetic force in their vicinity and hence are likely to cause the compass needle to deviate from the magnetic meridian. In certain localities, particularly in cities, its effect is so pronounced that the compass is practically useless.

To eliminate the effect of local attraction, use is made of the fact that the computed *angle* between two lines at any station is the correct angle even if local attraction from a fixed source exists, since the observed bearing of one line is affected by the same amount as that of the other line.

In a traverse—disregarding for the time being the accidental errors due to observing—it is probable that the terminal points of any line, as *AB*, are free from local attraction if the back bearing from *B* is the reverse of the forward bearing from *A*. Hence the bearing of such an unaffected line may be assumed correct. With the bearing of this line as a basis, the correct bearings of successive lines can be computed by means of the traverse angles.

Owing to errors of observation there are likely to be discrepancies between the observed forward and back bearings of lines, even though no local attraction exists. If the discrepancies are small and apparently not of a systematic character, it is reasonable to assume that the errors are due to causes other than local attraction.

80. Numerical Problems.

1. The magnetic bearing of a line is S42°40′W, and the magnetic declination is 6°10′E. What is the true bearing of the line?

2. In an old survey made when the declination was 1°40′W, the magnetic bearing of a given line was N16°15′W. The declination in the same locality is now 2°30′E. What are the true bearing and the present magnetic bearing of the line?

3. Following are the observed magnetic bearings of a compass traverse: *AB*, N28°15′E; *BC*, N68°10′E; *CD*, N45°30′W; *DE*, S79°45′W. Compute the deflection angles.

4. Following are the deflection angles of traverse *A* to *E*: *B*, 25°12′L; *C*, 16°46′L; *D*, 72°10′R. The true bearing of *AB* is N57°43′E. Compute the bearings of the remaining lines.

5. For the traverse of problem 3, the declination is 3°30′E. Compute the true azimuths reckoned from the north point.

6. For the traverse of problem 4, compute the true azimuths reckoned from the south point.

7. Following are the deflection angles of a closed traverse: *A*, 38°16′R; *B*, 59°42′R; *C*, 73°30′R; *D*, 84°57′R; *E*, 38°50′L; *F*, 76°21′R; *G*, 66°11′R. Compute the error of closure. Adjust the angular values on the assumption that the error is the same for each angle.

8. The following are bearings taken on a closed compass traverse. Compute the interior angles and correct them for observational errors.

Assuming the observed bearing of the line *AB* to be correct, adjust the bearings of the remaining sides.

Line	Forward bearing	Back bearing
AB	S37°30′E	N37°30′W
BC	S43°15′W	N44°15′E
CD	N72°45′W	S72°15′E
DA	N12°45′E	S13°15′W

9. In triangulating across a river a base line *AB* of the triangle *ABC* has a measured length of 620.43 ft., and the angles at *A* and *B* are respectively 61°22′ and 78°35′. Compute the distance *AC*.

81. Field Problem.

PROBLEM 7. SURVEY OF FIELD WITH SURVEYOR'S COMPASS AND TAPE

Object.—To find the true bearing and length of each side of an assigned field, using the surveyor's compass and 66-ft. chain tape. The data may be used in office problem 19 or 20 (p. 118).

Procedure.—(1) On the compass, set off the magnetic declination in order that true bearings may be read directly; if this cannot be done, observe the magnetic bearings and convert them to true bearings later. (2) Set up at one corner *A* of the field, and observe the forward bearing of the line *AB*. (3) Take a back bearing from *A*. (4) Chain the line *AB*, and record the distance in chains to the nearest link. (5) Set up the compass at *B*, and proceed in the same manner as at *A*. (6) Continue around the field, taking both back and forward bearings from each point and chaining the lines. (7) Compute the interior angles of the field and adjust them to make the sum equal $(n - 2)\ 180°$. (8) Assume the bearing of one line to be correct, and compute the correct bearings of the other lines. (9) If magnetic bearings have been observed, apply the magnetic declination to convert them to true bearings.

CHAPTER VI

THE ENGINEER'S TRANSIT

82. Description.—The transit may be employed for measuring and laying off horizontal angles, directions, vertical angles, differences in elevation, and distances.

Figures 82*a* and 82*b* show the type of transit in most common use. It is seen to consist of an *upper*, or *vernier*, *plate* to which are attached A-shaped standards supporting the telescope, and a *lower plate* to which is fixed a horizontal graduated circle. The upper and lower plates are fastened respectively to vertical inner and outer spindles, the two axes of rotation being coincident with and at the geometric center of the graduated circle. The outer spindle is seated in the tapered socket of the leveling head. Near the bottom of the leveling head is a ball-and-socket joint which secures the instrument to the foot plate yet permits rotation of the instrument about the joint as a center.

The outer spindle carrying the lower plate may be clamped in any position by means of the lower *clamp-screw*. Similarly, the inner spindle carrying the upper plate may be clamped to the outer spindle by means of the upper clamp-screw. After either clamp has been tightened, small movements of the spindle may be made by turning the corresponding *tangent-screw*. The axis about which the spindles revolve is called the *vertical axis* of the instrument.

Level tubes called *plate levels* are mounted at right angles to each other on the upper plate. Four *leveling screws*, or foot screws, are threaded into the leveling head and bear against the foot plate; when the screws are turned, the instrument is tilted about the ball-and-socket joint. When all four screws are loosened, pressure between the sliding plate and the foot plate is relieved, and the transit may then be shifted laterally with respect to the foot plate. From the end of the spindle is suspended a hook for the plumb line. The instrument is mounted on a tripod by screwing the foot plate on to the tripod head.

The *telescope* is fixed to a transverse *horizontal axis* which rests in bearings on the standards. The telescope may be rotated about this horizontal axis and may be fixed in any position in a vertical plane by means of the telescope clamp-screw; small movements may then be secured by turning the telescope tangent-screw. Fixed to the horizontal axis is the *vertical circle*, and attached to one of the standards is the vertical vernier. Some transits have a vertical vernier that may be rotated about the horizontal axis and is equipped with a control level tube. Beneath the telescope is the *telescope level tube.*

Attached to the upper plate is the *compass box*, the details of which are the same as for the surveyor's compass described in Art. 78.

FIG. 82*a*.—Engineer's transit.

82a. Telescope.—The telescope of the transit is similar to that of the engineer's level (Art. 41). When the transit is used as an instrument for direct or trigonometric leveling, any point on the horizontal cross-hair may be used in sighting; when the transit is used for establishing lines, measuring angles, or taking bearings, any point on the vertical cross-hair may be used. Most telescopes are equipped with two horizontal stadia hairs, one above and the other an equal distance below the horizontal cross-hair.

82b. Graduated Circles.—The vertical circle has two opposite zero points and is graduated usually in half degrees, the numbers increasing to 90° in both directions from the zero points as illustrated in Fig. 82*c*. When the telescope is level, the index of the vernier is at 0°.

The horizontal circle is graduated usually in half degrees but sometimes in third degrees or quarter degrees. It may be numbered

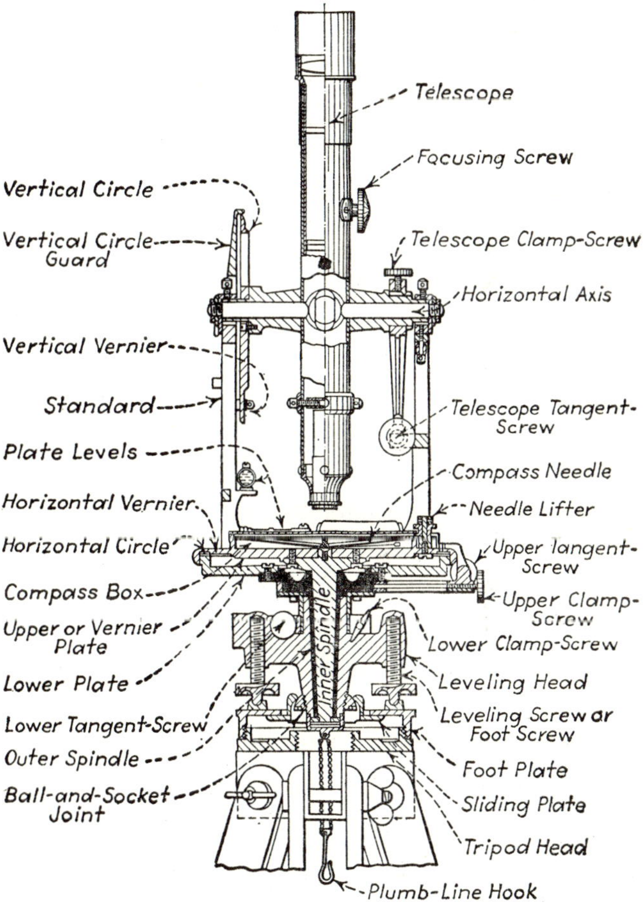

FIG. 82*b*.—Section of transit.

from 0° to 360° clockwise (Fig. 82*d*), 0° to 360° clockwise and 0° to 90° in quadrants, or 0° to 360° in both directions.

On the ordinary transit, both vertical and horizontal angles can be read to 01′ by means of the verniers.

82c. Verniers.—The verniers employed for reading the horizontal and vertical circles are identical in principle with those for the target rod described in Art. 47. Figure 82*e* shows the usual type of double direct vernier. The vernier on the left of the index is for reading clockwise angles, and that on the right for reading counter-clockwise angles. The circle is graduated in half degrees, or 30′. Each space on the vernier is 01′ less than a space on the circle, and 30 spaces on the vernier are equal to 29 spaces on the circle; the least count is 30′/30 = 01′. Considering counter-clockwise angles, the reading of the right vernier is seen to be 17°25′. Considering clockwise angles, the reading of the left vernier is 162°35′.

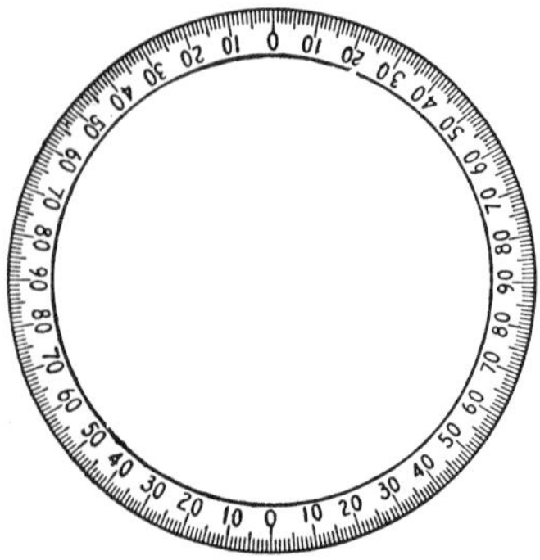

FIG. 82*c*.—Vertical circle numbered in quadrants.

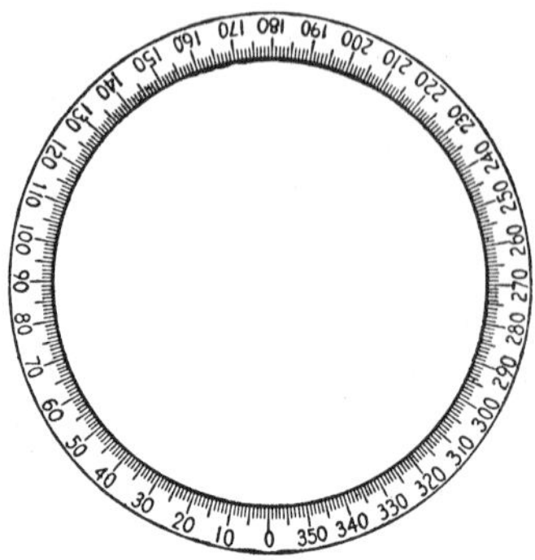

FIG. 82*d*.—Horizontal circle numbered 0 to 360.

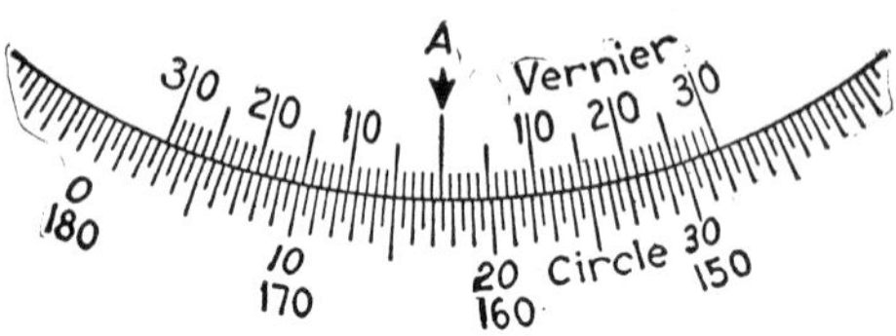

FIG. 82*e*.—Double direct vernier reading to minutes.

All transits have two verniers, called the *A* and *B* verniers, for reading the horizontal circle; their indices are 180° apart. If the two verniers fail to register readings exactly 180° apart on the circle, the error can be eliminated by taking the mean of the two vernier readings.

83. Use of the Transit.—The succeeding articles describe the elementary processes employed in running lines and in measuring horizontal and vertical angles with the transit. Transit surveys are considered in detail in the following chapter. Suggestions for the care and handling of the transit are given in Art. 5.

The process of taking magnetic bearings with the transit is the same as with the surveyor's compass. The transit may be employed for running direct levels in the same manner as with the engineer's level, the telescope level bubble being centered each time a rod reading is taken.

The operation of reversing the telescope by rotating it about the horizontal axis is called "plunging the telescope." When the telescope level tube is below, the telescope is said to be in the *normal* or *direct* position; when the level tube is above, the telescope is said to be in the *inverted* or *reversed* position.

84. Setting Up the Transit.—The transit is placed approximately over a definite point, such as a tack in a stake. Each tripod leg is then moved as required to bring the plumb bob within ¼ in. of the tack, with the foot plate nearly level and with the shoe of each tripod leg pressed firmly into the ground. The instrument is leveled approximately by means of the leveling screws. Two adjacent leveling screws are loosened, and the instrument is shifted laterally until the plumb bob is over the tack. The leveling screws are tightened to a firm, but not tight, bearing. The instrument is leveled by means of the leveling screws and the plate levels, each level tube being first brought parallel with a pair of opposite leveling screws. Both bubbles are brought approximately to center, and then each bubble is centered carefully. The telescope is tested for parallax (Art. 41) before observations are begun.

85. Measuring a Horizontal Angle.—If a horizontal angle, as *AOB*, is to be measured, the transit is set up over *O*. The upper motion is clamped, and by means of the upper tangent-screw one of the horizontal verniers is set at 0°. The telescope is sighted approximately to *A*; the lower motion is clamped; and by turning the lower tangent-screw the line of sight is set exactly on a range pole or other object marking the point. The upper clamp is loosened, and the telescope is turned until the line of sight cuts *B*. The upper clamp is tightened, and the line of sight is set exactly on *B* by turning the upper tangent-screw. The reading of the vernier which was initially set at 0° gives the value of the angle.

A horizontal angle is laid off in similar manner, except that after the sight is taken on *A* the angle is set off on the circle, and the point *B* is established on the line of sight.

Following is a list of suggestions:

1. Make reasonably close settings by hand so that the tangent-screws will not need to be turned through more than one or two revolutions.

2. Make the last movement of the tangent-screw clockwise, thus compressing the opposing spring. (See parts 20 to 26, Fig. 43.)

3. When reading the vernier, have the eye directly over the coinciding graduation, to avoid parallax. It is also helpful to observe that the

graduations on both sides of those coinciding fail to concur by the same amount.

4. As a check on the reading of one vernier, the other vernier may be read also.

5. The plate bubbles should be centered before measuring an angle, but between initial and final settings of the line of sight the leveling screws should not be disturbed. When an angle is being measured by repetition (Art. 87), the plate may be releveled after each turning of the angle before again sighting on the initial point.

6. When a number of angles are to be observed from one point without moving the horizontal circle, the instrumentman should sight at a clearly defined object that will serve as a reference mark and should observe the angle. If occasionally the angle to the reference mark is read again, any accidental movement of the horizontal circle will be detected.

86. Common Mistakes.—In measuring horizontal angles, mistakes often made are:

1. Turning wrong tangent-screw.
2. Failing to tighten clamp.
3. Reading numbers on the horizontal scale from the wrong row.
4. Reading angles in the wrong direction.
5. Dropping 30′, with a circle graduated to 30′; for example, calling an angle 21°14′ when it is actually 21°44′.
6. Reading the vernier in the wrong direction.
7. Reading the wrong vernier.

87. Measuring an Angle by Repetition.—By means of the transit, a horizontal angle may be mechanically multiplied, and the product can be read with the same precision as the single value. The precision increases directly with the number of repetitions up to six or eight; beyond this number the precision is not appreciably increased by further repetition, on account of accidental errors.

To repeat an angle, as *AOB*, the transit is set up at *O*, and the single value of the angle is observed as previously described. The vernier setting is left unaltered; the instrument is turned on its lower motion; and a second sight is taken to the first point, as *A*. The upper clamp is loosened, and the telescope is again sighted to *B*. In this way the process is continued until the angle has been multiplied the required number of times. The vernier is read, and the value of the angle is determined by dividing the difference between initial and final readings by the number of times the angle was turned.

Usually the angle is multiplied four to eight times, half the observations being made with the telescope normal and half with it inverted. Both verniers are read, and the mean values are used in the computations.

88. Measuring a Vertical Angle.—The vertical angle to a point is its angle of elevation (+) or depression (−) from the horizontal. The transit is set up and leveled as when measuring horizontal angles.

For a transit having a fixed vertical vernier, the plate bubbles should be centered carefully. The telescope is sighted approximately at the point, and the horizontal axis is clamped. The horizontal cross-hair is set exactly on the point by turning the telescope tangent-screw, and the angle is read by means of the vertical vernier.

For a transit having a movable vertical vernier with control level, the telescope is sighted on the point as described above, the vernier control bubble is centered, and the angle is read.

For leveling with the transit, for astronomical observations, or for measurement of horizontal angles requiring steeply inclined sights, usually it is desired to level the transit with greater accuracy than that which is possible through the use of the plate levels. In such cases the vertical axis is made truly vertical by means of the telescope level, the plate levels being disregarded.

If the axis of the telescope level is not parallel to the line of sight and if the vertical vernier reads zero when the bubble is centered, an error in vertical angle results. This error can be eliminated by double-sighting (Art. 88*b*), or it can be rendered negligible for ordinary work by careful adjustment of the instrument.

88a. Index Error.—Index error is the error in an observed vertical angle due either to displacement (lack of adjustment) of the vertical vernier or, for a transit having a fixed vertical vernier, to inclination of the vertical axis.

Assuming that the axis of the telescope level is parallel to the line of sight, the true vertical angle is determined by applying to the observed angle an *index correction* equal in amount but opposite in sign to the index error. Thus, if the observed angle is $+12°14'$ and the index error is determined to be $+02'$, the true value of the vertical angle is $+12°14' - 02' = +12°12'$.

Transit Having a Fixed Vertical Vernier.—Displacement of the vertical vernier introduces a constant index error which can be determined by leveling the transit carefully, leveling the telescope, and reading the vertical vernier. This error can be eliminated by adjustment or by double-sighting.

For this type of transit, any inclination of the vertical axis due to erroneous leveling of the instrument introduces an index error that varies with the direction in which the telescope is pointed. This index error can be rendered negligible by careful leveling of the instrument before each observation. It is not eliminated by double-sighting.

Transit Having a Movable Vertical Vernier with Control Level.—Displacement of the vertical vernier introduces a constant index error which can be determined by leveling both the telescope level tube and the vernier level tube and reading the vertical vernier. The error can be eliminated by double-sighting. Any moderate inclination of the vertical axis does not introduce an appreciable error in vertical angles, provided the instrument is in adjustment and provided the vernier control bubble is centered each time an observation is made.

88b. Double-sighting to Determine Correct Vertical Angle.—*If the instrument has been leveled carefully*, the correct vertical angle can be determined with a transit having a full vertical circle by reading once with the telescope normal and once with the telescope inverted, taking the mean of the two values thus obtained.

In traversing, a similar result is obtained by measuring the vertical angle of each traverse course from each end, with the telescope the *same side up*, and taking the mean of the two values thus obtained.

Double-sighting eliminates error in vertical angle due either to lack of parallelism between the line of sight and the axis of the level tube or to displacement of the vertical vernier; but in the case of a transit having a fixed vertical vernier, double-sighting does not eliminate the error due to lack of verticality of the vertical axis.

89. Prolonging a Straight Line.—A straight line, as AB (Fig. 89), may be prolonged as follows: The transit is set up at B, and a backsight is taken to A, with telescope normal. With both upper and lower motions clamped, the telescope is plunged, and a point C is set on line. The transit is moved to C; a backsight is taken to B; a point D is set in similar manner; and so on.

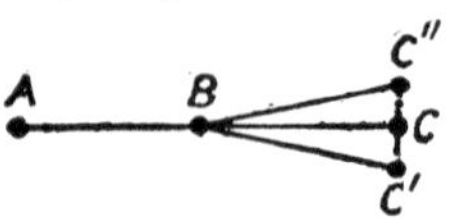

FIG. 89.—Prolonging a straight line.

If the line of sight is not perpendicular to the horizontal axis of the transit, C will not lie on the true prolongation of AB but to one side, as at C'. If the instrument is in poor adjustment or if it is desired to establish the line with high precision, the transit is revolved about its vertical axis, and a second backsight is taken to A with the telescope *inverted*. The telescope is plunged, and a point C'' is established on line beside C'. Midway between C' and C'' a point C is set defining a point on the correct prolongation of AB. This method is known as "double-sighting."

90. Prolonging a Line Past an Obstacle.—Figure 90 illustrates one method of prolonging a line AB past an obstacle where the offset space is limited. The transit is set up at A; a right angle is turned; and a point C is established at a convenient distance from A. Similarly the point D is established, the distance BD being made equal to

AC. The line *CD*, which is parallel to *AB*, is prolonged, and points *E* and *F* are established in convenient locations beyond the obstacle. From *E* and *F* right-angle offsets are made, and *G* and *H* are set as were *C* and *D*; *GH* then defines the prolongation of *AB*. If the line is to be prolonged accurately, *AB* and *EF* should be as long as practicable.

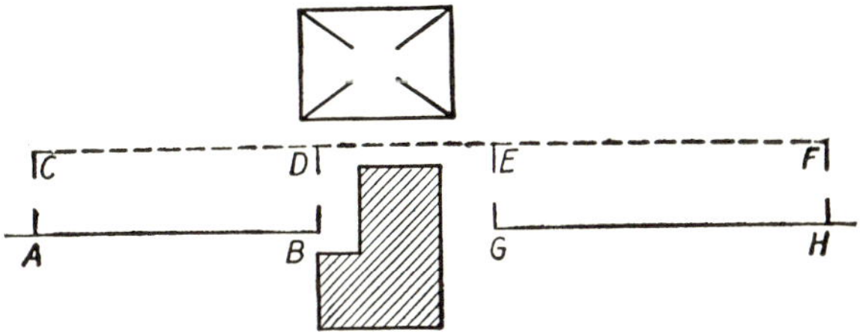

Fig. 90.—Prolonging a line past an obstacle.

91. Adjustment of the Transit.—Much of Arts. 50 and 51 concerning the adjustment of the level applies equally well to the transit.

The transit adjustments commonly made are described herein; it is assumed (1) that the objective slide is permanently fixed in the telescope tube with respect to transverse motion and (2) that the optical axis and the axis of the objective slide coincide and are perpendicular to the horizontal axis. Some general suggestions regarding adjustments are given in Art. 6.

1. *To Make the Vertical Cross-hair Lie in a Plane Perpendicular to the Horizontal Axis.*—Sight the vertical cross-hair on a well-defined point not less than 200 ft. away, and swing the telescope through a small vertical angle. If the point appears to depart from the vertical cross-hair, loosen two adjacent screws, and rotate the cross-hair ring in the telescope tube until the point traverses the entire length of the hair. Tighten the same two screws.

2. *To Make the Axis of Each Plate Level Lie in a Plane Perpendicular to the Vertical Axis.*—Rotate the instrument about the vertical axis until each level tube is parallel to a pair of opposite leveling screws. Center the bubbles by means of the leveling screws. Rotate the instrument end for end about the vertical axis. If the bubbles become displaced, bring them *halfway* back by means of the adjusting screws.

3. *To Make the Line of Sight Perpendicular to the Horizontal Axis.*—Level the instrument. Sight on a point *A* (see Fig. 91*a*) about 500 ft. away, with telescope normal. With both horizontal motions of the instrument clamped, plunge the telescope and set another point *B* on the line of sight and about the same distance away on the opposite side of the transit. Rotate the instrument end for end about

the vertical axis, and again sight at A (with telescope inverted). Plunge the telescope as before; if the line of sight does not fall on B, set a point C on the line of sight beside B. Mark a point D, *one quarter* of the distance from C to B, and adjust the cross-hair ring (by means of the two opposite horizontal screws) until the line of sight passes through D. The points sighted should be at about the same elevation as the transit.

4. *To Make the Horizontal Axis Perpendicular to the Vertical Axis.*—Set up the transit near a building or other object on which is some well-defined point A at a considerable vertical angle. Level the instrument very carefully, thus making the vertical axis truly vertical. Sight at the high point A (see Fig. 91*b*), and with the horizontal motions clamped, depress the telescope and set a point B on or near

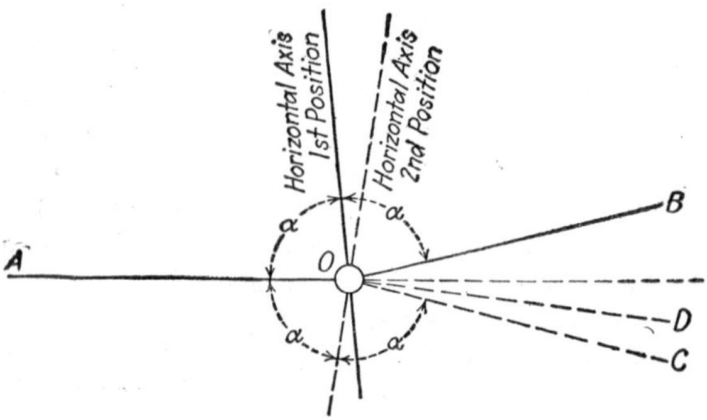

Fig. 91*a*.—Adjustment of line of sight.

the ground. If the horizontal axis is perpendicular to the vertical axis, A and B will be the same vertical plane. Plunge the telescope, rotate the instrument end for end about the vertical axis, and again sight on A. Depress the telescope as before; if the line of sight does not fall on B, set a point C on the line of sight beside B. A point D, halfway between B and C, will lie in the same vertical plane with A. Sight on D; elevate the telescope until the line of sight is beside A; loosen the screws of the bearing cap; and raise or lower the adjustable end of the horizontal axis until the line of sight is in the same vertical plane with A.

5. *To Make the Axis of the Telescope Level Parallel to the Line of Sight.*—Proceed the same as for the two-peg adjustment of the dumpy level (Art. 50, adjustment 3), except as follows: With the line of sight set on the rod reading established for a level line, the correction is made by raising or lowering one end of the telescope level tube until the bubble is centered.

6. (For Transit Having a Fixed Vertical Vernier) *To Make the Vertical Circle Read Zero When the Telescope Bubble Is Centered.*—With the plate bubbles centered, center the telescope bubble. If the vertical vernier does not read zero, loosen it and move it until it reads zero. Care should be taken that the vernier will not bind on the vertical circle.

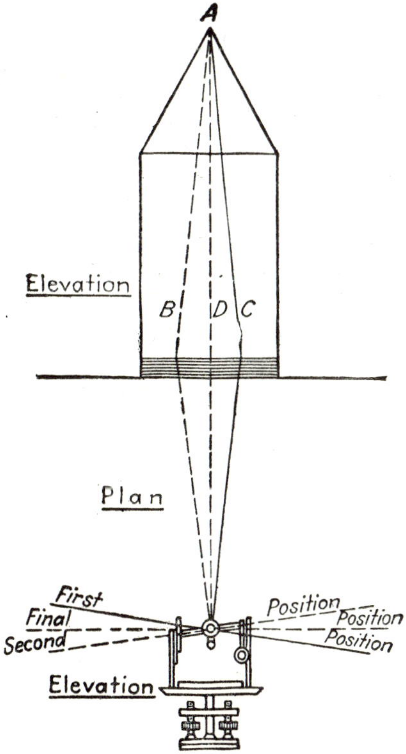

FIG. 91*b*.—Adjustment of horizontal axis.

6*a*. (For Transit Having a Movable Vertical Vernier with Control Level) *To Make the Axis of the Auxiliary Level Parallel to the Axis of the Telescope Level When the Vertical Vernier Reads Zero.*—Center the telescope bubble, and by means of the vernier tangent-screw move the vertical vernier until it reads zero. By means of the capstan screws adjust the level tube attached to the vertical vernier until the bubble is centered.

92. Errors in Transit Work.—Except in field astronomy, a measured angle is always closely related to a measured distance. On surveys of ordinary precision it usually requires much more care to keep *linear* errors within prescribed limits than to maintain a corresponding degree of *angular* precision. Often undue attention is paid to securing accuracy in angular measurements.

1. *Instrumental Errors.*—In work of ordinary precision, instrumental errors can be kept within permissible limits if the transit is maintained in good adjustment.

2. *Personal Errors.*—All the personal errors are accidental and hence cannot be eliminated. They can be kept within reasonable limits by care in leveling the instrument and observing.

3. *Natural Errors.*—In general, the errors resulting from natural causes are not large enough to affect appreciably the measurements of ordinary precision.

93. Precision of Angular Measurements.—Many factors influence the precision of transit work, and no rigid rules can be formulated to ensure a required accuracy. The values given below represent, in a general way, the *maximum* error likely to occur in measuring a horizontal angle under the average conditions of practice, instruments being in fair condition and in fair adjustment except as otherwise stated. The *average* angular error will be materially less. Also, as the errors are largely accidental, the resultant error in a series of measured angles may be expected to vary as the square root of the number of angles involved.

Case 1.—Short sights, point indicated by range pole obscured near ground. Range pole plumbed by eye. Single observation of angle. Error 02′ to 04′.

Case 2.—Long sights, but otherwise as stated for case 1. Error 01′ to 02′.

Case 3.—Unobscured sights on well-defined points; sights not steeply inclined. Single observation of angle, vernier reading to minutes. Error 30″ to 01′.

Case 4.—Unobscured sights on well-defined points. Sights not steeply inclined. Verniers reading to 30″. Single observation of angle represented by mean of readings of both verniers. Transit in excellent condition and in good adjustment. Error 15″ to 30″.

Case 5.—As for case 4, but verniers reading to 10″. Also instrument set up with great care. Error 10″ to 15″.

Case 6.—Unobscured sights on well-defined points. Instrument set up with great care. Sights not steeply inclined. Transit in good condition. Vernier reading to 30″. Angle repeated six times with telescope normal and six times with it inverted. Error 02″ to 04″.

Case 7.—As for case 6, but transit reading to 10″. Observations taken at favorable times. Error 01″ to 02″.

94. Numerical Problems.

1. Thirty spaces on a transit vernier are equal to 29 spaces on the graduated circle, and one space on the circle is 15′. What is the least count of the vernier?

2. A vertical angle measured by a single observation is −12°02′, and the index error is determined to be +06′. What is the true value of the angle?

3. A transit for which the circle is graduated 0° to 360° clockwise is used to measure an angle by 10 clockwise "repetitions," 5 with telescope normal and 5 with telescope inverted. Compute the most probable value of the angle from the accompanying data.

Telescope	Reading	Vernier A	Vernier B
Normal..................	Initial	48°46′	228°46′
Normal..................	After first turning	161°09′	
Inverted..................	After tenth turning	92°41′	272°42′

95. Field Problems.

PROBLEM 8. MEASUREMENT OF HORIZONTAL ANGLES WITH TRANSIT

Object.—To measure several angles about a point with the transit and to check the values by the use of magnetic bearings.

Procedure.—(1) Set up and level the instrument at any point O. (2) Set four chaining pins at about 150 ft. from the transit, forming four angles about the station O. The pins may be plumbed by the vertical cross-hair of the transit. (3) According to the procedure of Art. 85, measure each of the angles, using the A vernier only and resetting the vernier on each backsight. (4) The sum of the measured angles should not differ from 360° by more than ±02′. If this difference is exceeded, remeasure the angles until the sum falls within the stated limits. (5) Release the compass needle, sight on each point, and record the magnetic bearing to each pin. (6) Compute the angles by bearings, and compare with the transit angles. The discrepancy between any transit angle and the same angle by bearings should not exceed 30′.

PROBLEM 9. MEASUREMENT OF VERTICAL ANGLES WITH TRANSIT

Object.—To determine the height of a building above the water table, by measurement of vertical angles with the transit.

Procedure.—(1) Set up the transit at A, at a distance from the building approximately twice the height of the building. Level the telescope, and note the position of the horizontal hair on the building; mark the point sighted, and measure the distance h_a above or below the water table. (2) Determine the index error of the vertical circle. Sight on the high point T of the building, and record the vertical angle. (3) Depress the telescope, and set a point B in the same vertical plane with A and T, about halfway between A and the building. (4) Set up the transit at B, and measure h_b and the vertical angle to point T, as at A. (5) Measure the horizontal distance AB. (6) Draw a sketch, and compute the difference in elevation (a) between T and the horizontal line of sight from either transit station and (b) between T and the water table.

Hints and Precautions.—The index error may be read either before or after observing a vertical angle; while this reading is being taken, the line of sight should be in the same vertical plane as the point sighted.

Problem 10. Prolongation of Line Past Obstacle

Object.—To prolong a line *AB* past an obstacle where the conditions are such as to limit the lengths of the offsets.

Procedure.—(1) As outlined in Art. 90. (2) The lengths of offsets should be measured very carefully. If the instrument is in good adjustment, the points *E* and *F* may be set by a single reversal; or if a clear sight can be obtained, the transit may be set up at *C* and the points *E* and *F* established without reversal. (3) If the obstacle is imaginary, check the accuracy of the work by setting the transit at *A* and locating *G* and *H* by the direct prolongation of *AB*.

Problem 11. Measurement of Angles by Repetition

Object.—To obtain a more accurate determination of the horizontal angles between various stations about a point than would be possible by a single measurement (see Art. 87).

Procedure.—(1) Set up the transit very carefully over the point. (2) Set the *A* vernier at zero, read the *B* vernier, and record the readings. (3) With the telescope *normal,* measure one of the angles clockwise, and record both vernier readings to the least reading of the vernier. (4) Leaving the upper motion clamped, again set on the first point and again measure the angle clockwise (thus doubling the angle). (5) Continue until five "repetitions" (observations) have been secured. Record both vernier readings and the total angle turned. (6) In like manner, without resetting the vernier, measure the angle (five repetitions) with the telescope *inverted,* always measuring clockwise. (7) Go through the same process for all other angles about the point. (8) Compute the value of each of the angles for the 10 repetitions, and compare with the single measurement. (9) For a transit reading to single minutes, the error of horizon closure should not exceed $10'' \sqrt{\text{number of angles}}$. (10) Adjust the angles so that their sum will equal 360° by distributing the error equally among the mean values.

Problem 12. Adjustment of the Transit

Object.—To make the field adjustments of the engineer's transit.

Procedure.—As outlined in Art. 91.

CHAPTER VII

TRANSIT-TAPE SURVEYS

96. General.—The practices described in this chapter apply in general to perhaps nine-tenths of all surveying work of ordinary precision, excluding leveling.

Usually the transit party is composed of a *transitman*, a *head chainman*, and a *rear chainman*. The transitman directs the party, operates the transit, and keeps the notes. The chainmen give line, set stakes, and assist in clearing the line. In wooded country, axemen may be needed to clear the line. Where sights are long, a rear flagman may be employed to give backsights to the transitman. Where many observations are taken, the notes may be kept by a recorder, who may also act as the chief of party.

The equipment of the transit party usually consists of a transit, 100-ft. steel tape, two range poles, stake bag, stakes, tacks, axe or hammer, two or three plumb bobs, field notebook, chaining pins, and marking crayon.

Any temporary or permanent point of reference over which the transit is set up is called a *transit station*. Usually the transit station is a peg, called a *hub*, driven flush with the ground and having a tack driven in its top. The location of a hub is usually indicated by a flat *guard stake* extending above the ground and driven sloping so that its top is over the hub.

Lines connecting transit stations are called *transit lines*. Traverses and triangulation systems are formed by transit lines.

For many continuous traverses where lengths are measured with the tape, stakes or other marks are set every 100 ft. These 100-ft. points are called *full* stations, and intermediate points are called *plus* stations. A full station at, say, 1,200 ft. from the initial point would be numbered 12 + 00 or simply 12; and a plus station at, say, 1,927.2 ft. from the initial point would be numbered 19 + 27.2.

97. Transit Surveys.—The field work of surveying with the transit may be divided as follows:

1. Establishing transit stations and lines by angular and linear measurements. The transit lines may be said to form the skeleton of the survey and are called the *control* or *horizontal control*.

2. Locating objects and points with respect to the transit lines, thus furnishing the *detail* with which the transit lines are clothed (see Art. 107).

The details may be observed either while the transit lines are being run or after they have been established and checked. Instruments and methods employed in collecting details may differ from those employed in establishing the control.

Transit surveys may be made by the method of *radiation* (Art. 98), of *intersection* (Art. 99), of *traversing* (Arts. 100 to 105), or of *triangulation* (Art. 76). On surveys of ordinary precision, transit traverses in one form or another probably make up more than nine-tenths of the systems of control. The methods of radiation and intersection are employed on surveys of small areas and in the collection of details from transit traverses.

98. Radiation.—The transit is set up at any convenient station from which can be seen all points that it is desired to locate. The distance from the transit station to each of the points is measured, and the horizontal angle is observed. The angles between successive points may be measured; or the bearing or azimuth of each of the lines joining the points with the transit station may be observed. Where points only are to be located, the method is excellent. Where lengths and directions of lines other than the radial lines are to be determined, trigonometric computations are necessary.

99. Intersection.—In Fig. 99 let the points A, B, C, etc., represent objects that it is desired to locate, and let OP be a convenient line from both ends of which the unknown points are visible. The length of the base line OP is measured with the tape. The transit is set up at O, and angles to the unknown points are observed; these may be expressed as azimuths, as bearings, or as angles between successive points. A similar series of observations is made with the transit at P. In this manner each of the unknown points becomes the vertex of a triangle of which the base line OP is the side of measured length and in which the angles adjacent thereto are observed values; the locations of the unknown points are thus defined.

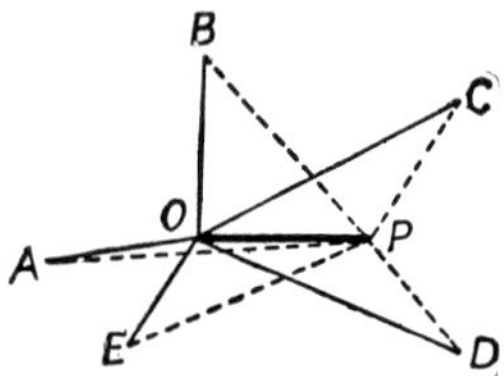

Fig. 99.—Intersection.

A variation of the method of intersection is the method of *resection*, by means of which the transit may be set up at a station of unknown location and its location determined by sighting on points of known location. The principle of resection employing the transit is as described for the plane table in Art. 168.

100. Traversing.—Following is a general description of the work of running a *closed* traverse, the transit stations being established in advantageous locations as the survey progresses, and distances being measured between successive transit stations. In Fig. 100 let *A* and *B* be selected locations for transit stations marking the first line of a traverse. The transit is set up at *B*, the horizontal vernier is set to a given angular value, a backsight is taken on *A*, and the lower motion is clamped. The line *AB* is then chained, the head chainman being lined in by the transitman. The distance *AB* is recorded. The position of *C* is selected, and the transit point is established. The transit is turned on the upper motion until a foresight is secured on *C*. The upper motion is clamped, and the angular value is read and recorded. The distance *BC* is chained and recorded in a manner similar to that for *AB*. The transit is moved forward to *C*; a backsight is taken to *B*; the point *D* is chosen; a foresight is taken to *D*; and the angle is read. The line *CD* is chained. The process is repeated for point *E*, etc., until the traverse is finally brought to closure on the initial point *A*. As a check, usually the initial station is occupied by the transit, and the angle between the first and last lines of the traverse is measured; in this way the angular error of closure is determined.

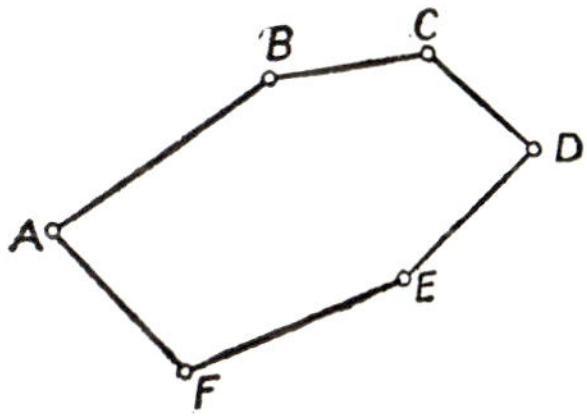

Fig. 100.—Traversing.

An *open* traverse may be run in exactly the same manner, except that of course there is no closure. The cumulative angular error cannot be determined except by astronomical observations.

The angles of the traverse may be measured by observing deflection angles, azimuths, or interior angles as desired. As a check against large errors, usually the magnetic bearings are observed on both backsights and foresights and are compared with bearings computed from the measured angles at each station.

One or more transit stations may be necessary between points at which angles are turned. In such cases the line is prolonged as described in Art. 89.

It is customary to show a sketch of the traverse on the right-hand page of the field notebook. For most short closed traverses the notes read down the page. For long traverses usually the center line of the right-hand page is taken to represent the traverse, and the notes and sketch read *up* the page (Fig. 101); thus objects on the sketch appear in the same relative position as they appear to the recorder proceeding along the line.

101. Deflection-angle Traverse.—Successive transit stations are occupied, and at each station a backsight is taken with the A vernier set at zero and the telescope inverted. The telescope is then plunged; the foresight is taken by turning the instrument about the vertical axis on its upper motion; and the deflection angle is observed. The angle is recorded as right R or left L, according to whether the upper motion is turned clockwise or counter-clockwise.

Figure 101 shows the notes for a portion of an open deflection-angle traverse where stakes are set every 100 ft. The notes are

PRELIMINARY SURVEY OF I.N. RY., C. TO ST. LEONARDS

(Defl. Ang. Traverse)

Sta.	Defl. Ang.	Mag. B.	Cal. B.	Remarks
12				
+82				Water Line
+17				" "
11				
10				
9				
⊙8+36.3	23°14'L	N 35°W	N34°47'W	
8				
+99.8				
+68				℄ Due E to St.L.
+33.1				
7				
6+00				Ctr. of Brook
5				
4				
3				
2				
+23.6				
1				
⊙0+00		N11°W	N11°33'W	on C.L. C.P.Ry. 523'N. M'l. Post—No. 26

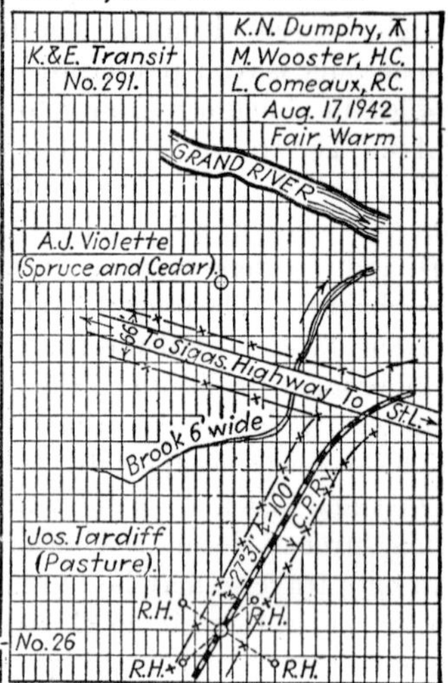

FIG. 101.—Notes for open deflection-angle traverse.

typical of the form used on highway, railroad, and other route surveys. Magnetic bearings have been observed as a check.

102. Azimuth Traverse.—The azimuth method of traversing is extensively employed on topographic and other surveys where a large number of details are located by angular and linear measurements from the transit stations. The reference meridian may be either true or assumed.

Successive stations are occupied beginning with the line of known azimuth. At each station the transit is "oriented" by setting the A vernier to read the back azimuth (forward azimuth $\pm 180°$) of the preceding line and then backsighting to the preceding transit station. The instrument is then turned on the upper motion, and a foresight on the following transit station is secured. The reading indicated by the A vernier is the azimuth of the forward line. Traverse angles are checked by magnetic bearings.

103. Interior-angle Traverse.—At each station the vernier is set at zero, and a backsight to the preceding transit station is taken. The instrument is then turned on its upper motion; a foresight is taken on the following station; and the interior angle is read. The notes may be kept in the form of a sketch, or the numerical values may be tabulated.

104. Checking Traverses.—Magnetic bearings offer an excellent means of checking observed angles against mistakes or large errors. On important traverses often the angular values are checked by doubling the angles, and the linear measurements are checked by chaining forward and backward over each line.

For a closed traverse it is possible by known geometrical relations to determine the angular error of closure. The linear error of closure

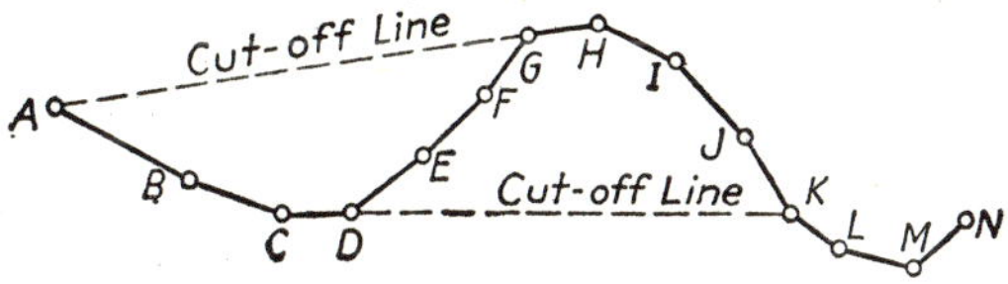

FIG. 104.—Cut-off lines.

may be determined by comparing the coordinates of the initial point as originally assumed and as computed by successive angles and distances around the traverse.

For an open traverse no such means are available for checking the measurements as a whole. Angular errors may be determined closely by taking astronomical observations at intervals. Occasionally conditions are favorable to establishing *cut-off lines* as the traverse progresses (Fig. 104). These lines together with the intermediate traverse lines form closed traverses covering parts of the survey. Another check is obtained by observing the angle to some distant landmark from each of several stations.

105. Specifications for Traversing with Transit and Tape.—The precision of transit traverses is affected by both linear and angular errors of measurement (see Arts. 28, 92, and 93). Many factors affect the precision, and it can be expressed only in very general terms.

The following specifications give approximately the *maximum* linear and angular errors to be expected when the methods stated are followed. If the surveys are executed by well-trained men, with instruments in good adjustment, and under average field conditions, in general the error of closure should not exceed *half* the specified amount. The specifications apply to traverses of considerable length. It is assumed that a standardized tape is used.

Class 1.—Precision sufficient for many preliminary surveys, for horizontal control of surveys plotted to intermediate scale, and for land surveys where the value of the land is low.

Transit angles read to the nearest minute. Sights taken on a range pole plumbed by eye. Distances measured with a 100-ft. steel tape. Pins or stakes set within 0.1 ft. of end of tape. Slopes under 3 per cent disregarded. On slopes over 3 per cent, distances either measured on the slope and corrections roughly applied or measured with the tape held level and using an estimated standard pull.

Angular error of closure not to exceed $1'30'' \sqrt{n}$, in which n is the number of observations. Total linear error of closure not to exceed 1/1,000.

Class 2.—Precision sufficient for most land surveys and for location of highways, railroads, etc. By far the greater number of transit traverses fall in this class.

Transit angles read carefully to the nearest minute. Sights taken on a range pole carefully plumbed. Pins or stakes set within 0.05 ft. of end of tape. Temperature corrections applied to the linear measurements if the temperature of air differs more than 15°F. from standard. Slopes under 2 per cent disregarded. On slopes over 2 per cent, distances either measured on the slope and corrections roughly applied or measured with the tape held level and using a carefully estimated standard pull.

Angular error of closure not to exceed $1' \sqrt{n}$. Total linear error of closure not to exceed 1/3,000.

Class 3.—Precision sufficient for much of the work of city surveying, for surveys of important boundaries, and for the control of extensive topographic surveys.

Transit angles read twice with the instrument plunged between observations. Sights taken on a plumb line or on a range pole carefully plumbed. Pins set within 0.05 ft. of end of tape. Temperature of air determined within 10°F., and corrections applied to the linear measurements. Slopes determined within 2 per cent, and corrections applied. If tape is held level, the pull kept within 5 lb. of standard, and corrections for sag applied.

Angular error of closure not to exceed $30'' \sqrt{n}$. Total linear error of closure not to exceed 1/5,000.

Class 4.—Precision sufficient for accurate surveying in cities and for other especially important surveys.

Transit angles read twice with the instrument plunged between readings, each reading being taken as the mean of both *A* and *B* vernier readings. Verniers reading to 30″. Instrument in excellent adjustment. Sights taken with special care. Pins set within 0.02 ft. of end of tape. Temperature of tape determined within 5°F., and corrections applied. Slopes determined within 1 per cent, and corrections applied.

If tape is held level, the pull kept within 3 lb. of standard, and corrections for sag applied.

Angular error of closure not to exceed $15'' \sqrt{n}$. Total linear error of closure not to exceed 1/10,000.

106. Referencing Transit Stations.—A transit station is said to be *referenced* when it is so tied by linear and/or angular measurements to near-by objects that it can be replaced readily. The reference points may be wooden *reference hubs* or may be objects of a more or less permanent character. Stations likely to be disturbed or destroyed should always be referenced, and the measurements recorded by means of a sketch.

The station may be referenced by linear measurements only (Fig. 106*a*) to two reference hubs which either may be on line with the station (as R.H.*A* and R.H.*B*) or may not be on line (as R.H.*A*

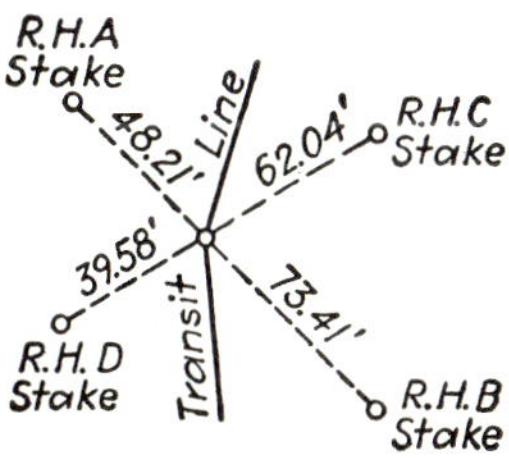

FIG. 106*a*.

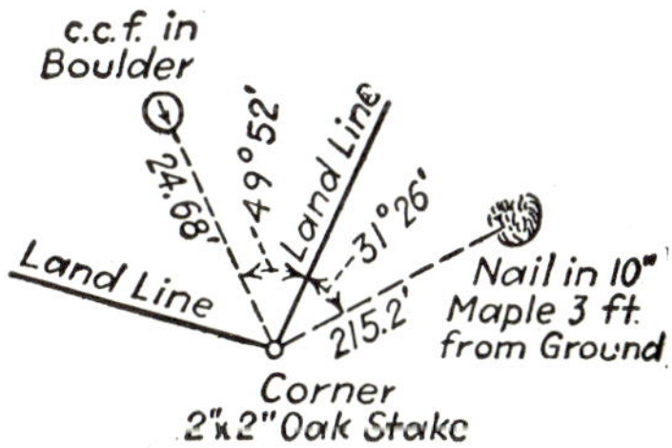

FIG. 106*b*.

and R.H.*D*). If either of the two reference hubs is disturbed, the station cannot be relocated.

If four reference hubs are placed as shown in Fig. 106*a*, with the transit station at the intersection of the line *AB* and the line *CD*, any two reference hubs may be destroyed, and still the station can be relocated.

Figure 106*b* is typical of the manner in which corners are referenced in land surveying. Both angular and linear ties are employed.

107. Details from Transit Lines.—On nearly all transit surveys certain *details*, or natural and artificial features of the terrain, are located with respect to the transit lines. The amount of detail required may be small, as on property-line surveys, or large, as on topographic surveys. If details are located solely for map-making purposes, generally the required precision of measurements to details is less than that for the transit lines.

Angular measurements to details are usually made with the transit. Linear measurements to details are made with the 100-ft. steel tape, with the 50-ft. metallic tape, with the stadia (Chap. VIII), or sometimes by pacing, according to the accuracy required and the convenience of measuring.

On most surveys the details are located as the traverse is run. The observations with the transit, called *side shots*, are made immediately after the foresight to the following station has been taken.

107a. Locating Details.—Following are descriptions of the common methods of locating details with the transit and tape. In general, the *dimensions* of an object (as a building) should be determined by direct measurement. Sketches are used freely. Where points are numerous, each point is given a number on the sketch, and the angles and distances are tabulated.

1. *By Angle and Distance from Transit Station.*—As illustrated by Fig. 107*a*, any given point on the object is located by an angle

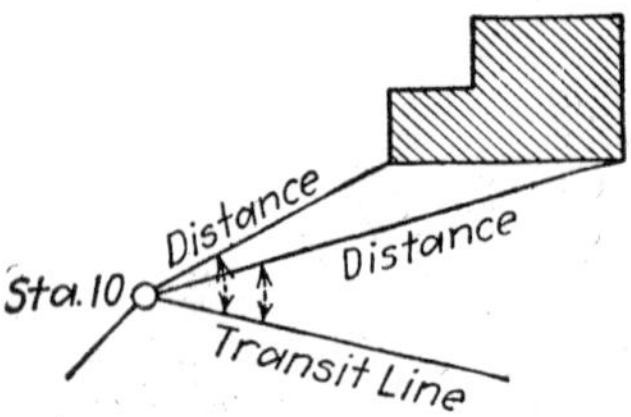

Fig. 107*a*.

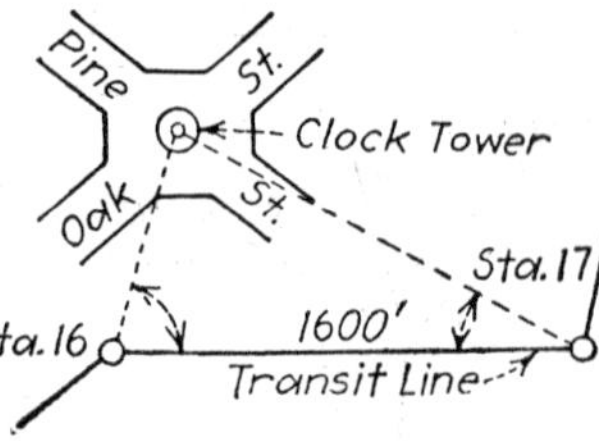

Fig. 107*b*.

and a distance. The method is widely used where details are close to transit stations.

2. *By Angles from Two Transit Stations.*—As shown by Fig. 107*b*, the point is located with respect to the transit line by angles taken from two stations, that is, by the method of intersection. No linear measurements are required. The method is useful for distant or inaccessible points.

3. *By Distances from Two Stations.*—If the details are close to the transit line yet distant from the nearest transit station, they are conveniently located by linear ties from two traverse stations. Thus the corners of the middle building of Fig. 107*c* are located by ties to stations 12 + 00 and 13 + 00.

4. *By Ranges, Range Ties, and Swing Offsets.*—If the features to be located are buildings, the sides of the building may be ranged in by eye, and the intersection of the lines thus defined with the transit line may be found. The station and plus of such a point of intersection is called a *range*, and a range together with the distance along the range line to a corner of the building is called a *range tie*. Thus in Fig. 107*c* the building at the right is completely located by two range ties from stations 12 + 18 and 13 + 26; a check tie to station

14 + 00 is also shown. A *swing offset*, or perpendicular distance from transit line to a point, is determined by swinging the tape about the point as a center and taping the least distance from the point to the transit line. The left building of Fig. 107*c* is located by the swing offset shown and by the range tie from station 11 + 11.

5. *By Perpendicular Offsets from the Transit Line.*—This method is adapted to the location of irregular or curved boundaries, streams, and roads that closely parallel the transit lines. As indicated by Fig. 107*d*, a point is located by measuring the distance along the

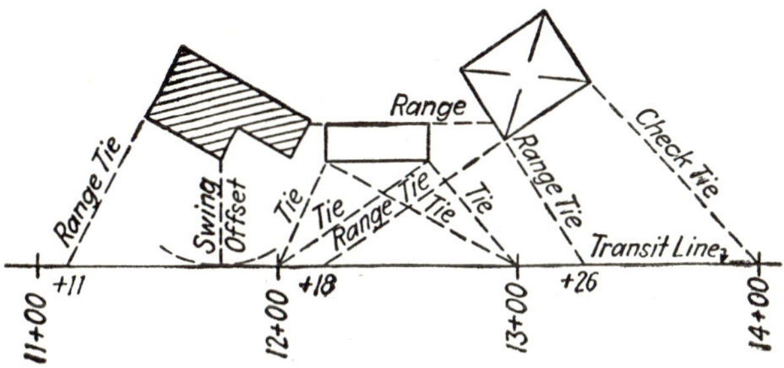

FIG. 107*c*.

FIG. 107*d*.

transit line to the foot of a perpendicular offset through the point and then measuring the length of the perpendicular offset. Usually the direction of the perpendiculars is estimated by eye except where offsets are long; in this case the tape or one of the sighting instruments is employed.

108. Field Problems.

PROBLEM 13. OPEN DEFLECTION-ANGLE TRAVERSE WITH TRANSIT AND TAPE

Object.—To locate a section of an assigned route, setting stakes at full stations. This problem may be employed in connection with field problems 5 (p. 53) and 27 (p. 177). For plotting, see office problems 19 and 20 (p. 118).

Procedure.—(1) Stake out a route perhaps 1,500 ft. long with three or four changes in direction. (2) Set up the transit over the first stake marking a change in direction. With the *A* vernier set at zero and with the telescope inverted, sight on the stake at the beginning of the line. Read and record the magnetic bearing of this backsight. (3) Chain the line from the beginning (station 0 + 00) to the transit station, and record the length to the nearest 0.01 ft. The transitman lines in the head chainman. As the rear chainman pulls each pin, he replaces it with a stake on which he has written the station number. (4) Plunge the telescope, unclamp the upper motion, and sight on the next forward transit station. Record the reading of the *A* vernier and *R* or *L* to indicate whether the deflection is right or left from the prolongation of the preceding line. Also record the magnetic bearing of the forward line. (5) Before the transit is moved, compute the deflection angle from the magnetic bearings taken at the station, and compare with the deflection angle indicated on the horizontal circle, as a rough check. (6) Chain the forward line. (7) Set up the transit at succeeding stations, and observe the deflection angle at each. Give line for the chainmen on each forward line. (8) Keep notes in the form of the sample notes (Fig. 101). Include sketches of streams, roadways, fences, etc., crossed by the line. (9) Assume as correct the magnetic bearing of some course the back and forward bearings of which have the same angular value, and compute the forward bearing of each of the other courses.

PROBLEM 14. CLOSED AZIMUTH TRAVERSE WITH TRANSIT AND TAPE

Object.—To collect data sufficient for plotting the boundaries and determining the area of a field, employing the azimuth method of traversing. The data may be used later, as in field problem 15 and in office problem 19 or 20 (p. 117).

Procedure.—(1) Stake out an irregular field having perhaps five sides and containing an acre or more. (2) Measure the sides with the steel tape, to the nearest 0.01 ft. (3) Set up the transit at one corner. Set the *A* vernier at 0°, and turn the instrument on the lower motion to sight along the magnetic meridian either north or south according as azimuths are to be reckoned from north or south. Clamp the lower motion in this position. (4) Unclamp the upper motion, and turn the instrument to sight the next corner forward. The angle turned off in a clockwise direction and read from the *A* vernier is the azimuth of the forward line. Record the azimuth. (5) Record the magnetic bearing. Compare this with the bearing computed from the azimuth of the line. (6) Compute the back azimuth of the line by adding 180° to the forward azimuth. (7) Set up the instrument on the next corner forward, and with the *A* vernier set on the back azimuth of the line, backsight on the corner previously occupied. The instrument is now oriented. (8) Turn the instrument on the upper motion, and sight on the next corner forward; the azimuth of the forward line is then indicated by the *A* vernier. (9) Proceed in this manner until each corner has been occupied. Also set up again at the first corner, and take readings as for the other corners. (10) Note the angular error of closure, which should not exceed $30'' \sqrt{\text{number of sides}}$. Distribute the error equally among the angles.

PROBLEM 15. DETAILS WITH TRANSIT AND TAPE

Object.—To obtain sufficient data for plotting a detailed map of a portion of the campus. The data may be used in office problem 19 or 20 (p. 117).

Procedure.—(1) Run a closed azimuth traverse, as described in the preceding problem, through the area to be mapped. Locate the lines and corners of the traverse so that linear and angular measurements necessary to fix the position of details with respect to the traverse may be taken with the least labor. Reference the corner hubs. (2) Obtain the location of buildings, streets, walks, hydrants, fences, etc., by the several methods of Arts. 107 and 107*a*. (3) Sketch indefinite details such as trees, streams, and shrubbery without measurements other than by pacing. (4) When the traverse is plotted (office problem 19 or 20), plot the details according to the manner in which they were secured.

Hints and Precautions.—(1) Details may be taken as the traverse is run. (2) Determine the angular error of closure of the traverse. If this exceeds $30'' \sqrt{\text{number of sides}}$, rerun the traverse until the mistake is discovered. (3) The location of important details should be checked, preferably by a different method. (4) Sketches should not be overcrowded with measurements. Many measurements can be tabulated on the left-hand page and referred to on the sketch by a single letter or number.

CHAPTER VIII

STADIA SURVEYING

109. The Stadia Method.—The stadia method of measuring distances is far more rapid than chaining and under certain conditions is as accurate. It is a useful means of checking more precise measurements. It is employed extensively in transit surveying, plane-table surveying, and leveling.

The equipment for stadia measurements consists of (1) a telescope with two horizontal *stadia hairs*, one above and the other an equal distance below the horizontal cross-hair; and (2) a *stadia rod* graduated usually in feet and decimals. Any self-reading leveling rod may be used as a stadia rod, but for long sights special graduations are employed.

The process of taking a stadia measurement consists in observing through the telescope the apparent locations of the two stadia hairs on the rod, which is held vertical. The interval between the two rod readings, called the *stadia interval* or *stadia reading*, is a direct function of the distance from instrument to rod. For most instruments the ratio of distance to stadia interval is 100.

For convenience and to lessen the chance of mistake, on transit or plane-table surveys the stadia interval is determined by setting the lower stadia hair on a foot mark and reading the position of the upper hair.

110. Principle of the Stadia.—In Fig. 110 the line of sight is horizontal, and the rod is vertical. The stadia hairs are indicated by the points a and b; the distance between the stadia hairs is i. The stadia interval is s.

In optics it is shown that a ray of light passing through the optical center of a lens remains undeviated in direction and, further, that rays which are parallel on one side of the lens are all brought to a focus at a fixed point on the optical axis. This point is called the *principal focus*, and its distance from the optical center is called the *focal length* of the lens.

Imagine that aa' and bb' in the figure are parallel rays emanating from the stadia hairs a and b. Then F is the principal focus; f is the focal length of the objective; and the emerging rays take the positions $a'FA$ and $b'FB$. Also imagine that aOA and bOB are rays emanating respectively from a and b that pass undeviated through the optical center O.

As $ab = a'b'$, by similar triangles

$$\frac{f}{i} = \frac{d}{s}$$

Hence the horizontal distance from the principal focus to the rod is $d = \frac{f}{i}s = Ks$, in which $K = \frac{f}{i}$ is a coefficient called the *stadia interval factor* which for a particular instrument is a constant so long as conditions remain unchanged. Thus for a horizontal sight the distance from principal focus to rod is obtained by multiplying the stadia

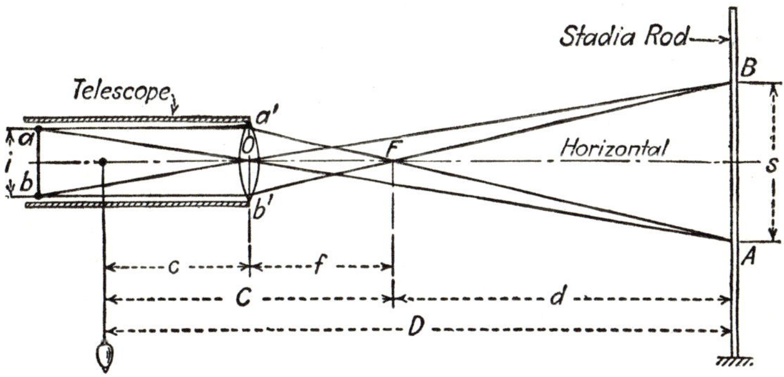

FIG. 110.—Horizontal stadia sight.

interval factor by the stadia interval. The horizontal distance from center of instrument to rod is then

$$D = Ks + (f + c) = Ks + C \tag{1}$$

in which C is the distance from the center of the instrument to the principal focus. This formula is employed in computing horizontal distances from stadia intervals when sights are horizontal.

111. Stadia Constants.—The focal distance f is a constant for a given instrument. It can be determined with all necessary accuracy by focusing the objective on a distant point and then measuring the distance from the cross-hair ring to the objective. The distance c, though a variable depending on the position of the objective, may for all practical purposes be considered a constant. Its mean value can be determined by measuring the distance from the vertical axis to the objective when the objective is focused for an average length of sight.

Usually the value of $C = f + c$ is determined by the manufacturer and is stated on the inside of the instrument box. Under ordinary conditions C may be considered as 1 ft. without error of consequence. Internal-focusing telescopes are so constructed that C is zero or nearly so.

111a. Stadia Interval Factor.—The nominal value of the stadia interval factor $K = f/i$ is usually 100. The interval factor can be determined by observation. The usual procedure is to set up the instrument in a location where a horizontal sight can be obtained and with a tape to lay off, from a point distant $C = (f + c)$ in front of the center of the instrument, distances of 100 ft., 200 ft., etc., up to perhaps 1,000 ft., stakes being set at the points thus established. The stadia rod is then held on each of the stakes, and the stadia interval is read. The stadia interval factor is computed for each sight, and the mean is taken as the most probable value. To overcome prejudicial tendencies in sighting, the stakes may be set at random distances which are measured later.

112. Inclined Sights.—In stadia surveying most sights are inclined, and usually it is desired to find both the horizontal and the vertical distances from instrument to rod. For convenience, the rod is always held vertical.

Figure 112 illustrates an inclined line of sight, AB being the stadia interval on the vertical rod and $A'B'$ being the corresponding projection normal to the line of sight. The length of the inclined line of sight from center of instrument is

$$D_i = \frac{f}{i} A'B' + C \tag{2}$$

For all practical purposes the angles at A' and B' may be assumed to be 90°. Let $AB = s$; then $A'B' = s \cos \alpha$. Making this substitution in Eq. (2) and letting $K = f/i$, the inclined distance is

$$D_i = Ks \cos \alpha + C \tag{3}$$

The horizontal component of this inclined distance is

$$H = Ks \cos^2 \alpha + C \cos \alpha \tag{4}$$

which is the general equation for determining the horizontal distance from center of instrument to rod, when the line of sight is inclined.

The vertical component of the inclined distance is

$$V = Ks \cos \alpha \sin \alpha + C \sin \alpha$$

or

$$V = \tfrac{1}{2} Ks \sin 2\alpha + C \sin \alpha \tag{5}$$

which is the general equation for determining the difference in elevation between the center of the instrument and the point where the line of sight cuts the rod. To determine the difference in ground elevations, the height of instrument and the rod reading of the line of sight must be considered.

Equations (4) and (5) are known as the *stadia formulas for inclined sights.*

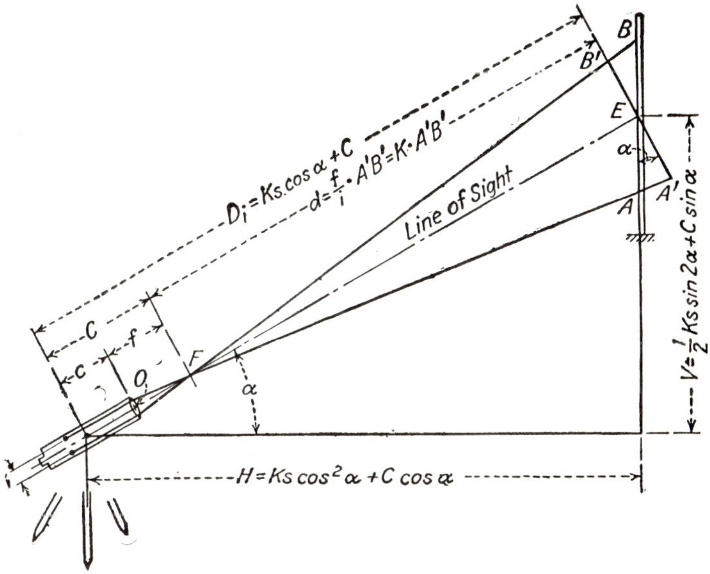

Fig. 112.—Inclined stadia sight.

113. Permissible Approximations.—More approximate forms of the stadia formulas are sufficiently precise for most stadia work. Generally distances are computed only to feet and elevations to tenths of feet. For side shots where vertical angles are less than 3°, Eq. (4) may properly be reduced to the form

$$H = Ks + C \tag{6}$$

which is the same as for horizontal sights (Art. 110).

Owing to unequal refraction and to accidental inclination of the rod, observed stadia intervals are in general slightly too large. To offset the systematic errors from these sources, frequently on surveys of ordinary precision the constant C is neglected. Hence in any ordinary case Eq. (4) may with sufficient accuracy be expressed in the form

$$H = Ks \cos^2 \alpha \text{ (approx.)} \tag{7}$$

Also Eq. (5) may often be expressed with sufficient accuracy for ordinary work in the form

$$V = \tfrac{1}{2}Ks \sin 2\alpha \text{ (approx.)} \qquad (8)$$

except where the vertical angle is large.

Equations (7) and (8) are simple in form and are most generally employed.

When K is 100, the common practice is to multiply mentally the stadia interval by 100 at the time of observation and to record this value Ks, called the *stadia distance*, in the field notebook. Thus, if the stadia interval were 7.37 ft., the stadia distance recorded would be 737 ft.

For external-focusing telescopes, the degree of approximation in using Eqs. (7) and (8) may be greatly reduced either by adding 0.01 ft. to the observed stadia interval s or—when K is 100—by adding 1 ft. to the observed stadia distance Ks. It is convenient to add the correction mentally and to record the corrected value in the field notebook. The notes should state that corrected values are recorded.

114. Stadia Reductions.—Ordinarily in practice the horizontal distances and the differences in elevation are not computed by actually solving the stadia formulas but are obtained by the use of a table, diagram, stadia slide rule, or stadia arc on the vertical circle of the transit, all these devices being based on the formulas.

Table VI gives, for each 02′ of vertical angle up to 30°, the horizontal distances (from principal focus to rod) and differences in elevation for Ks = 100 ft., computed from the equations $H = Ks \cos^2 \alpha$ and $V = \tfrac{1}{2}Ks \sin 2\alpha$ [see Eqs. (4), (5), (7), and (8)]. For any other value of Ks, the tabular quantities are to be multiplied by the value of Ks in hundreds of feet. The table also gives the horizontal distances and differences in elevation for three values of $C = (f + c)$, indicated as C in the table. More elaborate tables may be found in various other publications. If Table VI or a similar table is used, the necessary multiplications may be carried out with sufficient accuracy with the ordinary slide rule.

Diagrams showing graphically the quantities $Ks \cos^2 \alpha$ and $\tfrac{1}{2}Ks \sin 2\alpha$ for all ordinary distances are published in a variety of forms. It is a simple matter to prepare such a diagram.

The stadia slide rule is constructed like the ordinary slide rule, except that on the slide are given values of $\cos^2 \alpha$ and $\tfrac{1}{2} \sin 2\alpha$.

The Beaman stadia arc, in modified form known also as the *stadia circle*, is a specially graduated arc on the vertical circle of the transit or the plane-table alidade. The stadia arc has no vernier, but settings are read by means of an index mark "H" for horizontal distances and another index mark "V" for differences in elevation.

For any given vertical angle of sighting, the reading of each index mark is a correction factor, in per cent, to be applied to the observed stadia distance. In one type of stadia arc, the factor is multiplied by the stadia distance. In another type, the product of the factor and the stadia distance is subtracted from the stadia distance. The ordinary slide rule is sufficiently accurate for the multiplications.

115. Transit-stadia Surveying.—On transit surveys of low precision where only horizontal angles and distances are required, the stadia is more rapid than chaining. In general, the surveying procedure parallels that when the tape is used. Stadia intervals are observed as each point is sighted. Horizontal angles are measured, but vertical angles are observed only if large enough to make the horizontal distance appreciably different from the stadia distance (say, when greater than 3°) and then are estimated without reading the vernier. The field party consists of a transitman, one or more rodmen, and usually a recorder.

On topographic and similar surveys, both the elevation of each point and its location in a horizontal plane are desired. As each point is sighted, both the horizontal and the vertical angle are measured, and the stadia interval is observed. This method may be employed merely for the location of details, the horizontal and vertical control being established by other means; or it may be employed for establishing control as well as for details.

If details only are to be located, the transit is set up at a traverse or triangulation station the elevation and location of which are known. The height of the instrument (H.I.) above the station is measured with a rod or a tape. The transit is oriented by sighting along a line whose azimuth is known, this azimuth having been set off on the horizontal circle. The upper motion is unclamped, and sights to desired points are taken.

Where the required precision is not high, the control may be established by transit-stadia traverse, and the details may be located at the same time. For the traverse it is customary to observe the stadia interval and the vertical angle both forward and back from each set-up of the transit, employing the mean value in computations. In measuring vertical angles it is customary, wherever practicable, to sight at a rod reading equal to the height of instrument above the station over which the transit is set. Horizontal sights are taken wherever possible.

Figure 115 shows a page of notes for a stadia traverse for which side shots are taken as the work of running the traverse progresses. The elevation of station *P* 49 has previously been determined as 785.1. Directions of the traverse lines are determined by azimuths and roughly checked by observed magnetic bearings, and stadia distances are recorded, rather than the rod intervals. The backsight from station *P* 50 to *P* 49 checks reasonably close with the foresight from station *P* 49 to *P* 50. The sights to points 502 and *P* 51 are horizontal; the rod reading is shown in the notes, and the difference in elevation is determined by direct leveling. In topographic surveying a sketch is included in the notes, the points to which sights are taken being numbered in the sketch.

PRELIMINARY (STADIA) SURVEY OF I.N.RY., BRIGHTON TO CAMBY

Obj.	Az.	Mag.B.	Stad.Dist.	Vert.Ang.	Hor.Dist.	Diff.Elev.	Elev.	Nov. 27, 1912 J.C.Clark, ⊼
	Inst. at Sta. P49; H.I.=4.7						785.1	Cold T.N.Tillman, Notes
P48	169°34'	S10°30'E	637	-2°27'	636	-27.2	757.9	W.W. & H.H., Rods
P50	38°21'	N38°15'E	681	+1°14'	681	+14.6	799.7	On slope.
491	151°10'		366	-7°21'	360	-45.6	739.5	West bank Green River.
492	126°35'		418	-5°59'	413	-43.3	741.8	East " " "
493	78°05'		385	-5°36'	381	-37.4	747.7	West " " "
494	81°20'		387	-5°40'	383	-38.0	747.1	East " " "
495	298°55'		214	+6°34'	211	+24.3	809.4	Top slope.
	Inst. at Sta. P50; H.I.=4.9						799.7	
P49	218°21'	S38°30'W	683	-1°13'	683	-14.5	785.2	
501	294°40'		415	+4°38'	412	+33.4	833.1	Top slope.
502	16°00'		308	0° on 2.1	308	+2.8	802.5	On slope.
503	137°35'		374	-6°36'	369	-42.8	756.9	West bank Green River.
504	136°10'		486	-5°52'	481	-49.4	750.3	East " " "
505	5°45'		322	+7°36'	316	+41.8	841.5	Top slope.
P51	59°38'	N59°30'E	529	0° on 10.1	529	-5.2	794.5	On slope.
506	94°25'		487	-3°36'	485	-30.5	769.2	West bank Green River.

FIG. 115.—Stadia notes for preliminary route survey.

To maintain a given precision in computed values of difference in elevation, stadia intervals must be observed with much greater refinement where vertical angles are large than where they are small.

116. Precision of Stadia Surveying.—An important advantage of transit-stadia surveying over transit-tape surveying is that, in determining distances and differences in elevation by stadia, the important errors can be made accidental whereas, in chaining, the important errors are systematic.

Many factors influence the precision of stadia surveying, and no definite statement of the precision for a given procedure can be made. Following are estimates which are believed to be fairly representative of several classes of stadia work, these estimates being based upon experience.

1. For side shots where a single observation is taken with sights steeply inclined and with no particular care taken to ensure the rod's being plumb, horizontal distances may have a precision lower than $1/100$, and individual differences in elevation may be in error 2 ft. or more per 1,000 ft. of horizontal distance.

2. Under the same conditions as in (1) but with small vertical angles and reasonable care used in approximately plumbing the rod and with lengths of sight between 200 and 1,500 ft., the precision of horizontal distances should not be lower than $1/200$; differences in elevation per 1,000 ft. of horizontal distance need not be in error more than 0.3 ft. if vertical angles are observed to 01′ or more than 1 ft. if vertical angles are estimated to 05′.

3. For a rapid stadia traverse of considerable length run through rough country with numerous long sights, angles being measured to minutes but without special precaution to eliminate systematic errors, the error of closure may be as low as 25 ft. per mile in plan and 3 ft. per mile in elevation.

4. For conditions as in (3) but for country fairly level so that all vertical angles are small, the error of closure ought not to exceed 15 ft. per mile in plan and 0.5 ft. $\sqrt{\text{distance in miles}}$ in elevation.

5. For rough country with vertical angles up to 15°, angles to minutes, rod standardized, rod plumbed with level, sights limited to 1,500 ft. and taken forward and back from each transit station, and interval factor carefully determined, the error of closure may be less than 15 ft. $\sqrt{\text{distance in miles}}$ in plan and 1 ft. $\sqrt{\text{distance in miles}}$ in elevation.

6. For conditions as in (5) but for level country so that all vertical angles are small, the error of closure may be as small as 6 ft. $\sqrt{\text{distance in miles}}$ in plan and 0.3 ft. $\sqrt{\text{distance in miles}}$ in elevation.

7. For conditions as in (5) but stadia intervals determined by use of a target rod with two targets and observations made during cloudy days, the error of closure in plan should not exceed 4 ft. $\sqrt{\text{distance in miles}}$.

117. Numerical Problems.

1. With line of sight horizontal, a stadia reading is taken on a rod held at a chained distance of 600.0 + C ft. from the transit station. The rod reading of the lower stadia hair is 0.82 ft. and of the upper stadia hair is 6.77 ft. What stadia interval factor is indicated by this observation?

2. A stadia interval of 6.31 ft. is observed with a transit for which the stadia interval factor is 98.5 and C is 1.0 ft. The vertical angle is +7°42′. Determine the horizontal distance and difference in elevation by means of (*a*) the exact stadia formulas for inclined sights, (*b*) the approximate formulas, and (*c*) Table VI.

3. Following are stadia intervals and vertical angles for a transit-stadia traverse. The elevation of station A is 418.6 ft. The stadia interval factor is 100.0, and $C = 1.0$ ft. Compute the horizontal lengths of the courses and the elevations of the transit stations.

Station	Object	Stadia interval, feet	Vertical angle
B	A	8.50	$+0°48'$
	C	4.37	$+8°13'$
C	B	4.34	$-8°14'$
	D	12.45	$-2°22'$
D	C	12.41	$+2°21'$
	E	7.18	$-1°30'$

118. Field Problems.

Problem 16. Determination of Stadia Interval Factor

Object.—To determine the stadia interval factor $K = f/i$ of transit or level.

Procedure.—(1) As described in Art. 111a, employing a line about 800 ft. long. Determine f and c by measurement (Art. 111). (2) For each observation, read the rod for lower, middle, and upper hairs. (3) Compute K for each distance for lower half interval, upper half interval, and full interval; take the mean of all computed values as the factor for the instrument. Discard any readings that differ widely from the others.

Problem 17. Preliminary Traverse of Route with Transit and Stadia

Object.—To obtain data for plotting a topographic map of a proposed highway route between two governing points. The data may be used in office problem 21 (p. 132).

Procedure.—(1) Run a stadia azimuth traverse between the two points, establishing stadia stations at advantageous points near where it appears that the line will eventually be placed. (2) Determine the distance between adjacent stations by observing the stadia interval on both backsights and foresights. (3) Observe the vertical angle between instrument stations on both backsights and foresights. Record the H.I. at each set-up. (4) Make the available checks before moving the transit. (5) While running the traverse, take side shots 200 to 600 ft. on each side of the traverse line, as necessary to define the configuration of the land and the location of objects that might affect the proposed line. (6) Note the type of soil, any indications of rock near the surface, and the type of cover. (7) Keep notes in the form of the sample notes (Fig. 115).

Hints and Precautions.—(1) In determining the differences in elevation and the horizontal distances between traverse points, use the mean of the two vertical angles and the mean of the two stadia readings taken along the line joining these points. (2) Before taking side shots about a station occupied, set the next stadia station in advance. (3) In running the traverse the magnetic bearing of each line should be recorded and immediately compared with the bearing computed from the azimuth of the line. (4) Inclined distances with vertical angles of less than 3° may be considered as horizontal without appreciable error. (5) Observe vertical angles to the nearest minute. Observe azimuths of traverse lines to the nearest minute and azimuths of sights to details to the nearest 05′ without the use of the vernier. Read the rod intercept to the nearest 0.01 ft. (6) Many shots can be taken with the telescope leveled as in direct leveling. (7) The observer should form the habit of judging distances by eye, in order to avoid large mistakes. The middle cross-hair should not be mistaken for one of the stadia hairs.

Problem 18. Closed Traverse and Location of Details with Transit and Stadia

Object.—To collect sufficient data for making a topographic map of an assigned tract. The data may be used in office problem 21 (p. 132).

Procedure.—(1) Make a rapid reconnaissance of the tract, selecting the most advantageous points for instrument stations from which areas comprising the entire area can be observed. (2) Run a closed azimuth traverse through the selected points, observing the stadia intervals and vertical angles. The allowable angular error of closure should not exceed $01' \sqrt{\text{number of sides}}$. The error of closure in elevation should not exceed $0.3 \text{ ft.} \sqrt{\text{distance in thousands of feet}}$. (3) Occupy each of the traverse stations, and with the instrument correctly oriented observe the azimuth, stadia distance, and vertical angle to all changes in ground slope and to other natural and artificial features which are within range of the instrument. (4) Include in the notes a sketch drawn approximately to scale. (5) By means of Table VI, a stadia slide rule, or a stadia reduction diagram, determine the horizontal distance and the elevation of each side shot.

Hints and Precautions.—(1) See "Hints and Precautions" of the preceding field problem.

CHAPTER IX

MAP DRAFTING

119. The Drawings of Surveying.—It is assumed that the student is familiar with the use of the ordinary drafting instruments and with the elements of mechanical drawing. The drawings of surveying consist of maps, profiles, cross-sections, and (to some extent) graphical calculations; the usefulness of these drawings is dependent largely upon the accuracy with which points and lines are projected on paper. For the most part, few dimensions are shown, and the person who makes use of the drawings must rely either upon distances as measured with a scale or upon angles as measured with a protractor. A consistent relation between the field measurements and the map requires great care in plotting.

Profiles and cross-sections are plotted as described in Arts. 65 and 202.

If the relief of the ground surface is not to be indicated, the map is called a *plan*, a *plot*, or a *planimetric map*. If the relief is to be shown, the map is called a *topographic map*. In general, the following information should appear on any map:

a. The direction of the meridian.
b. A graphical scale of the map with a corresponding note stating the scale at which the map was drawn.
c. A legend or key to symbols other than the common conventional signs (Figs. 124*a–d*).
d. An appropriate title.

In addition, a map that is to become a part of a public record of land division should contain the following information:

1. The length of each line.
2. The bearing of each line or the angle between intersecting lines.
3. The location and kind of each monument set, with distances to reference marks.
4. The location of the tract with reference to established coordinate axes.
5. The number of each formal subdivision such as a section, block, or lot.
6. The location and name of each road, stream, landmark, etc.
7. The names of all property owners, including owners of property adjacent to the tract mapped.

8. A full and continuous description of the boundaries of the tract by bearing and length of sides; and the area of the tract.

9. The witnessed signatures of those possessing title to the tract mapped; and if the tract is to be an addition to town or city, a dedication of all streets and alleys to the use of the public.

10. A certification by the surveyor that the plot is correct to the best of his knowledge.

Explanatory notes may be employed to give such information as the sources of data for the map, the precision of the survey, or the datum to which the elevations are referred.

120. Scales.—The scale of a map is the fixed relation that every distance on the map bears to the corresponding distance on the

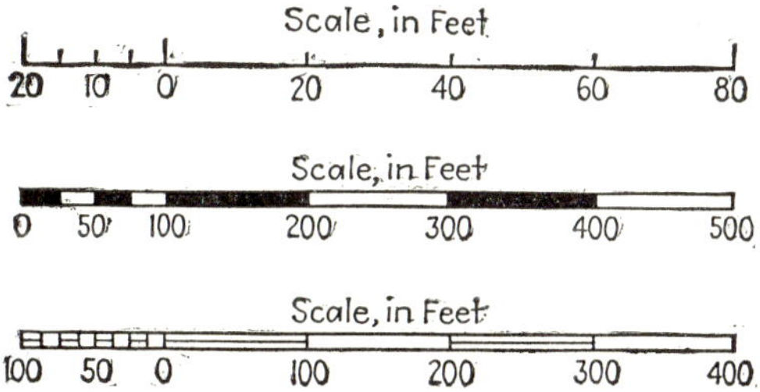

FIG. 120.—Graphical scales.

ground. It may be stated either by numerical relations or graphically, as follows:

1. One inch on the drawing represents some whole number of tens, hundreds, or thousands of feet on the ground, as 1 in. = 200 ft. This type of scale is used for most maps for construction purposes.

2. One unit of length on the drawing represents a stated number of the *same* units of length on the ground, as 1/62,500. This ratio of map distance to corresponding ground distance is called the *representative fraction.* The scale is independent of the units of measurement. It is used extensively for geographic and military maps.

3. A *graphical scale* is a line subdivided into map distances corresponding to convenient units of length on the ground. Various forms of subdivision are shown in Fig. 120, in which the top line represents a scale of 1 in. = 40 ft. and the two lower scales represent 1 in. = 200 ft. On a graphical scale the units of measurement should always be stated.

The numerical scales described above are subject to error if the drawing paper shrinks or swells or if the map is reproduced in other sizes by photographic means. If distances are to be determined accurately from the map, *a graphical scale should always be shown.*

In general, the scale of a map should be no larger than is necessary to represent the location of details with the required precision. For convenience in discussion, maps are herein divided arbitrarily into those of

Large scale: 1 in. = 100 ft. or less.
Intermediate scale: 1 in. = 100 ft. to 1 in. = 1,000 ft.
Small scale: 1 in. = 1,000 ft. or more.

121. Meridian Arrows.—The direction of the meridian is indicated by a needle or feathered arrow pointing north, of sufficient length to be transferred with reasonable accuracy to any part of the map. The true meridian is usually represented by an arrow with *full* head or by a star; the magnetic meridian by an arrow with *half* head. When both are shown, the angle between them should be indicated. A simple design is shown in Fig. 121.

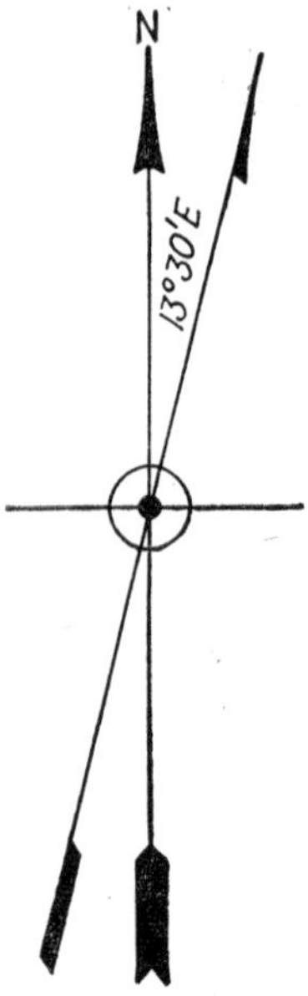

FIG. 121.—Meridian arrows.

122. Lettering.—In general, lettering should be freehand, although guide lines and slope lines should be employed. The *Reinhardt* style of single-stroke lettering is employed almost entirely. The letters are constructed rapidly and are easy to read. Reinhardt letters are made either vertical (Fig. 122*b*) or inclined (Fig. 122*a*.) Detailed information concerning lettering is to be found in textbooks on drawing.

123. Titles.—The best position for the title is the lower right-hand corner of the sheet. Each line should be centered, and the title as a whole balanced. The parts should be weighted in order of their importance, beginning with the principal object of the drawing or with the name of the area. Generally the title should state the purpose or kind of map, the name and location of the tract or project, the name of the owner, the scale of the drawing (unless this is shown elsewhere), the contour interval, the name of the engineer and the draftsman, and the date. Revisions of the map should be shown by dated notes at the left of the title.

124. Conventional Signs.—Objects are represented on a map by signs or symbols, many of which are conventional. Some of these are shown in Figs. 124*a*–*d*. The size of the symbols should be proportioned somewhat to the scale of the map.

In drawing the symbols for culture, marsh, etc. (Figs. 124*c–d*) the separate symbols are distributed evenly over the area but are irregularly spaced so as not to give the appearance of rows. The water-surface lines for marsh are drawn with a ruling pen, but all other parts of the symbols are drawn freehand.

Tree symbols are usually drawn in plan. It is common practice to differentiate between deciduous and evergreen trees (Fig. 124*c*).

To draw the water-line symbol, the draftsman begins by sketching at the shore line two hair lines spaced closely. Each succeeding line is drawn in close conformity with the one preceding, and the spacing

Fig. 122*a*.—Reinhardt letters, slope form.

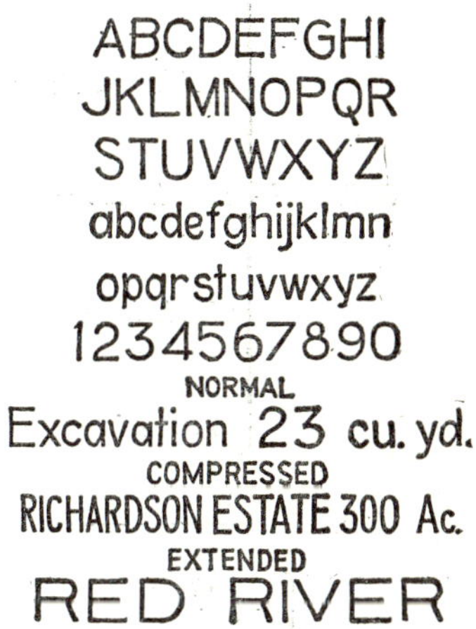

Fig. 122*b*.—Reinhardt letters, vertical form.

between lines is increased uniformly outward from the shore. Each line is composed of a series of smooth intersecting arcs, each concave toward the shore. If the space is large it may be left without lines except near the shore line.

Contour lines are drawn as fine, smooth, freehand lines of uniform width (Chap. XII). Preferably a contour pen (Fig. 126*c*) is used.

A sand bar or flat is represented by dots spaced evenly over the area. The boundary of the area is formed by two or three rows of closely and evenly spaced dots somewhat larger than the rest.

125. Drawing and Tracing Papers.—Pencil drawings and temporary drawings are often made on a smooth manila *detail paper*. For general map work a fairly smooth, tough *drawing paper* of uniform texture is desirable. The paper should take ink well and should stand erasures without its surface becoming fibrous.

A *tracing* is a drawing in ink or pencil on a transparent sheet of paper or cloth, for the purpose of reproduction. *Tracing paper* is economical and is satisfactory for many purposes, but it will not

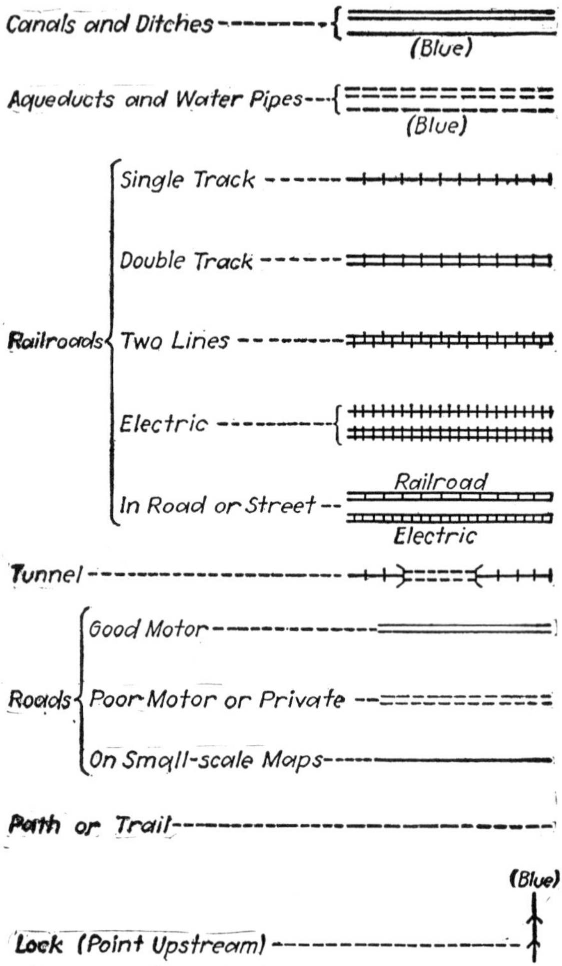

FIG. 124*a*.—Conventional signs: works and structures.

stand repeated erasures well, and it will become torn and cracked unless it is handled carefully. *Tracing cloth* is made from fine linen cloth specially treated to render it transparent. It must be kept dry. The unglazed side is preferred by most draftsmen. Preparatory

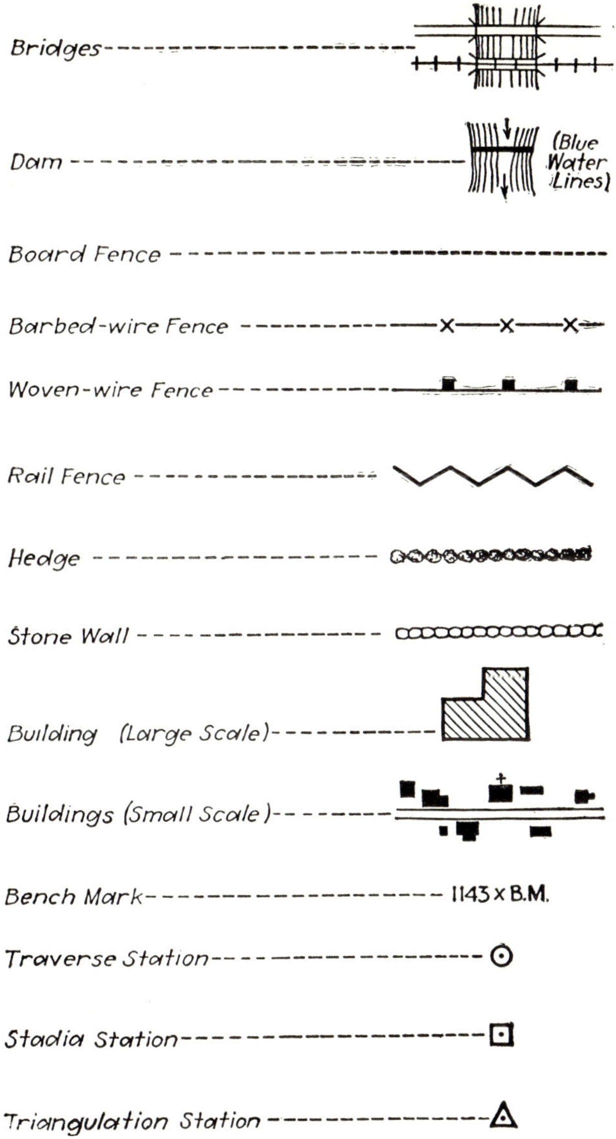

FIG. 124b.—Conventional signs: works, structures, stations.

to tracing in ink on cloth, the surface is dusted with powdered talc or chalk and is rubbed with a dry cloth.

Erasures of ink on cloth are made with least damage to the surface by rubbing lightly with a soft pencil eraser, using an erasing shield. Ordinary ink erasers are too abrasive and produce a fibrous surface.

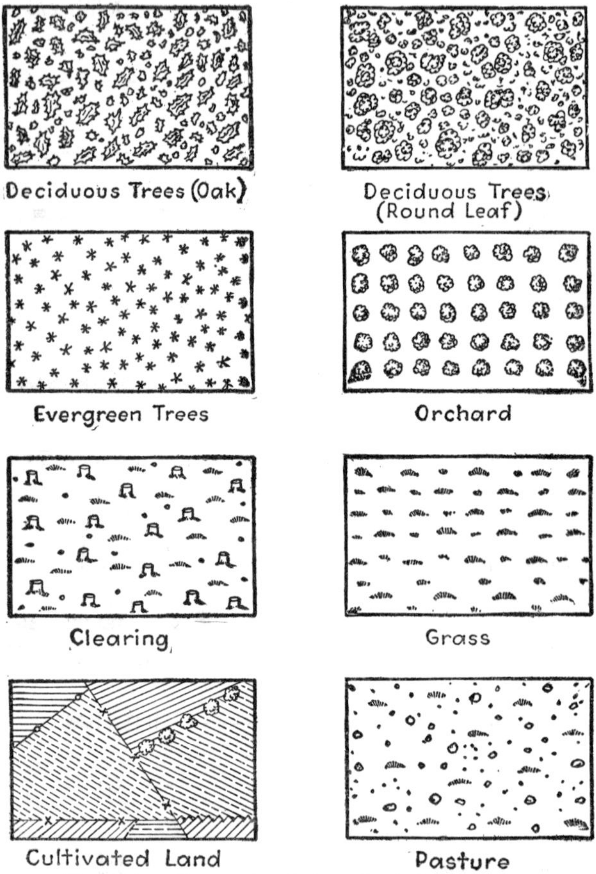

Fig. 124*c*.—Conventional signs: land classification.

Pencil lines are removed and the tracing is cleaned by rubbing either with art gum or with a cloth saturated in gasoline, cleaner's naphtha, benzine, or carbon tetrachloride.

126. Drawing Instruments.—*Engineer's scales* are divided into 10, 20, 30, 40, 50, 60, 80, or 100 parts to the inch. The 12-in. triangular boxwood rule (Fig. 126*a*) is most commonly used in mapping. A flat rule with two scales on edges of opposite bevel is also used.

Probably $\frac{1}{100}$ in. is as close as distances can be plotted by the ordinary methods of drafting. For accurate drawings, points should

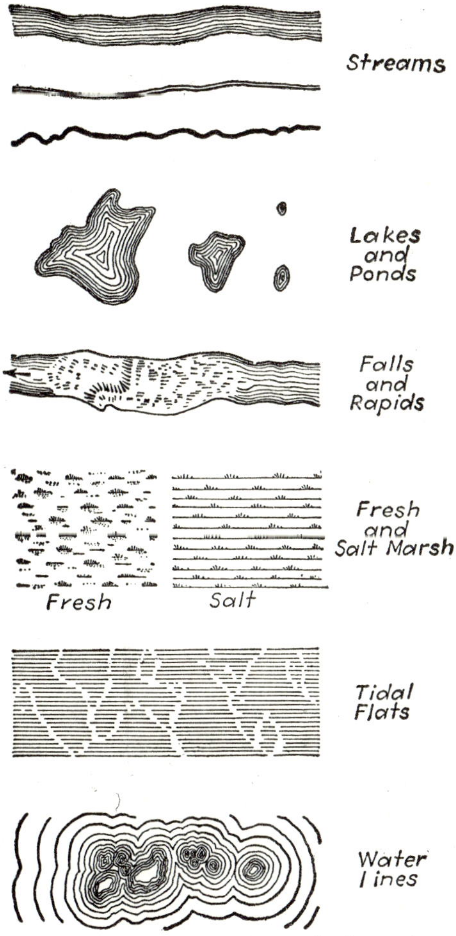

FIG. 124*d*.—Conventional signs: hydrography.

be pricked with a fine needle, and distances should be plotted with the eye directly above the graduation to which the distance is measured.

The *protractor* is a device for laying off and measuring angles on drawings. A *beam compass* is used for drawing the arcs of large circles, say of radius greater than 6 in. Arcs of large circles can also be drawn by means of *railroad curves*, which are thin curved strips of cardboard, wood, metal, hard rubber, or celluloid. The *road pen*

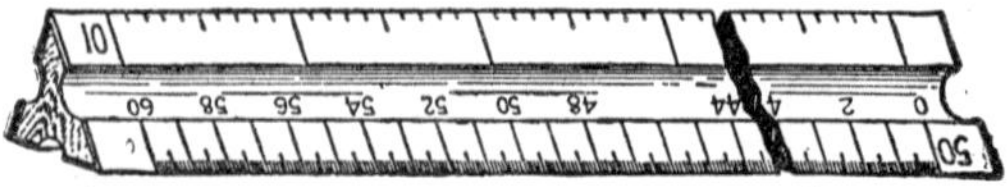

FIG. 126*a*.—Engineer's scale, triangular.

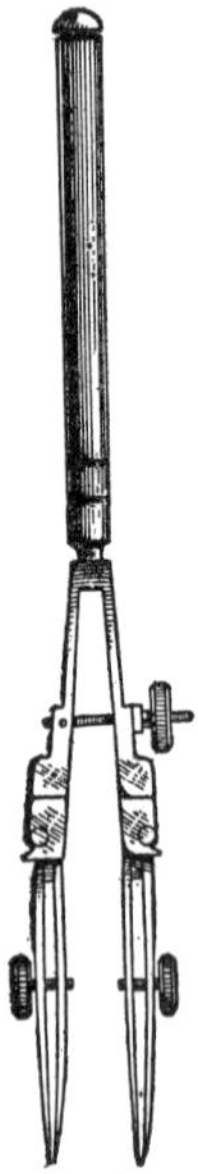

FIG. 126*b*.—Road pen.

FIG. 126*c*.—Contour pen.

or *railroad pen* (Fig. 126*b*) is used principally for drawing two parallel lines. The *contour pen* (Fig. 126*c*) is useful for drawing contours or other freehand curves. For transferring distances from one map to another at a different scale, *proportional dividers* are useful.

CHAPTER X

MAP PLOTTING

127. Process of Making a Map.—The operations of plotting are in a sense the reverse of the operations of surveying. The general aim is to plot the more definite features within proper limits of error. The process of mapping involves (1) the plotting, by more precise methods, of points of horizontal *control* which are generally transit stations and which may be traverse points, triangulation points, or both and (2) the plotting, by less precise methods, of the map *details*, employing angular and linear measurements from the lines and points in the horizontal control system.

Horizontal control may be plotted (1) by use of the protractor, (2) by the tangent method, or (3) by the coordinate method. In any case distances are measured with the engineer's scale and points are pricked with a needle. Control lines are drawn carefully with a hard pencil having a fine point. Usually, particularly for maps for general purposes, the control is not shown on the finished map.

For methods of plotting details, see Art. 137. For methods of plotting profiles and cross-sections, see Arts. 65 and 202.

128. Protractor Method.—When a deflection-angle traverse is to be plotted by protractor, the common practice is as follows: The position of the first line is fixed by estimation, and its length (as AB) is laid off by measurement. The protractor is oriented at the forward point (as B), the deflection angle to the succeeding line is laid off, and a line of indefinite length is drawn. Along this line is laid off the given distance (as BC) to the succeeding traverse point (as C); and so on.

If azimuths or bearings of the lines of a traverse are given or computed, a meridian line is drawn through each station, and the direction of the succeeding line is laid off with respect to the meridian.

129. Tangent Method.—In principle, this method is similar to the protractor method, with the difference that an angle is laid off by a linear measurement which is a constant times the natural tangent of the angle. In Fig. 129, AB represents the plotted position of the initial line of a traverse, and α is the deflection angle at B which it is desired to lay off in order to determine the direction of BC. By the tangent method, the line last established (in this case

AB) is prolonged some convenient distance, usually 10 in., to form a base line *Bb*, at the end of which a perpendicular *bb′* is erected; and the distance *bb′* is laid off equal to the length of the base line *Bb* multiplied by the natural tangent of α. A line drawn from *B* through *b′* defines the direction of *BC*.

Generally if the deflection angle is greater than 45°, the base line is established not as a prolongation of the preceding line but as a perpendicular to the preceding line at the point last plotted. Thus in the figure *Cc* is the base line perpendicular to *BC*; and the perpendicular offset *cc′* is the length of the base line *Cc* times the natural cotangent of the angle β.

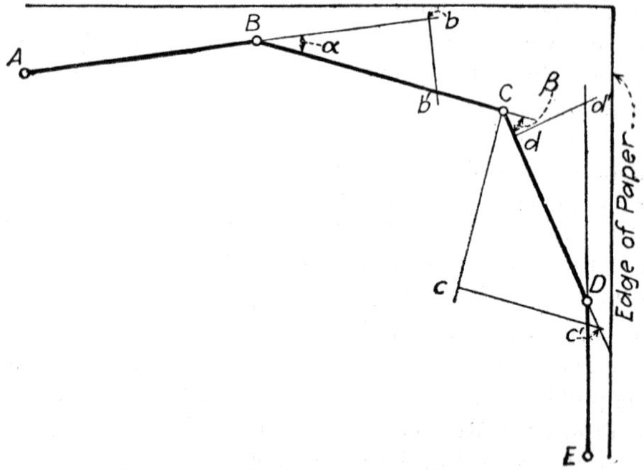

FIG. 129.—Plotting by tangent offsets.

130. Checking.—The methods of checking are essentially the same for the protractor and the tangent methods of plotting just described. A traverse that has been brought to closure in the field should likewise close on paper. If the field work is assumed as correct, any error of paper closure is due to inaccuracies in laying off angles and distances.

If the error of closure is small, it is usually assumed to have accumulated gradually, and the traverse is made to close by an arbitrary progressive change in the position of the plotted lines, as shown in Fig. 130*a*. Another method of closure is illustrated in Fig. 130*b*; each traverse point is moved in a direction parallel to the side of closure, by an amount proportional to the distance along the traverse from the initial point to the given point. For example, $C'C = AB'C'/AB'C'D'E'A' \times A'A$.

If the error of closure is large, a mistake in laying off an angle or a distance has occurred; sometimes the mistake may be located quickly by observing the simplest way of bringing the traverse to a closure. Thus in Fig. 130*b*, with the error of closure *A'A* as shown, one might expect that a mistake had been made in the length of the side *C'D'* which is nearly parallel to the side of closure.

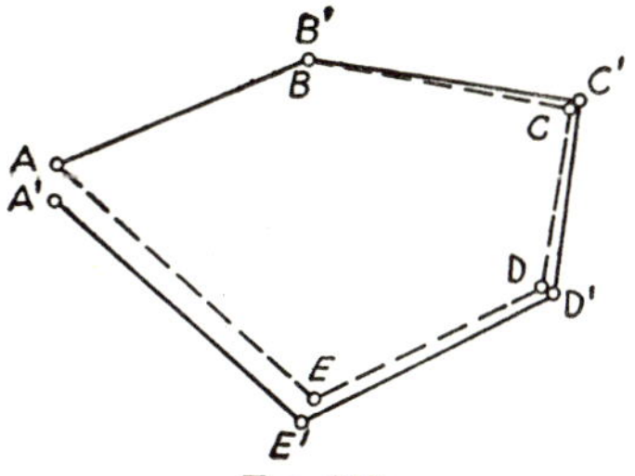

Fig. 130*a*.

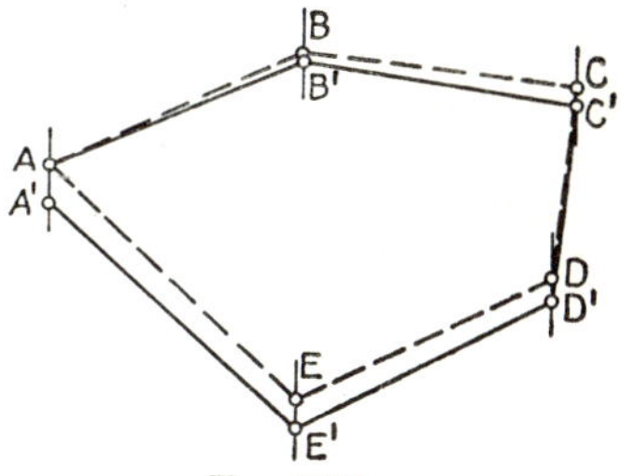

Fig. 130*b*.

131. Method of Rectangular Coordinates.—This method, also known as the *method of total latitudes and departures*, is recognized as the most reliable method of plotting. It is the only practical method for plotting extensive systems of horizontal control, whether they be established by traversing or by triangulation. Rectangular coordinates are employed not only for plotting maps but also frequently for calculating areas, as described in Art. 142.

The coordinate axes are a *reference meridian* (true, magnetic, or assumed) and a line at right angles thereto called a *reference parallel*. The intersection of these lines, marking the origin, may be any point in the survey or may be a point entirely outside the survey. With the direction of each survey line determined and its length known, computations are made to determine its *latitude* and *departure*, or the lengths of its orthographic projection upon the meridian and the parallel respectively. The origin having been chosen, the coordinates for the several control points are computed by using the latitudes and departures. The coordinate of a point measured normal to the {parallel / meridian} is called the {*total latitude* or the *parallel distance* / *total departure* or the *meridian distance*} of the point. {Total latitudes / Total departures} are positive or negative according to whether the corresponding points lie {north or south of the reference parallel / east or west of the reference meridian}. With the coordinate axes established on paper, a point is plotted by laying off its total latitude and total departure to the required scale. Articles 132 to 136 are devoted to a more detailed discussion of the processes involved.

Before the latitudes and departures of the traverse lines are computed, the angular error of closure of the traverse is determined by the known geometrical conditions, and the angles or bearings are so adjusted or corrected that the known geometrical conditions will be fulfilled (see Art. 134*a*).

132. Latitudes and Departures.—In Fig. 132*a*, AB represents any line the latitude and departure of which it is desired to determine, and the lines NS and EW represent a meridian and a parallel. The line AB makes the angle α with the meridian.

As the latitude of a line is the orthographic projection of the line upon a meridian, the latitude of AB is $A'B' = Ab = AB \cos \alpha$.

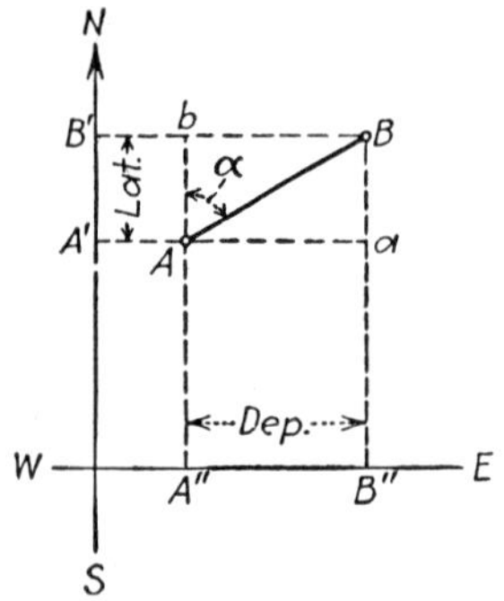

Fig. 132*a*.

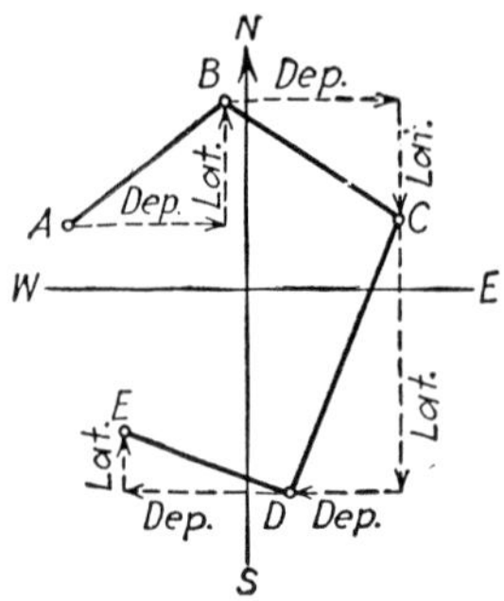

Fig. 132*b*.

And as the departure of a line is its orthographic projection upon a parallel, the departure of AB is $A''B'' = Aa = AB \sin \alpha$. Stated in the form of a general rule applicable to any line:

$$\text{Latitude} = \text{length} \times \text{cosine bearing angle} \quad (1)$$
$$\text{Departure} = \text{length} \times \text{sine bearing angle} \quad (2)$$

Latitudes are designated as *North* or *positive* for all lines having a northerly bearing, and *South* or *negative* for all lines having a southerly bearing. Departures are designated as *East* or *positive* for lines having an easterly bearing, and *West* or *negative* for lines having a westerly bearing. Thus, in Fig. 132*b*, for the line AB the latitude is North or + and the departure is East or +. For BC the latitude is South or − and the departure is East or +. For CD the latitude is South or − and the departure is West or −. For DE the latitude is North or + and the departure is West or −.

Computations are made either with a computing machine or by logarithms. If the computing machine is employed, the data may be kept in the following form:

Line	Bearing	Length	cos bearing	sin bearing	N latitude	S latitude	E departure	W departure

First the bearings and lengths of the lines are tabulated. Then the natural cosines and sines of the bearing angles are recorded. And finally for each line the latitude and departure are computed, the length being set on the machine as the multiplicand.

If logarithms are used, the form of the following example is convenient:

Example: For a given course in a traverse the computed bearing is N34°21′W and the observed length is 1,215.3 ft. The traverse is to be plotted to the scale of 1 in. = 100 ft. It is desired to compute the latitude and departure.

At the given scale, distances may be plotted within 1 or 2 ft., and therefore latitudes and departures should be correct to perhaps the nearest quarter foot if the traverse has many sides. Five-place logarithms will be used. The computations are given in the accompanying tabulation. Beginning with "log distance" in the tabulation, computations for latitude are made reading upward, and computations for departure are made reading downward.

Latitude....................................	1,003.3 ft.
log latitude....................................	3.00145
log cos bearing....................................	9.91677
log distance....................................	3.08468
log sin bearing....................................	9.75147
log departure....................................	2.83615
Departure....................................	685.7 ft.

133. Error of Closure.—In any closed traverse it is obvious that the sum of the north latitudes should equal the sum of the south latitudes and that the sum of the east departures should equal the sum of the west departures. In other words, for any closed traverse the algebraic sum of the latitudes (ΣL) should be equal to zero, and the algebraic sum of the departures (ΣD) should be equal to zero. However, owing to errors in field measurements of both angles and distances, in general an unadjusted traverse will not close on paper even though the plotting be without error. The conditions stated above make it possible to determine the error of closure by means of the computed latitudes and departures.

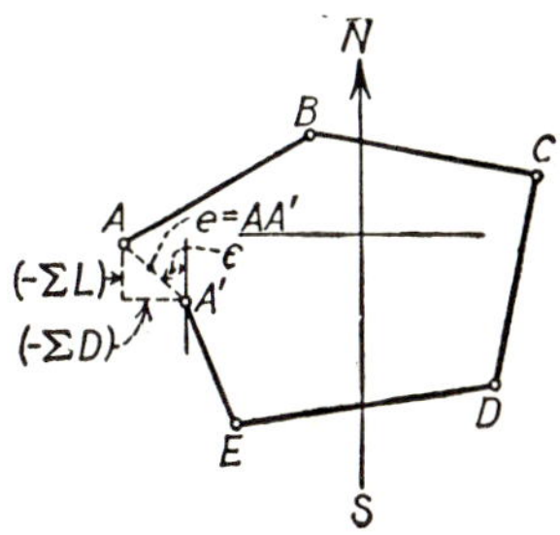

FIG. 133.—Error of closure.

Figure 133 shows a traverse that does not close, the line $A'A = e$ being the side of error. In order that the algebraic sum of the latitudes and the algebraic sum of the departures shall each be equal to zero, the latitude of the side of error must be $-\Sigma L$, and its departure must be $-\Sigma D$, considered algebraically. These two quantities form the base and altitude of the right-angle triangle of which the side of error is the hypotenuse; hence the linear error of closure is

$$e = \sqrt{(\Sigma L)^2 + (\Sigma D)^2} \tag{3}$$

The direction of the side of error is given by the relation

$$\tan \epsilon = \frac{-\Sigma D}{-\Sigma L}$$

with due regard to sign. The data will make it apparent in which quadrant the bearing lies.

134. Balancing the Survey.—When the error of closure has been determined as described in the preceding article, usually corrections are made so that the traverse will form a mathematically closed figure, and the corrections are applied to the latitudes and departures in such manner as to make their algebraic sum equal zero. This operation is called *balancing the survey* or *balancing the traverse.* There are several rules for distribution of errors; however, many surveyors rely upon their own judgment and distribute the error arbitrarily in accordance with their estimation of the field conditions.

134a. Adjustment of Angular Error.—The total angular error determined from the known geometrical conditions is distributed among the angles or computed bearings of the traverse before computations of the latitudes and departures are made. Often this adjustment is arbitrary, based upon a knowledge of the field conditions; but if all angles have been measured under like conditions, the error is distributed equally to each angle in the traverse.

The angular adjustment does not yield true values, but it meets the known geometrical conditions; and the adjusted values are the most probable values that can be assigned.

134b. Compass Rule for Balancing a Survey.—The compass rule to balance a traverse is the one most commonly employed. It states that the correction to be applied to the $\left\{\begin{matrix}\text{latitude}\\ \text{departure}\end{matrix}\right\}$ of any course is to the total correction in $\left\{\begin{matrix}\text{latitude}\\ \text{departure}\end{matrix}\right\}$ as the length of the course is to the length of the traverse.

Example: The accompanying tabulation lists the bearings, lengths, latitudes, and departures for a closed traverse of six sides. The survey is to be balanced by the compass rule.

The error of closure in latitude is 15.0 ft., and in departure is 32.0 ft The sum of the lengths of the sides is 4,997.6 ft. or practically 5,000 ft. The length of the side of closure is $\sqrt{15^2 + 32^2} = 35.3$, and the linear error of closure is $35.3/5{,}000 = 1/142$. (This error is too large to be permitted in practice but was made large for the purpose of this example.)

Line	Bearing	Length, ft.	Latitude		Departure		Correction, ft.	
			N	S	E	W	Latitude	Departure
AB	N	500.0	500.0	....	0.0	....	+1.5	+3.2
BC	N45°00′E	848.6	600.0	....	600.0	....	+2.6	+5.4
CD	S69°27′E	854.4	....	300.0	800.0	....	−2.6	+5.4
DE	S11°19′E	1,019.8	....	1,000.0	200.0	....	−3.1	+6.6
EF	S79°42′W	1,118.0	....	200.0	....	1,100.0	−3.3	−7.2
FA	N54°06′W	656.8	385.0	....	....	532.0	+1.9	−4.2
Sum.......	....	4,997.6	1,485.0	1,500.0	1,600.0	1,632.0	15.0	32.0

Sta.	Adjusted lat.		Adjusted dep.		Total lat.		Total dep.		Adjusted bearing	Adjusted length, feet
	N	S	E	W	N	S	E	W		
A					0	0	0	0		
	501.5	....	3.2	....					N0°22′E	501.5
B					501.5	.....	3.2	..		
	602.6	....	605.4	....					N45°08′E	854.2
C					1,104.1		608.6			
	....	297.4	805.4	....					S69°44′E	858.6
D					806.7	.....	1,414.0	..		
	....	996.9	206.6	....					S11°42′E	1,018.1
E					....	190.2	1,620.6	..		
	....	196.7	....	1,092.8					S79°48′W	1,110.4
F					....	386.9	527.8	..		
	386.9	....	....	527.8					N53°45′W	654.4
A					0	0	0	0		
Sum	1,491.0	1,491.0	1,620.6	1,620.6	....	.....	....	...	....	4,997.2

For example, the correction in latitude of the course *AB* is $15/5{,}000 \times 500 = 1.5$ ft. The north latitudes are too small, and hence the correction is to be added; that is, the correction is +1.5 ft. The correction to the departure of *AB* is $32/5{,}000 \times 500 = 3.2$ ft. As the east departures are too small, the corrected departure will be east, and the correction is to be added.

In the eighth and ninth columns of the tabulation are given the corrections in feet. As a check, the arithmetical sum of the corrections in latitude or departure (given in the last line of each column) should equal the total error in latitude or departure. When the corrections have been applied, the algebraic sum of the latitudes and of the departures must be zero, and hence the survey is balanced.

135. Computation of Coordinates.—When it is desired to plot points of horizontal control by the method of coordinates, the latitudes and departures of the control lines are computed, and in the case of a closed traverse the survey is balanced. The origin is chosen, and the coordinates, or total latitudes and departures, of the several points in the survey are computed by summing algebraically the latitudes and departures of lines between that point and the origin.

Example: In the second tabulation are given (1) the adjusted latitudes and departures of the lines in the traverse of the preceding example, (2) the computed total latitudes and departures, or coordinates, referred to station *A* as the origin, and (3) the adjusted bearings and lengths of the traverse lines.

In a closed traverse the additions are verified if on making the complete circuit the total latitude and total departure of the initial point check these quantities as assumed at the beginning. In the tabulation this check has been performed.

136. Plotting Control by Coordinates.—When a system of horizontal control is to be plotted by rectangular coordinates, the size of the enclosing rectangle is determined from an examination of the total latitudes and departures. In order to ensure the proper location of points on the map sheet, usually the enclosing rectangle and the principal points of control are first plotted roughly to small scale. On the drawing paper the enclosing rectangle is drawn to the required scale, its position being fixed by estimation from the small-scale sketch.

Let *HJKL* (Fig. 136) be the enclosing rectangle for the traverse of the preceding article. The rectangle should be drawn with great care, and it should be checked by scaling the lengths of the diagonals. The locations of the reference meridian and the reference parallel are determined by scaling along the sides of the enclosing rectangle. For example, in Fig. 136 the reference parallel is located by scaling *JA* and *KM* equal to 387 ft., the total latitude of the most southerly point of the survey. The location is checked by scaling *AH* and *ML* equal to 1,104 ft., the total latitude of the most northerly point of the survey.

If the drawing is large and many control points are to be plotted, other meridians and parallels are constructed to divide the area into squares which represent some whole number of hundreds of feet at the given scale. Thus in the figure, a meridian is drawn 1,000 ft. east of the reference meridian, and a parallel is drawn 1,000 ft. north of the reference parallel. If the scale of the map is 1 in. = 100 ft., the actual length of the sides of the resulting squares is 10 in.

Each point of horizontal control is located on the sheet by plotting its total latitude and departure, and its location is verified by scaling the length of the preceding course. For example, point *D* (Fig. 136) has a north total latitude of 807 ft. and an east total departure of 1,414 ft. The point D is plotted by laying off to scale above the reference parallel the distances ab and Mc each equal to 807 ft., then drawing the line bc, and finally laying off to scale the distance, $bD = 1{,}414 - 1{,}000 = 414$ ft. As a check, the distance CD as scaled from the drawing should agree with the length of this line as computed from the adjusted latitude and departure ($CD = 858.6$ ft. by the tabulation in Art. 135).

For long open traverses the enclosing rectangle is not constructed. The reference meridian and reference parallel are drawn as accurately

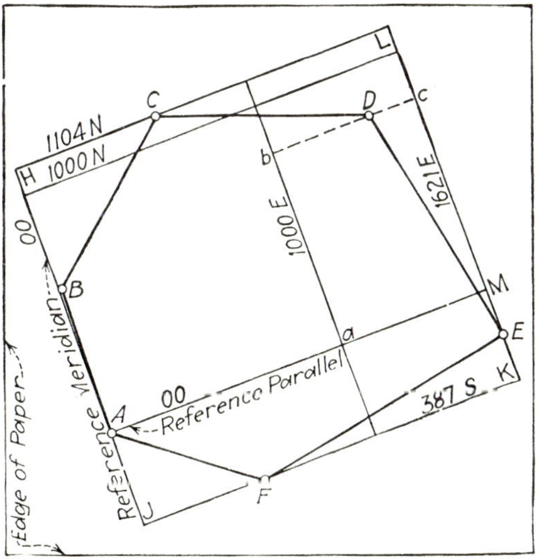

FIG. 136.—Plotting control by coordinates.

as possible; and with these as a basis other meridians and parallels are drawn as necessary for plotting the traverse points.

137. Plotting Details.—If an object has been located in the field by the method of radiation, on the map a line is drawn from the plotted location of the instrument station in a direction corresponding to that on the ground; the distance from station to object is then scaled off along this line. If an object has been located by intersection, the corresponding angles are laid off from the two control stations, and the intersection of the two lines thus defined marks the plotted location of the object. If an object has been located by linear measurements from two stations, its location is plotted by intersecting arcs of the given radii drawn with the dividers. If objects of somewhat indefinite form have been located by perpendicular offsets from a traverse line, usually these offsets may be plotted by estimating the perpendiculars with the eye. The angles are laid off with a protractor.

The work of plotting details is less refined than that of plotting the horizontal control, but the aim is to plot objects of definite size and shape within the allowable limit of error. The location of all important details should be checked by measurement, and that of less important objects by inspection.

138. Omitted Measurements.—When for any reason it is impossible or impractical to determine by field observations the length and bearing of every side of a closed traverse, the missing data may generally be calculated provided that not more than two quantities (lengths and/or bearings) are omitted. It must be assumed that the observed values are without error, and hence all errors of measurement are thrown into the calculated lengths or bearings.

There are three general cases: (1) length and bearing of one side unknown, (2) omitted measurements in adjoining sides of the traverse, and (3) omitted measurements in nonadjoining sides. In case 3, the solution involves changing the order of sides in the traverse in such a way as to make the two partly unknown sides adjoin.

139. Numerical Problems.

1. With the protractor, plot to the scale of 1 in. = 200 ft. the closed deflection-angle traverse for which the accompanying notes are given. Measure the linear error of closure, and record it on the sheet. Distribute the error as suggested in Art. 130.

Station	Deflection angle	Length, feet
A	. . . .	
		338
B	9°00′*L*	
		307
C	76°45′*R*	
		792
D	74°45′*R*	
		822
E	102°00′*R*	
		624
F	23°30′*R*	
		620
A	92°00′*R*	

2. Plot the following open deflection-angle traverse to the scale of 1 in. = 400 ft. by the method of tangents, using a 10-in. base. Lay off successive lines by deflection angles as described in Art. 129. Assume the direction of the first course to be north, and compute the bearings

of the other courses. Check the accuracy of the plotting by methods described in Art. 130. (See also office problem 19.)

Station	Deflection Angle
114 + 92.3	....
96 + 47.7	18°43′*L*
75 + 20.2	36°22′*L*
68 + 00	9°00′*R*
50 + 33.1	52°51′*L*
41 + 89.0	12°08′*L*
32 + 01.4	20°17′*R*
15 + 65.3	6°42′*R*
7 + 18.6	10°24′*L*
0 + 00	....

3. Given the notes of problem 1. Assume the original direction of *AB* to be north, and compute the bearings of the several courses. Compute the latitudes and departures, using four-place logarithms. Compute the error of closure, and balance the survey by the compass rule (Art. 134*b*).

4. Given the following data of an open azimuth traverse. Compute the coordinates of the several points in the survey, assuming that the origin is at *A*. Plot the traverse by the coordinate method, using a scale of 1 in. = 400 ft. (See also office problem 20.)

Course	Azimuth	Distance, feet
AB	124°20′	704.2
BC	153°16′	1,257.1
CD	267°00′	410.7
DE	294°53′	927.6
EF	220°44′	615.3
FG	178°09′	1,973.1
GH	166°28′	2,684.3
HJ	150°13′	1,574.8
JK	112°47′	475.3

140. Office Problems.

Problem 19. Plotting by Tangents; Map Construction

Object.—To plot a given traverse by the method of tangents. The data of field problems 7 (p. 62), 13 (p. 85), or 14 and 15 (p. 86) may be used.

Procedure.—(1) Tabulate angles and distances of the given traverse in the computation book. If angles are given as azimuths or bearings, change them to deflection angles. (2) Tabulate and check the natural tangent of each deflection angle less than 45° and the cotangent of each angle greater than 45°. (3) Plot the traverse roughly to small scale, using the protractor for angles; note its general form. (4) Carefully plot the first line of the traverse on drawing paper to the required scale, estimating the position of the line by means of the small-scale sketch. The line should lie so that the drawing, when finished, will be symmetrical with the sheet. (5) Plot the remaining lines of the traverse by the method described in Art. 129. Verify all measurements as soon as they have been plotted. (6) Check the traverse by the methods of Art. 130. If it is a closed traverse, distribute the error of closure as indicated by the conditions of the problem. (7) Plot the details by methods corresponding to those used in the field. Use conventional signs wherever applicable. Show the meridian. Make a title.

Hints and Precautions.—(1) Particular care should be taken in scaling the lengths of the lines of the traverse and the tangent distances. Points should be pricked with a needle. The eye should be above each point as it is plotted. (2) Lengths of lines and tangent distances plotted by estimation using the 10 scale are not sufficiently accurate; the 50 scale should be used. (3) A perpendicular should not be erected with the corner of the triangle at the point of intersection of base line and perpendicular, as the triangle corner may be rounded. The hypotenuse of the right triangle should be fitted to the base line, and a straightedge (or another triangle) placed in contact with the base of the triangle. Then the triangle should be turned through 90°, and its third edge placed in contact with the straightedge and moved along it until the hypotenuse passes through the point of intersection. (4) It is well to test each perpendicular by measuring the hypotenuse of a 45° right triangle having sides perhaps 8 in. long; in this case the length of the hypotenuse is 11.31 in. (5) If the construction lines would otherwise fall off the edge of the paper, measure back along the line as shown for the plotting of *DE* from *CD* in Fig. 129.

Problem 20. Plotting by Coordinates; Map Construction

Object.—To plot a given traverse by the method of coordinates. The data of field problems 7 (p. 62), 13 (p. 85), or 14 and 15 (p. 86) may be used.

Procedure.—(1) Transcribe the given data to a computation book in form shown by the first three columns of the tabulation in Art. 134*b*. If directions of lines are given as azimuths or deflection angles, compute either the true or the assumed bearings. (2) Compute the latitudes and departures of the traverse lines; and if it is a closed traverse, balance the survey as described in Art. 134*b*. (3) Assume one of the traverse points as the origin of coordinates, and compute the total latitudes and departures as described in Art. 135. (4) Check all computations. (5) To small scale, plot roughly the traverse and enclosing rectangle (Art. 136), and note their relative positions. (6) On drawing paper plot the enclosing rectangle to the required scale, estimating its position by means of the small-scale sketch. The traverse, not the rectangle, should be symmetrical with the sheet. (7) Test the accuracy of the plotting by scaling the length of the diagonals. (8) Plot the reference meridian, reference parallel, and any supplementary meridians and parallels as described in Art. 136. Check the plotting. (9) Locate each traverse point by plotting its total latitude and departure. Check by scaling the length of the preceding traverse line. (10) Plot the details by methods corresponding to those used in the field. Use conventional signs wherever applicable. Finish the map without erasing construction lines. Label each traverse line with its corrected length and its corrected bearing. Show the meridian. Make a title.

Hints and Precautions.—For details of plotting, see "Hints and Precautions" 1 and 2 of the preceding office problem.

CHAPTER XI

CALCULATION OF AREAS OF LAND

141. Methods of Determining Area.—One of the primary objects of most land surveys is to determine the area of the tract. The area may be determined by any of the following methods:

1. By plotting the boundaries to scale as described in Chap. X; the area of the tract may then be found by use of the planimeter or it may be calculated by dividing the tract into triangles and rectangles, scaling the dimensions of these figures, and computing their areas mathematically. This method is useful in roughly determining areas or in checking those that have been calculated by more exact methods.

2. By computing mathematically the areas of the individual triangles into which the tract may be divided, the computations being based upon angular and linear measurements made in the field. Table XI gives the relations between the area of a triangle and its angles and lengths of sides.

3. By calculating the area from the coordinates, or meridian distances and parallel distances, of the *corners* of the tract (Art. 142).

4. By calculating the area from the double meridian distances and the latitudes of the *sides* of the tract. This method is not described herein.

For computation of areas of cross-sections, see Arts. 203 and 204.

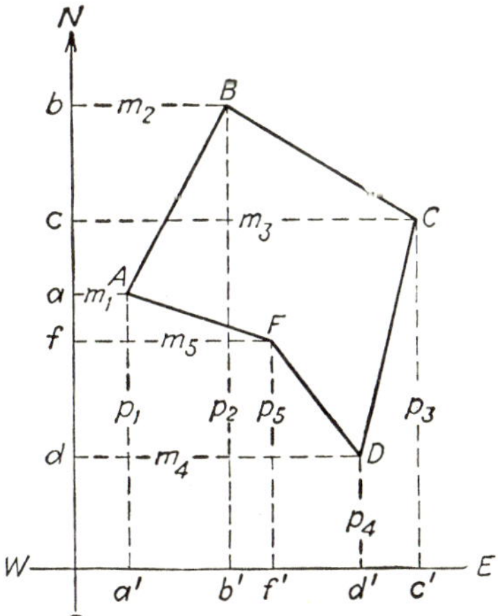

Fig. 142.—Area by coordinates.

142. Area by Coordinates.—If the coordinates of the corners of a tract have been found as described in Art. 135, the area can be calculated expeditiously. Essentially the calculation is that of finding the areas of trapezoids formed by projecting the lines upon one of a pair of coordinate axes.

In Fig. 142, *ABCDF* represents a tract the area of which is to be determined, *SN* a reference meridian, and *WE* a reference parallel. The coordinates of *A*, *B* . . . *F* are known; for any point (as *A*) these coordinates are the perpendicular distance from the reference meridian (as $aA = m_1$), defined as the *total departure* or the *meridian distance;* and the perpendicular distance from the reference parallel (as $a'A = p_1$), defined as the *total latitude* or the *parallel distance.* Meridian distances are regarded as positive or negative according to whether they lie east or west of the reference meridian; parallel distances are regarded as positive or negative according to whether they lie north or south of the reference parallel. In the figure all meridian and parallel distances are positive.

The area of the tract can be computed by summing algebraically the areas of the trapezoids formed by projecting the lines upon the reference meridian; thus

$$\text{Area } ABCDF = \text{area } BCcb + \text{area } CDdc - \text{area } DFfd - \text{area } FAaf - \text{area } ABba \quad (1)$$

or

$$\text{Area} = \tfrac{1}{2}(m_2 + m_3)(p_2 - p_3) + \tfrac{1}{2}(m_3 + m_4)(p_3 - p_4) - \tfrac{1}{2}(m_4 + m_5)(p_5 - p_4) - \tfrac{1}{2}(m_5 + m_1)(p_1 - p_5) - \tfrac{1}{2}(m_1 + m_2)(p_2 - p_1) \quad (2)$$

By multiplication and a rearrangement of terms in Eq. (2) there is obtained

$$2 \cdot \text{area} = -[p_1(m_5 - m_2) + p_2(m_1 - m_3) + p_3(m_2 - m_4) + p_4(m_3 - m_5) + p_5(m_4 - m_1)] \quad (3)$$

Rule.—*To determine the area of a tract of land when the coordinates of its corners are known, multiply the parallel distance, or ordinate, of each corner by the difference between the meridian distances, or abscissas, of the following and the preceding corners, always algebraically subtracting the following from the preceding. One half of the algebraic sum of the resulting products is the required area.*

A result identical except for sign would be obtained by always subtracting the preceding from the following. The sign of the area is not significant.

Example: Given the following data. Find the required area by applying the foregoing rule.

Corner	1	2	3	4	5
Meridian distance, ft.........	300	400	600	1,000	1,200
Parallel distance, ft..........	300	800	1,200	1,000	400

$$
\begin{aligned}
2 \cdot \text{area} &= -[300(800) + 800(-300) + 1{,}200(-600) + 1{,}000(-600) \\
&\qquad + 400(700)] \\
&= -240{,}000 + 240{,}000 + 720{,}000 + 600{,}000 - 280{,}000 \\
&= 1{,}040{,}000 \text{ sq. ft.}
\end{aligned}
$$

$$
\text{Area} = \frac{1{,}040{,}000}{2} = 520{,}000 \text{ sq. ft.}
$$

143. Area of Tract with Irregular or Curved Boundaries.—If the boundary of a tract of land follows some irregular or curved line,

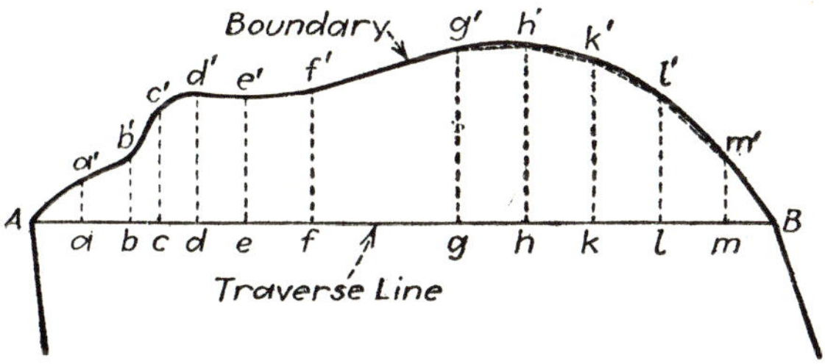

Fig. 143*a*.

it is customary to run a traverse in a convenient location near the boundary and to locate the boundary by offsets from the traverse line. Figure 143*a* represents a typical case, *AB* being one of the traverse lines. The offset distances are aa', bb', etc., and the corresponding distances along the traverse line are *Aa*, *Ab*, etc. Where the boundary is irregular, as from a' to f', it is necessary to take offsets at points of change and hence generally at irregular intervals. Where a segment of the boundary is straight, as from f' to g', offsets are taken only at the ends. Where the boundary is a gradual curve, as from g' to m', ordinarily the offsets are taken at regular intervals.

If the offsets are taken sufficiently close together, the error involved in considering the boundary as straight between offsets is small as compared with the inaccuracies of the measured offsets. The assumed boundary takes some such form as that illustrated by the dotted lines $g'h'$, $h'k'$, etc., in Fig. 143*a*, and the areas between offsets are of trapezoidal shape. Under such an assumption, irregular areas are said to be calculated by the *trapezoidal rule*.

143a. Offsets at Regular Intervals.—Let Fig. 143*b* represent a portion of a tract lying between a traverse line *AB* and an irregular boundary CD, offsets $h_1, h_2 \ldots h_n$ having been taken at the regular intervals d. The summation of the areas of the trapezoids comprising the total area is

$$\text{Area} = \frac{h_1 + h_2}{2} \cdot d + \frac{h_2 + h_3}{2} \cdot d \cdots + \frac{h_{(n-1)} + h_n}{2} \cdot d$$

$$\text{Area} = d\left(\frac{h_1 + h_n}{2} + h_2 + h_3 \cdots + h_{(n-1)}\right) \qquad (4)$$

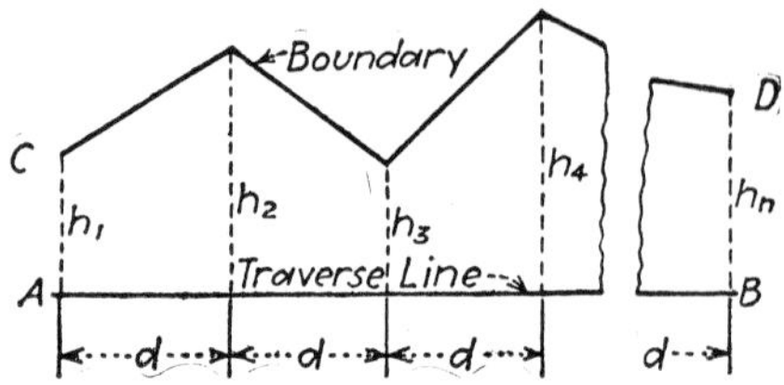

FIG. 143b.

Equation (4) may be expressed conveniently in the form of the following rule:

Trapezoidal Rule.—*Add the average of the end offsets to the sum of the intermediate offsets. The product of the quantity thus determined and the common interval between offsets is the required area.*

Example: Find the area between a traverse line and a curved boundary, rectangular offsets being taken at intervals of 20 ft., and the values of the offsets in feet being $h_1 = 3.2$, $h_2 = 10.4$, $h_3 = 12.8$, $h_4 = 11.2$, and $h_5 = 4.4$. By the trapezoidal rule,

$$\text{Area} = 20\left(\frac{3.2 + 4.4}{2} + 10.4 + 12.8 + 11.2\right) = 764 \text{ sq. ft.}$$

143b. Offsets at Irregular Intervals.—The method of coordinates described in Art. 142 may be applied to this problem by assuming the origin as being on the traverse line and at the point where the first offset is taken. The coordinate axes are then the traverse line and a line at right angles thereto. The rule of Art. 142 may then be modified to the following:

Rule.—*Multiply the distance (along the traverse) of each intermediate offset from the first by the difference between the two adjacent offsets, always subtracting the following from the preceding. Also multiply the distance of the last offset from the first by the sum of the last two offsets. The algebraic sum of these products, divided by two, is the required area.*

144. Numerical Problems.

1. What is the area of a triangle having sides of length 386.1, 337.0, and 415.9 ft.? Of a triangle having two sides of length 672.4 and 318.5 ft. and an included angle of 29°06′?

2. Given the notes tabulated below, for a closed traverse. Balance the survey by the compass rule, and calculate the area in acres by the coordinate method, using four-place logarithms.

Course	Bearing	Length, feet
AB	S45°45′E	294.4
BC	N65°30′E	263.4
CD	N35°15′E	313.6
DE	N64°15′W	392.0
EF	S59°00′W	197.2
FA	S26°15′W	240.0

3. Plot the traverse of problem 2 to scale; divide the field into rectangles and right-angle triangles; and compute the approximate area by slide rule.

4. Given the following offsets from traverse line to irregular boundary, measured at points 25 ft. apart.

Distance, feet	Offset, feet	Distance, feet	Offset, feet
0	0.0	125	28.2
25	16.6	150	11.9
50	35.1	175	30.7
75	39.3	200	43.4
100	42.0		

By the trapezoidal rule, calculate the area between traverse line and boundary.

5. Following are offsets from a traverse line to an irregular boundary, taken at irregular intervals:

Distance, feet	Offset, feet	Distance, feet	Offset, feet
0	18.5	100	44.1
25	37.7	170	53.9
60	58.2	200	46.0
70	40.5	220	34.2

Calculate the area between traverse line and boundary by means of the rule based on coordinates.

CHAPTER XII

TOPOGRAPHIC MAPS

145. Relief.—A topographic map shows not only the natural and artificial features of the terrain but also the *relief*, or configuration of the earth's surface. Topographic maps are a necessary aid in the design of engineering projects which require a consideration of land forms, elevations, or gradients. Relief is usually shown by *contour lines.*

146. Contour Lines.—A *contour* is an imaginary line of constant elevation on the ground surface. It may be thought of as the trace formed by the intersection of a level surface with the ground surface, for example, the shore line of a still body of water. Contours are represented on the map by contour lines; the terms *contour* and *contour line* are often used interchangeably. On a given map, successive contour lines represent elevations differing by a fixed vertical distance called the *contour interval.*

The principal characteristics of contour lines may be illustrated by reference to Fig. 146. For the purpose of this discussion the slope of the river surface is disregarded, and it is considered that the shore line on the map marks the position of the 510-ft. contour line.

1. The horizontal distance between contour lines is inversely proportional to the slope; hence on steep slopes (as along the river banks in Fig. 146) the contour lines are spaced closely.

2. On uniform slopes the contour lines are spaced uniformly.

3. Along plane surfaces (such as those of the railroad cuts and fills in Fig. 146) the contour lines are straight and parallel to one another.

4. Since contour lines represent level lines, they are perpendicular to the lines of steepest slope. They are perpendicular to ridge and valley lines where they cross such lines.

5. Since all land areas may be regarded as summits or islands above sea level, all contour lines must close upon themselves either within or without the borders of the map. It follows that a closed contour line on a map always indicates either a summit or a depression. If water lines or the elevations of adjacent contour lines do not indicate which condition is represented, a depression is shown by a hachured contour line, called a *depression contour*, as shown at *M* in Fig. 146.

6. Since contour lines represent contours of different elevation on the ground, they cannot merge or cross one another on the map, except in the rare cases of vertical surfaces (see bridge abutments in Fig. 146) or overhanging ground surfaces as at a cliff or a cave.

7. A single contour line cannot lie between two contour lines of higher or lower elevation.

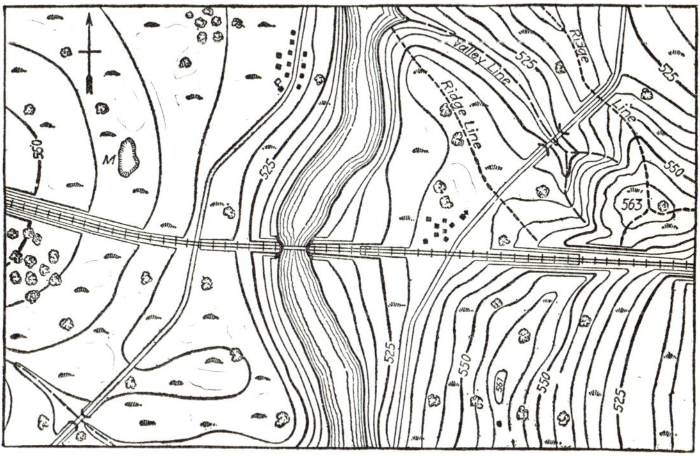

Fig. 146.—Contour lines.

147. Contour Interval.—The contour interval may be thought of as the scale by which vertical distances or elevations are measured on a map. The choice of contour interval depends upon the purpose

Table 147.—Relation between Scale of Map, Slope of Ground, and Contour Interval

Scale of map	Slope of ground	Interval, ft.
Large (1 in. = 100 ft. or less)	Flat Rolling Hilly	0.5 or 1 1 or 2 2 or 5
Intermediate (1 in. = 100 ft. to 1,000 ft.)	Flat Rolling Hilly	1, 2, or 5 2 or 5 5 or 10
Small (1 in. = 1,000 ft. or more)	Flat Rolling Hilly Mountainous	2, 5, or 10 10 or 20 20 or 50 50, 100, or 200

and scale of the map and upon the character of terrain represented. Table 147 represents good practice for usual conditions.

148. Contour-map Construction.—Normally the construction of a topographic map consists of three operations: (1) the plotting of the horizontal control (Chap. X); (2) the plotting of details, including the map location of points of known ground elevation, called *ground points*, by means of which the relief is to be indicated; and (3) the construction of contour lines at a given contour interval, the ground points being employed as guides in the proper location

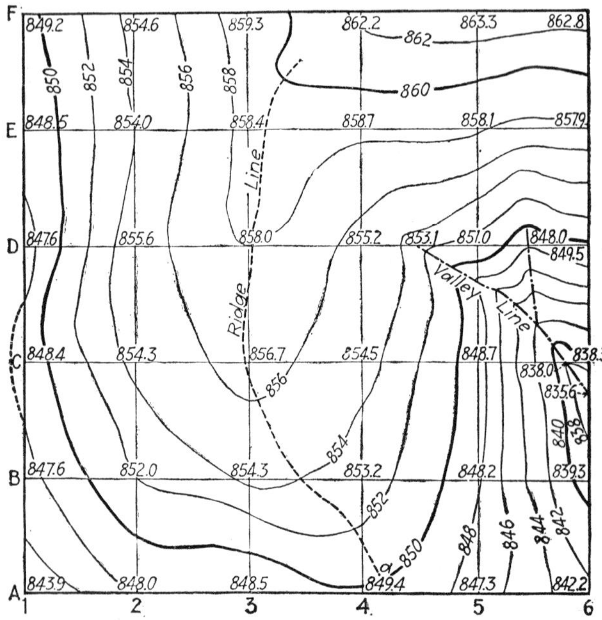

FIG. 148.

of the contour lines. A ground point on a contour is called a *contour point*.

Ridge and valley lines are important aids to the correct drawing of contour lines (Fig. 148). Frequently they are drawn first, and the contour crossings are spaced along them before the contour lines are drawn.

Contour lines are shown for elevations which are multiples of the contour interval. They are drawn as fine, smooth, freehand lines of uniform width, preferably by means of a contour pen (Fig. 126*c*). Usually each fifth line is made heavier than the rest. Elevations of contours are indicated by numbers placed at appropriate intervals;

usually only the fifth or heavier contour lines are numbered. The line is broken to leave a space for the number.

Any contour line must be drawn, to some degree, by estimation. Skill and judgment are required to the end that the contour lines may best represent the actual configuration of the ground surface.

149. Interpolation.—The process of spacing the contour lines proportionately between plotted points is called *interpolation.* For example, in Fig. 148 the elevations of points *A*-2 and *B*-2 are respectively 848.0 and 852.0 ft., and the 848 and 852-ft. contour lines pass through these points; if it is assumed that the slope is uniform, the 850-ft. contour line passes through a point midway between *A*-2 and *B*-2.

On intermediate- and small-scale maps, often the desired precision can be obtained if the interpolation is made by careful estimation supplemented by approximate mental computations.

Where considerable accuracy is desired in the map, the computations for interpolation may be made with the aid of a slide rule. For example, the elevations of points *E*-6 and *F*-6 (Fig. 148) are 857.9 and 862.8 ft., respectively. The difference in elevation between *E*-6 and the 858-ft. contour is 0.1 ft. Then, since the total difference in elevation is 4.9 ft., the proportional part of the distance from *E*-6 to *F*-6 to locate the 858-ft. contour line is 0.1/4.9 of the map distance between these points.

A convenient method of interpolation is by means of a rubber band graduated with lines forming a scale of equal parts. The scale divisions are not numbered, but numbers are assigned mentally for each interpolation. The band is stretched along the line between two plotted points in such manner that these points fall on the scale at points corresponding to their elevations. The intermediate contour points are then marked on the map. For example, with the band stretched so that *E*-6 (Fig. 148) is at 7.9 and *F*-6 is at 12.8 on the scale, the contour points representing elevations 858, 860, and 862 are marked on the map at scale divisions 8, 10, and 12.

150. Systems of Ground Points.—The typical systems of ground points commonly used in the preparation of topographic maps are the *trace-contour*, *checkerboard*, *controlling-point*, and *cross-profile* systems.

1. *Tracing Contours.*—A number of points on a given contour are located on the ground, and their corresponding locations are plotted on the map. The contour line is then drawn through these plotted points.

2. *Checkerboard.*—A system of squares or rectangles for which the corner elevations have been determined is plotted as in Fig. 148,

and near each corner is written its elevation. Also, the locations of valley and ridge lines are shown. The contour crossings are interpolated on the valley and ridge lines and on the sides of the squares, and the contour lines are drawn.

3. *Controlling Points.*—Ground points are chosen as necessary to define the summits, depressions, valley and ridge lines, and all important changes in slope. This irregular system is plotted, and the contour lines are located by interpolation.

4. *Cross-profiles.*—The cross-profile method is most frequently used in connection with route surveys. The field surveys determine the location either of all contour points or of all points of change in slope, along selected lines perpendicular to the route traverse line. The traverse, cross-profile lines, and ground points are plotted, and the contour lines are drawn.

151. Tests for Accuracy.—A topographic map can be tested for accuracy, both in plan and in elevation. In this discussion it is assumed that the errors in field measurement may be disregarded and that a graphical scale is provided on the map to render negligible any effect of shrinkage of the paper.

The test for horizontal dimensions consists in comparing distances scaled from the map and distances measured on the ground between the corresponding points. The precision with which distances may be scaled from a map depends on the scale of the map and on the size of the plotting errors. Thus, if for a map scale of 1 in. = 100 ft. it is known that the error in location of any one point with respect to any other on the map is $\frac{1}{40}$ in., then the error represents 2.5 ft. on the ground.

One test for elevations consists in comparing, for selected points, the elevations determined by field levels and the corresponding elevations taken from the map. Usually the points are taken at 100-ft. stations along traverse lines crossing typical features of the terrain. A more searching test is to plot selected profiles of the ground surface as determined by the field levels and the corresponding profiles taken from the map. The presence of systematic errors will be evidenced if the map profile is above or below the ground profile for an undue proportion of its length.

152. Choice of Map Scale.—From a consideration of the tests described in the preceding article it is possible to choose a map scale consistent with the purpose of the survey if the approximate size of the plotting error is known. For example, if it is known that (with reasonable care in plotting) the average error in distance between any two definite points on the map is $\frac{1}{40}$ in., and if it is known that the purpose of the survey will be met if the average error in scaled distances is 10 ft., these conditions are satisfied by a map scale of 1 in. = 400 ft.

USES OF TOPOGRAPHIC MAPS

153. Cross-sections and Profiles.—Figure 153 shows (below) a portion of a contour map for use in connection with roadway construction and (above) a vertical section on which the profile of the ground line and of the grade line are plotted. In the case illustrated the center line of the roadway is straight, and the profile of the

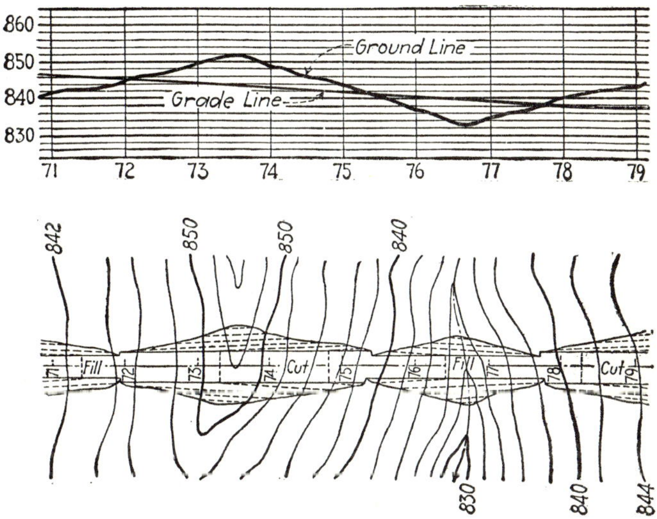

FIG. 153.—Map and profile for proposed roadway.

ground along this center line is constructed by projecting points (either full stations or contour points) to corresponding points on a base line of the vertical section and then by plotting the elevations of these points to scale on the vertical section. If the horizontal alinement of the roadway either is curved or is not parallel to the base line of the vertical section, usually the distances from full stations to contour crossings are scaled from the map, and then these distances are plotted on profile paper as if the stations and elevations were being taken from profile notes (Art. 63).

Vertical cross-sections for estimating earthwork for other purposes are constructed in a manner similar to that just described for the profile of a roadway. In practice, parallel lines along which the cross-sections are to be taken are drawn on the map, and the distance between them is scaled. One person scales horizontal distances to contour crossings and reads contour elevations, while a second person plots the data on regular cross-section paper.

154. Earthwork for Roadway.—In Fig. 153 the contour lines for the proposed roadway (the grade line of which has been fixed on the profile) are drawn dotted over the existing contour map. The contour crossings of the roadway are located on the map by projecting the contour crossings of the grade-line profile from the vertical section to the map. The side slopes of the earthwork are 1½ to 1. The width of the roadway is 36 ft. in cut and 30 ft. in fill. From a study of the figure the following observations may be made:

1. On the side slopes of the earthwork at any station the distance out from the edge of the roadway to a contour line is given by the difference in elevation multiplied by the side-slope ratio. Thus at station 76 + 40 the elevation of grade is 840.0 ft., and the elevation of the first contour line out from the edge of the fill is 838.0 ft.; hence the distance out is 2 ft. × 1½ = 3 ft. (For clearness, in the illustration the lateral scale is exaggerated.)
2. Since the grade line is not level, the contour lines on the earthwork slopes are not parallel to the roadway. Thus the 844-ft. dotted contour line which crosses the roadway at station 73 + 30 is so inclined in direction that at station 74 + 80 where the elevation of grade is 842 ft. the 844-ft. contour line is out from the edge of the roadway a distance of 2 × 1½ = 3 ft.
3. The toe of a slope is drawn on the contour map by connecting the points where the dotted lines intersect the corresponding full lines.

154a. If a line is drawn across a plotted roadway, normal to the center line, a cross-section of the proposed roadway (Art. 202) can be plotted from the data of the contour map. Between adjacent cross-sections, the quantity of earthwork can be calculated as explained in Arts. 207 and 208.

155. Earthwork for Grading Areas.—Earthwork for grading areas may be estimated by vertical cross-sections, as follows: A cross-section is plotted for each of a number of selected parallel lines on the contour map. On each cross-section is shown the profile of (*a*) the original line and (*b*) the ground line after the proposed grading. To obtain the volume of a given cut or fill between adjacent cross-sections, the area between these ground lines is averaged for the two cross-sections and is multiplied by the distance between them.

156. Reservoir Areas and Volumes.—A contour map may be employed to determine the capacity of a reservoir, the position of the flow line, the area of the reservoir, and the area of the drainage basin. The procedure may be illustrated by reference to the fill across the valley in Fig. 153, the fill being considered as a dam. If water is imagined to stand at the elevation of 834 ft., the water surface is represented by that within the full and dotted 834-ft. contour line. If the water were to rise through a 2-ft. stage to the elevation of 836 ft., the water surface would be represented by that within the full and dotted 836-ft. contour line. The volume of water that caused the 2-ft. rise is given by the average of the two surface areas multiplied by the vertical distance of 2 ft. By a similar procedure the volume of successive layers of the reservoir may be estimated.

The flow line is given by the contour line representing the maximum stage of the impounded water. The drainage area may be estimated by sketching on the map the watershed line and measuring the extent of the watershed with a planimeter.

157. Route Location.—In roadway location it is desired to fix the center line of the proposed construction so that the subgrade will

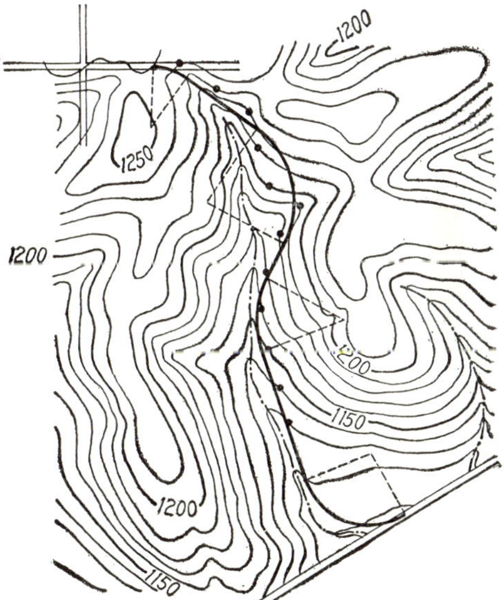

Fig. 157.—Route location.

conform as nearly as practicable to the original ground surface (Chap. XV). Suppose that a proposed highway is to be located in the valley shown in Fig. 157, joining the existing highways at opposite corners of the map; and let the maximum permissible gradient be 4 per cent. Beginning at the lower highway, the proposed route is projected on the map up the valley until the steep slopes require a careful study of the ground, say to the 1,130-ft. contour line. Then a pair of dividers is set at a map distance equal to the contour interval divided by the desired gradient, in this case 10 ft. ÷ 0.04 = 250 ft. One foot of the dividers is placed on the intersection of the proposed roadway and the 1,130-ft. contour line, and the other foot of the dividers is placed on the 1,140-ft. contour line, as indicated in the figure by a heavy dot; and so on for successive contour lines. The series of dots marks on the plotted ground surface the location of a 4 per cent grade line up the valley. Hence, the route is made to follow this line as closely as other limitations, such as the radius of curves, will permit.

158. Office Problem.

PROBLEM 21. TOPOGRAPHIC-MAP CONSTRUCTION

Object.—To construct a complete topographic map from field notes, relief being represented by contours. The data of field problem 17, 18, 25, 26, or 27 may be used.

Procedure.—(1) If the skeleton of the survey is a traverse, plot the traverse either by the method of tangents or by the method of coordinates; if the horizontal control is in the form of rectangles or squares, plot the control by the method of coordinates. (2) Plot the details of the map by methods corresponding to those used in the field. Use conventional signs wherever applicable. (3) Mark each ground point by a dot, and mark each elevation in such position that there will be no doubt to which point it refers, by letting the decimal point of the elevation represent the ground point. (4) Interpolate the contour crossings. (5) Place necessary notes so that they will not interfere with the map. (6) Draw a meridian arrow, and make an appropriate title. (7) Ink in the map.

CHAPTER XIII

THE PLANE TABLE

159. General.—The plane table consists essentially of (1) a *drawing board* mounted on a tripod and (2) an *alidade* having the vertical plane of the line of sight fixed parallel to a straightedge which rests upon but is not attached to the board (Fig. 159). A sheet of drawing paper, called a *plane-table sheet*, is fastened to the board.

The location of any object is determined as follows (Fig. 165): With the straightedge through the plotted point o representing the station O occupied by the instrument, the line of sight is directed to the object A, and a line oa is drawn along the straightedge on the plane-table sheet; this line represents the direction from station to object. The measured distance OA between station and object is then plotted to scale, thus locating A on the map at a.

The term *plane table* is somewhat ambiguous, being used sometimes to designate only the board with its supporting tripod and sometimes (more generally) both the table proper and its accompanying alidade.

In the construction of maps the plane table has wide application; but as no record of numerical values is secured, the instrument is useful for mapping only.

160. Tables.—The *Coast Survey table* (Fig. 159) is the most stable of the three types in common use. The drawing board, 24 by 31 in., is attached to a metal casting which is so arranged that the board can be leveled accurately by means of three leveling screws. By means of a clamp and tangent-screw the board can be fixed in any position in azimuth. The plane-table sheet is held in position by metal spring clamps.

The drawing board of the *Johnson table* is either 18 by 24 in. or 24 by 31 in. It is attached to the tripod head by means of a ball-and-socket joint having two clamp-screws. When one clamp is loosened, the ball is free to rotate in the socket and the table can be leveled. When the other clamp is loosened, the table can be rotated about the vertical axis and thus can be oriented. The plane-table sheet is held in position by countersunk screws in the top of the board.

The *traverse table* consists of a small drawing board, usually 15 by 15 in., mounted on a light tripod. A compass is fixed into a recess

FIG. 159.—Coast Survey table and alidade.

in the board. Ordinarily a peep-sight alidade is used with the traverse table. The traverse table is suitable for (1) military reconnaissance sketches, (2) traverses for small-scale maps, and (3) the mapping of relatively inaccessible areas to fill in a topographic map being drawn on a larger sheet.

161. Alidades.—The *telescopic alidade* commonly used consists of (1) a 3 by 18-in. brass straightedge beveled on one edge and (2) a telescope similar to that of the transit, mounted on the straightedge as shown in Fig. 159 and equipped with a level, a vertical arc, and stadia hairs.

In the obsolete *fixed-tube* type of telescopic alidade, the telescope tube is rigidly attached to or is an integral part of the horizontal axis, as in the engineer's transit. In the *tube-in-sleeve* type now in common use, the telescope tube is fitted into a cylindrical sleeve which is rigidly attached to the horizontal axis. In this type the telescope can be turned about its axis in the sleeve much as the telescope of the wye level can be turned in its wyes. On the telescope are turned two shoulders perhaps 5 in. apart, upon which rests a striding level.

A *peep-sight alidade* consists of two vertical sight vanes similar to those of a surveyor's compass (Fig. 78), mounted on a graduated straightedge ruler usually 6 to 10 in. long. It is convenient for sighting details while sketches are being made and is often employed by topographers as an auxiliary to the telescopic alidade.

162. Compasses and Levels.—For some instruments, upon one end of the straightedge is mounted a *trough compass* consisting of a magnetic needle mounted in a narrow box with a short graduated arc at the end; upon the other end of the straightedge is mounted either a circular level or a pair of level tubes at right angles to each other. For other instruments the compass and level form a separate movable unit. For still other instruments a compass is fixed into a recess in the drawing board.

163. Plane-table Sheet.—Only the best drawing papers should be used for plane-table work. To prevent undue expansion or shrinkage, plane-table sheets may be seasoned by alternate exposures to moist and dry air or may be mounted on aluminum. If the plotting is to be continued for several days, the map sheet is protected by a cover sheet of tough paper which is torn away in pieces to expose the map sheet as the work progresses. In wet weather a sheet of celluloid with roughened surface may be used temporarily.

Sharply pointed, hard (6H to 9H) pencils are used for drawing lines, and a fine needle is used for plotting station points.

164. Setting Up and Orienting the Table.—The plane table is set up approximately waist-high, so that the topographer may bend over the board without resting against it. The tripod legs are spread well apart and planted firmly in the ground. The board is leveled, but no special attempt is made to see that it is perfectly level each time an observation is made.

For map scales smaller than perhaps 1 in. = 50 ft. the plane table is set over the station without any attempt to place the plotted point vertically above the station point. For maps of larger scale the table is set up roughly and oriented approximately, and then it is shifted bodily until the point on the paper is practically over the station point, as indicated by plumbing. In any case the aim is to set up with sufficient care so that the plotted position of lines drawn from the station will be shown correctly within the scale of the map.

The table may be oriented (1) by use of the *compass*, (2) by *backsighting*, or (3) by solving the *three-point problem* as described in Art. 170. As soon as the table is oriented, it is clamped in position, and all mapping at the station is carried on without disturbing the board.

Orientation by Compass.—For rough mapping at small scale, often orientation by the magnetic compass is sufficiently exact. If the compass is fixed to the drawing board, the board is oriented by rotating it about the vertical axis until the fixed bearing (usually magnetic north) is observed. If the compass is attached to the alidade or to a movable plate, the edge of the ruler or the plate is alined with a meridian previously established on the plane-table sheet, and the board is turned until the needle reads north.

Orientation by Backsighting.—For mapping at intermediate or large scale, usually the board is oriented by backsighting along an established line the direction of which has been plotted previously but the length of which need not be known. The plane table is set up as at B on the line AB which has been plotted previously as ab, the straightedge of the alidade is placed along the line ba, and the board is oriented by rotating it until the line of sight falls at A.

164a. When a station or object is sighted and a ray is drawn through the plotted location of the station occupied toward the station or object, the sight is called a *foresight.* When the straightedge of the alidade is placed along a previously plotted line passing through the plotted location of the station occupied and that of another station or object and then the board is turned until the line of sight cuts the station or object, the sight is called a *backsight.* When a known station is sighted and a line is drawn through the plotted location of that station toward the station occupied, the sight is called a *resection.* Resection is also a general term applied to the process of determining the location of the station occupied (Art. 168).

165. Radiation.—When the table has been oriented, the direction to any object in the landscape may be drawn on the map by pivoting the alidade around the plotted location of the plane-table station, pointing the alidade toward the distant object, and drawing a line along the straightedge.

Thus in Fig. 165 the plane table is shown in position over station *O* in the center of the field. The plotted location of the plane-table station is indicated at *o* on the plane-table sheet. The alidade is pivoted about this point; and as sights are taken to points as *A*, *B*,

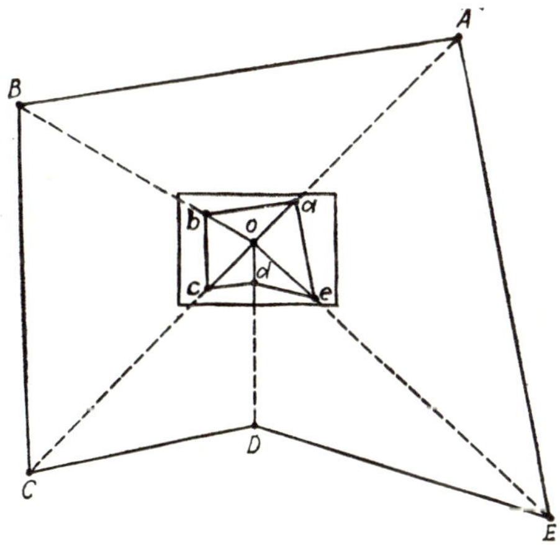

FIG. 165.—Radiation with plane table.

and *C*, rays are drawn along the edge of the ruler. The distances are measured and are then plotted to scale along the corresponding rays, thus locating the points *A*, *B*, and *C* on the map at *a*, *b*, and *c*. This procedure is called *radiation*.

166. Traversing.—Traversing with the plane table involves the same principles as traversing with the transit. As each successive station is occupied, the table is oriented, usually by backsighting on the preceding station. A foresight is then taken to the following station, and its location is plotted as in the radiation method just described. The distances between successive instrument stations must be measured.

Thus in Fig. 166 a series of traverse stations *A*, *B*, *C*, etc., is represented. The plane table is set up as at *A*, and *a* representing *A* is plotted in such location that the other stations will fall within the limits of the sheet. With the straightedge passing through *a* a foresight is taken to station B, and b, its location on the map, is plotted by radiation as just described. The instrument is then set up at station B and oriented by backsighting on station A as described in Art. 164. A foresight is taken to station C, and its location is plotted at c. By a similar procedure the locations of the remaining stations are plotted. If the traverse forms a closed figure, any error of closure will become apparent on the plane-table sheet when the initial station is again plotted at the end of the traverse.

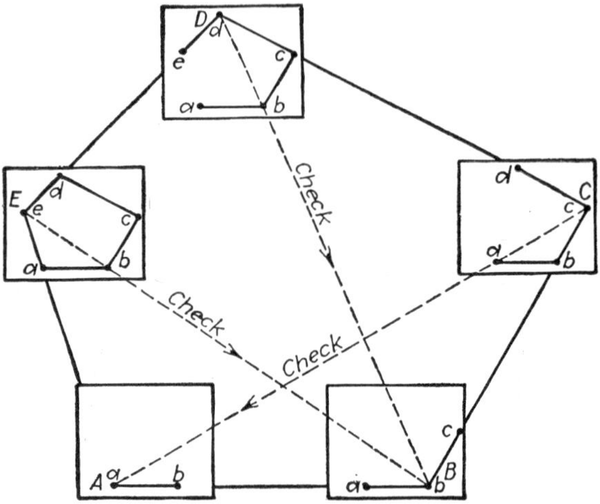

FIG. 166.—Traverse with plane table.

167. Intersection.—This method is similar to that described for the transit in Art. 99. It is useful for locating objects otherwise not conveniently accessible. The location of an object is determined by sighting at the object from each of two plane-table stations (previously plotted) and by drawing rays as in the method of radiation; the intersection of the two rays thus drawn marks the plotted location of the object. No linear measurements are required except to determine the length of the line joining the two plane-table stations.

Thus the locations of the objects A, B, C, etc. (Fig. 167), may be plotted as follows: The plane table is set up at station M, a sight to station N is taken as in the method of traversing, and the line mn representing the line MN is drawn to scale. Rays of indefinite length are drawn from m toward the objects A, B, C, etc. The plane table is then set up at station N and is oriented by sighting to station M. Rays are drawn from n toward the same objects. The intersections of these rays with the corresponding rays drawn from m mark the plotted locations of the objects at a, b, c, etc. Distances to the objects are not measured but may be scaled from the map. If the angle between the intersecting rays is small, the location will be indefinite.

167a. Graphical Triangulation.—Graphical triangulation is similar to triangulation with the transit (Art. 76), but the procedure differs in that the plotted locations of the distant signals are determined graphically on the plane-table sheet. The plotted location of each triangulation station, other than the station at each end of the base line, is determined by intersection or, wherever necessary, by resection. The method is advantageous where the terrain offers unobstructed sights, considerable relief, and many well-defined objects such as spires and lone trees.

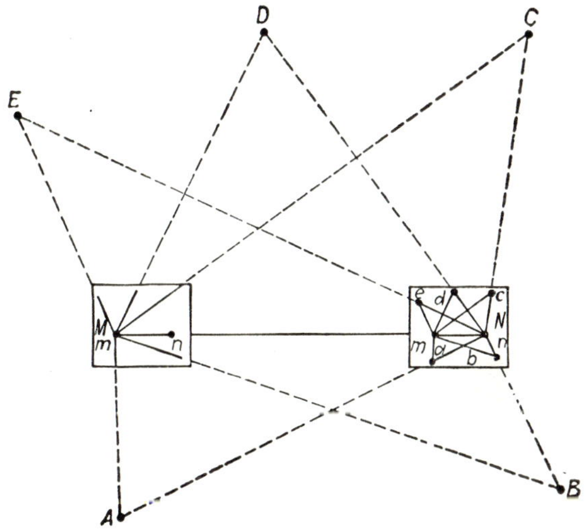

FIG. 167.—Intersection with plane table.

168. Resection.—Resection is the process of determining the plotted location of a station occupied by the instrument, by means of sights taken toward known points the locations of which have been plotted. Resection enables the topographer to select advantageous plane-table stations which have not been plotted previously.

With the plane table oriented over the desired station of unknown location, two or more objects of known location are sighted; as each object is sighted, a line of indefinite length is drawn through the plotted location of that object on the map. The intersection of these lines marks the plotted location of the plane-table station.

The table may be oriented by any of the methods stated in Art. 164. It is emphasized that, for the methods of orientation by compass and by backsighting, resection can be accomplished only *after the board has been oriented.* If resection is by the three-point problem, orientation and resection are accomplished in the same operation.

169. Resection after Orientation.—If the plane table has been oriented either by means of the compass or by backsighting, the method of resection is as follows: Let P be the station of unknown location occupied by the plane table, and let A and B be two visible stations which have been plotted on the sheet at a and b. Then the plotted location of P is determined by drawing a line, or resecting, through a in the direction of A and resecting through b in the direction of B. The point p where the two (or more) lines cross, or resect, marks the plotted location of the instrument station P.

170. Resection and Orientation: Three-point Problem.—Frequently the topographer wishes to occupy an advantageous station

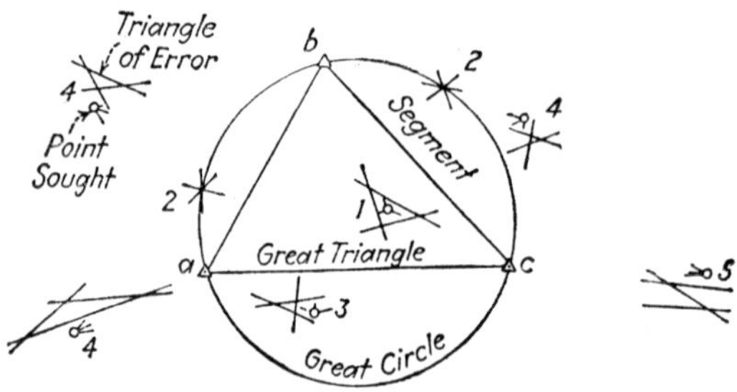

FIG. 170.—Three-point problem, solution by trial.

that has not been located on the map and toward which no ray from located stations has been drawn, and at the same time orientation by use of the compass is not sufficiently accurate. If three located stations are visible, the three-point problem offers a convenient method of orienting and resecting in the same operation. There are several solutions of the three-point problem. In the United States, experienced topographers commonly employ a method of direct trial, guided by published rules, as follows:

The plane table is set up over the station of unknown location and is oriented approximately either by compass or by estimation. Resection lines from the three stations of known location are drawn through the corresponding plotted points. These lines will not intersect at a common point unless the trial orientation happens to be correct. (An exception to this statement is discussed later in this article.) Usually a small triangle called the *triangle of error* is formed by the three lines (Fig. 170). The true plotted location of the plane-table station, called the *point sought*, is then estimated more closely by means of Rules 1 and 2, given below. The board is then reoriented by backsighting through this trial point toward one of the known stations (preferably the most distant); and the orientation is checked by resecting from the other two known stations. If the three lines still do not meet at a point, the process is repeated until they do; the orientation is then correct, and the common intersection of the three lines is the correct plotted location of the plane-table station.

Rule 1.—*The point sought is on the same side of all resection lines.* That is, it lies either to the right of each line (as the observer faces the corresponding station) or to the left of each line.

Rule 2.—*The distance from each resection line to the point sought is proportional to the length of that line.* By "length" is meant either the actual distance from plane-table station to known station or the corresponding plotted distance.

Rules 1 and 2 are general and apply to any location of the plane-table station except on the *great circle* passing through the three known stations (Fig. 170). In this case, regardless of the orientation of the table, the lines will meet in a common point (see points 2 in the figure) which will not necessarily be the point sought. If it is suspected but not known that the plane-table station is on the great circle, either the great circle should be plotted on the plane-table sheet or the orientation of the board should be changed slightly and a second trial made. If it is found that the station is on the great circle, one (or more) of the three known stations must be replaced by a known station (or stations) suitably located.

Rules 1 and 2 are supplemented by several auxiliary rules which apply to particular locations of the plane-table station. For example, if the station happens to be within the great triangle, the point sought is within the triangle of error.

The strength of the determination varies with the location of the plane-table station. The strength of determination should be considered not only in selecting the most favorable of the available known stations but also in deciding if the three-point problem can satisfactorily be used with the plane table at a given location.

171. Measurement of Difference in Elevation.—The methods of plotting the *horizontal projection* of ground points have been described. As the plane table is used principally in the work of topographic mapping, many *elevations* of ground points are determined by methods closely similar to those for leveling by means of the transit, namely, by direct leveling and trigonometric leveling including stadia leveling.

The plane table is not so precise or so stable as the horizontal plate of the transit; accordingly in measuring vertical angles it is necessary either to determine the index error of each sight or preferably to employ a control level attached to the vernier arm.

The elevation of a distant point the location of which has been plotted by the method of intersection may be secured readily by the method of trigonometric leveling (Art. 36), the horizontal distance being determined usually by scaling from the map. The conditions as to distance and as to the accuracy required will determine whether or not corrections for earth's curvature and atmospheric refraction should be applied.

172. Field Checks.—The previously plotted location of any visible object is verified in part if, during a subsequent set-up, a ray drawn toward the object passes through the plotted point. The plotted location of the plane table itself may be verified by resecting from distant visible objects the plotted locations of which are known to be correct. Checks of this sort should be applied at each station.

173. Adjustments of Alidade.—The adjustments of the telescopic alidade are similar to those previously described for the wye level and the transit. In general, the adjustments need not be so refined, and in one or two cases they are omitted.

1. *To Make the Axis of Each Plate Level Parallel to the Plate.*—Center the bubble (or bubbles) of the plate level (or levels) by manipulating the board. On the plane-table sheet, mark a guide line along one edge of the straightedge. Turn the alidade end for end, and again place the straightedge along the guide line. If the bubble is off center, bring it back *halfway* by means of the adjusting screws. Again center the bubble by manipulating the board, and repeat the test.

2. *To Make the Vertical Cross-hair Lie in a Plane Perpendicular to the Horizontal Axis.*—Sight the vertical cross-hair on a well-defined point about 200 ft. away, and swing the telescope through a small vertical angle. If the point appears to depart from the vertical cross-hair, loosen two adjacent screws of the cross-hair ring, and rotate the ring in the telescope tube until by further trial the point sighted traverses the entire length of the hair. Tighten the same two screws.

3. (For Alidade of Tube-in-sleeve Type)[1] *To Make the Line of Sight Coincide with the Axis of the Telescope Sleeve.*—Sight the intersection of the cross-hairs on some well-defined point. Rotate the telescope in the sleeve through 180° (usually the limits of rotation are fixed by a shoulder and a lug). If the cross-hairs have apparently

[1] For the fixed-tube type of alidade, adjustments 3 and 4 are replaced by the "two-peg" adjustment to make the axis of the telescope level tube parallel to the line of sight, as for the transit.

moved away from the point, bring each hair halfway back to its original position by means of the capstan screws holding the cross-hair ring. The adjustment is made by manipulating opposite screws, bringing first one cross-hair and then the other to its estimated correct position. Again sight on the point, and repeat the test.

4. (For Alidade of Tube-in-sleeve Type)[1] *To Make the Axis of the Striding Level Parallel to the Axis of the Telescope Sleeve (and Hence Parallel to the Line of Sight).*—Place the striding level on the telescope, and center the bubble. Remove the level; turn it end for end; and replace it on the telescope tube. If the bubble is off center, bring it back halfway by means of the adjusting screw at one end of the level tube. Again center the bubble (by means of the tangent-screw), and repeat the test.

5. (For Alidade Having a Fixed Vertical Vernier) *To Make the Vertical Vernier Read Zero When the Line of Sight Is Horizontal.*—With the board level, center the bubble of the telescope level. If the vertical vernier does not read zero, loosen it and move it until it reads zero.

5*a*. (For Alidade Having a Movable Vertical Vernier with Control Level) *To Make the Axis of the Vernier Control Level Parallel to the Axis of the Telescope Level When the Vertical Vernier Reads Zero.*—Center the bubble of the telescope level, and move the vernier by means of its tangent-screw until it reads zero. If the control-level bubble is off center, bring it to center by means of the capstan screws at the end of the control-level tube.

174. Field Problems.

PROBLEM 22. ADJUSTMENT OF THE PLANE-TABLE ALIDADE

Object.—To make the field adjustments of the telescopic alidade.
Procedure.—As outlined in Art. 173.

PROBLEM 23. PLANE-TABLE SURVEY OF FIELD

Object.—To make a plane-table survey of an assigned portion of the campus by a combination of the methods of radiation, intersection, and traversing. Either a plain map or a topographic map may be made.

Procedure.—(1) Stake out an irregular field having perhaps five sides. (2) Set up the plane table at one corner of the field, and so locate the plotted position of the station that the area to be covered will fall within the limits of the plane-table sheet. (3) Draw a magnetic meridian on the plane-table sheet. (4) Locate and sketch all conveniently accessible objects by the method of radiation (Art. 165), employing the stadia. Also, draw and identify rays toward relatively inaccessible objects which can also be seen from another point on the traverse. (5) Move to the next corner of the field; orient the board by backsighting; check the

[1] See footnote. p. 142.

orientation by the compass; and similarly locate objects by radiation and intersection. (6) Continue in this manner around the field to form a closed traverse, and plot sufficient data to construct a complete map. At each station, check the location of the plane table by resection (Art. 172). (7) If a marked error becomes evident, the orientation of the board should be checked, and, if necessary, one or more of the previous stations should be reoccupied. (8) If it is convenient to occupy some station other than a traverse station, locate the station occupied by resection, employing the method of backsighting (Art. 169). A ray must have been drawn previously toward the station to be occupied. (9) If a topographic map is to be made, locate a sufficient number of controlling ground points to enable contours to be drawn in the field. (10) Note the error of closure of the traverse; if the error is considerable, the work should be repeated. (11) Finish the map.

PROBLEM 24. ORIENTATION AND RESECTION BY THREE-POINT PROBLEM

Object.—With three stations (or objects) of known location assigned, to plot the location of a fourth station occupied by the plane table and simultaneously to orient the board.

Procedure.—(1) Plot the three known points on the plane-table sheet. (2) Set up the plane table at another assigned station and solve the three-point problem by the trial method as described in Art. 170. (3) If other known points are visible, plot their locations and check the location and orientation of the plane table by sighting on these points.

Hints and Precautions.—If the station occupied is at or near the circumference of a circle passing through the three known points, the solution is indeterminate.

CHAPTER XIV

TOPOGRAPHIC SURVEYING

175. General.—The distinguishing feature of a topographic survey is the determination of the location, both in plan and in elevation, of selected ground points which are necessary to the plotting of the contour lines and to the construction of the topographic map. The topographic survey of a tract consists in (1) establishing over the area a system called the *control*, which consists of key stations connected by measurements of relatively high precision, and (2) locating the *details* (Art. 107), including the selected ground points, by measurements of lower precision from the control stations.

Topographic surveys fall roughly into three classes, according to the map scale to be employed, as follows:

Large scale: 1 in. = 100 ft. or less.
Intermediate scale: 1 in. = 100 ft. to 1,000 ft.
Small scale: 1 in. = 1,000 ft. or more.

In general, this chapter deals with ground surveys for areas of moderate size and for maps of intermediate and large scale. Aerial surveys are discussed briefly in Art. 182. The choice of a suitable contour interval is discussed in Art. 147.

176. General Field Methods.—In the choice of field methods for topographic surveying, the items to be considered are (1) the *intended use of the map* (surveys for detailed maps should be made by more refined methods than surveys for maps of a general character), (2) the *area of the tract* (control measurements for a large area should be more precise than those for a small area), (3) the *map scale* (the probable errors in the field measurements should be considerably less than the probable errors in plotting at the given scale), and (4) the *contour interval* (the smaller the contour interval the more refined should be the field methods).

The principal instruments used are the engineer's transit, the plane table, the engineer's level, the hand level, and the clinometer. The use of the transit has advantages over the use of the plane table where there are many definite points to be located or where the ground cover limits the visibility and requires many set-ups. Conditions favorable to the use of the plane table are open country and many irregular lines to be mapped. Sometimes the transit and the plane table, or the transit and the engineer's level, may be used together to advantage. Through dense woods, elevations of details are determined most advantageously by means of the hand level or the clinometer, and distances are usually determined by chaining.

The horizontal control is established by traversing or by triangulation, and the vertical control is established by leveling.

The details are located by methods described in Chaps. VII, XII, and XIII and Arts. 178–178*d*. The locations of the selected ground points in plan are determined in the same manner as for definite details, usually by radiation. The ground points may or may not be contour points. The elevations of ground points are determined usually by trigonometric leveling or, where the terrain is flat, by direct leveling. The stadia is used extensively except on large-scale surveys, for which the errors in stadia distances are large compared with the errors of plotting; on large-scale surveys usually the distances to definite details are measured with the tape. The details may be located either at the time of establishing the control or later.

177. Control.—Control consists of (1) *horizontal control*, for which by traversing or triangulation the control stations are located in plan; and (2) *vertical control*, for which by leveling the bench marks are established and the control stations are located in elevation.

On surveys of wide extent a few stations distributed over the tract are connected by more precise measurements, forming the *primary control;* within this control system other control stations are located by less precise measurements, forming the *secondary control.* For small areas only one control system is necessary, corresponding in precision to the secondary control for larger areas.

177a. Horizontal Control.—The horizontal control may consist of a traverse system, a triangulation system, or a combination of the two. The required precision of horizontal control depends upon the size and shape of the tract and upon the scale of the map.

Traversing.—The primary traverse for an intermediate-scale map is usually run with transit and tape. Convenient routes are chosen which will result in the advantageous location of stations. Wherever secondary traverses are required to establish the instrument stations from which the details are located, an area within a closed primary traverse may be divided by means of the secondary traverses into a series of roughly parallel strips.

Triangulation.—The field conditions favorable to the use of triangulation (Art. 76) to establish the horizontal control are (1) a fairly extended area in an open, hilly region; (2) a city where accurate traversing is difficult because of street traffic; or (3) a rugged, mountainous region where traversing would be slow and laborious. A general layout of the scheme of primary triangulation is planned on an existing small-scale map; the field stations are established on summits where visibility is good; and signals are erected. One or more base lines are established and measured, and their true azimuths are determined by astronomical observations (Chap. XVII). Observations of angles are made on (1) major stations, which are marked by signals and which are to be occupied by the instrument, and (2) minor stations marked by such objects as trees, spires, and chimneys. When the field measurements have been completed, the necessary computations and adjustments are made; then the coordinates of each station are determined for use in plotting.

177b. Vertical Control.—Bench marks are established at convenient intervals over the area, to serve as points of departure and closure for the leveling operations of the topographic parties when locating details.

Primary and secondary level routes are required in about the same amount, and bear about the same relation to each other, as do the primary and secondary traverses or triangulation systems. Often the level routes follow the traverse lines, the traverse stations being used as bench marks.

178. Location of Details.—The veracity with which contour lines represent the terrain depends upon (1) the accuracy of the observations, (2) the number of observations, and (3) the distribution of the points located. Ground points are definite, but as the contour lines must necessarily be generalized to some extent it would be inappropriate to locate the points with refined measurements. The error of field measurement in plan should be consistent with the error in elevation, which in general should not exceed one fifth of a contour interval; thus generally the error in plan should not exceed one fifth of the horizontal distance between contours. The purpose of the survey will be better served by locating a greater number of points with less accuracy, within reasonable limits, than by locating fewer points with greater accuracy.

178a. Controlling-point Method.—In this method, the ground points form an irregular system along ridge and valley lines and at other artificial features of the terrain. The ground points are located in plan by radiation or intersection, and their elevations are determined commonly by trigonometric leveling or sometimes by direct leveling.

The transit, the plane table, or the transit and plane table together may be employed. Details in the vicinity of each control station are located by angle and distance measurements (Art. 115). The distances are usually measured by stadia. If the transit alone is employed, notes may be kept in the form shown in Fig. 115, perhaps supplemented by sketches.

The rodmen choose ground points along valley and ridge lines and at summits, depressions, important changes in slope, and definite details. The selection of points is important, and the rodmen should be instructed and trained for their work. They should follow a systematic arrangement of routes such that the entire area is covered. They should observe the terrain carefully and report important features which cannot be seen from the control station.

178b. Cross-profile Method.—For the cross-profile method of locating details, the party consists of a topographer and usually two men, herein called "chainmen," who act either as chainmen or as rodmen. The equipment consists of a leveling rod, a steel or metallic tape, a hand level or a clinometer, and usually a cross-ruled, wide-page sketchbook.

The control points are the 100-ft. stations of the transit traverse; these points have been marked on the ground by stakes, and their elevations have been determined by profile leveling and have been furnished to the topography party.

The ground points are points on relatively short crosslines transverse to the traverse line. They are either contour points or more commonly points of change in slope; in the latter case the intermediate contour points are located by interpolation.

The party proceeds from station to station along the traverse. At each station, the topographer notifies the chainmen of the elevation of the station. The head chainman carrying the rod moves out on a line estimated to be at right angles with the traverse line until the rod is on the next contour (either higher or lower) from the station, as determined by the rear chainman employing the hand level; the distance out to the contour is then measured with the tape. The rear chainman then goes out to the point occupied by the rod, and the head chainman again moves out until the next contour is reached; and so the process is repeated until all contour points are located out to the edge of the strip being surveyed. A similar procedure is followed on the other side of the traverse line.

For relatively small-scale maps, sometimes the clinometer is employed to determine the elevations of controlling points on the crosslines by rough indirect leveling, and the distances are measured by pacing.

For large-scale surveys or in flat country where the contour interval is small, usually elevations are determined either by direct leveling with the engineer's level or by indirect stadia leveling with the transit or the plane table.

178c. Checkerboard Method.—The checkerboard method of locating details is well adapted for large-scale surveys, as the points are located in plan by tape measurements. It is also useful where the tract is wooded, where the topography is smooth, and on urban surveys where blocks and lots are rectangular.

The tract is staked off into squares or rectangles—usually 50- or 100-ft. squares. The ground points and other details are then located with reference to the stakes and connecting lines. The usual procedure is first to run a rectangular transit-tape traverse near the perimeter of the tract, with stakes set at each 100-ft. station. The error of closure becomes apparent in the field; if this is greater than the permissible error, the stakes are reset. The interior of the figure bounded by this rectangular traverse is then filled in with stakes set at the corners of the 100-ft. squares; each stake is marked usually with a letter and a number indicating its position with respect to a pair of coordinate axes, as illustrated in Fig. 148. By direct leveling, the ground elevation at each stake is determined. Sketch sheets are prepared, on which are shown the elevations of the corners of the squares. The location of ground irregularities or other details inside the squares is determined by measurements either from adjacent coordinate points or from the sides of the squares. The elevation of such details is determined by use of the hand level. The map is constructed in the office.

178d. Trace-contour Method.—In this method, the contours are traced out on the ground. The various contour points occupied by the rod are located by radiation. The method is commonly used on large-scale surveys, and sometimes on intermediate-scale surveys where the ground is irregular. Although the transit may be used in this work, either alone or with the engineer's level, the plane table is commonly used. Often the plane table and the engineer's level are used together. In this case the levelman sets up the level at a convenient location and directs the rodman up or down the slope until a point on a given contour is located. This point is immediately sighted by the plane-table man and is plotted on the plane-table sheet. The rodman then moves to another contour point, usually on the same contour. The distances from plane-table station to contour points are measured by stadia; if the scale is large, definite objects may be located by taped distances.

179. Site and Construction Surveys.—Surveys for construction generally involve (1) a topographic survey of the site, to be used in the preparation of plans for the structure; (2) the establishment on the ground of a system of stakes or other markers, both in plan and in elevation, from which measurements of earthwork and structures can be taken conveniently by the construction force; (3) the giving of line and grade as needed either to replace stakes disturbed by construction or to reach additional points on the structure itself, and (4) the making of measurements necessary to determine the volume of work actually performed up to a given date (usually each month), as a basis of payment to the contractor. In connection with construction, often it is necessary to make property-line surveys as a basis for the acquisition of lands or rights of way (Chap. XVI).

The detailed methods employed on construction surveys vary greatly with the type, location, and size of structure and with the preference of the engineering and construction organizations. Much depends on the ingenuity of the surveyor to the end that the correct information is given without confusion or needless effort. In the following paragraphs are given some general methods applicable to all types of structures. Surveys for buildings are described in Arts. 180 and 181, and surveys for route structures (including bridges) in Chap. XV.

Alinement.—At the corners and along the sides of the proposed structure, outside the limits of excavation or probable disturbance but close enough to be convenient, transit stations are set and well referenced (Art. 106). Permanent targets or marks called *foresights* may be erected as convenient means of orienting the transit on the principal lines of the structure and for sighting along these lines by eye. Stakes or markers are set on all important lines in order to mark clearly the limits of the work.

In many cases, line and grade are given more conveniently by means of batter boards than by means of stakes. A batter board is a board (usually 1 by 6 in.) nailed to two substantial posts (usually 2 by 4 in.) with the board horizontal and its top edge preferably either at grade or at some whole number of feet above or below grade. The alinement is fixed by a nail driven in the top edge of the board. Between two such batter boards a stout cord or wire is stretched to define the line and grade.

Often it is impracticable to establish permanent markers on the line of the structure. In such cases the survey line is established parallel to the structure line, as close as practicable and with the offset distance some whole number of feet.

Grade.—A system of bench marks is established near the structure in convenient locations that will probably not be subject to disturbance. From time to time these bench marks should be checked against one another to detect any disturbance.

The various grades and elevations are defined on the ground by means of pegs and/or batter boards, as a guide to the workmen. The grade pegs may or may not be the same as the stakes used in giving line. When stakes are used, the vertical measurements may be taken from the top of the stake, from a keel mark or a nail on the side of the stake, or (for excavation) from the ground surface at the stake; in order to avoid mistakes, only one of these bases for measurement should be used for a given kind of work. When batter boards are used, the vertical measurements are taken from the top edge of the board.

When a stake is to be driven with its top at a given elevation, the rodman starts the stake and then holds the rod on the stake. The levelman reads the rod and calls out the approximate distance that the stake must be driven to reach grade. The rodman drives the stake nearly the desired amount, and a second rod reading is taken; and so the process is continued until the rod reading is made equal to the difference between the height of instrument and the desired elevation. A mark or nail on the side of the stake may be used instead of the top of the stake. In some cases the stake is sawed off at the desired elevation. If the grade elevation is only a short distance below the ground elevation, often a hole is dug in order that the stake may be driven to grade.

Precision.—For purposes of excavation only, usually elevations are given to the nearest 0.1 ft. For points on the structure, usually elevations to 0.01 ft. are sufficiently exact. Alinement to the nearest 0.01 ft. will serve the purposes of most construction, but greater precision may be required for prefabricated steel structures or members

It is desirable to give dimensions to the workmen in feet, inches, and fractions of an inch. Ordinarily measurements to the nearest $\frac{1}{4}$ or $\frac{1}{8}$ in. are sufficiently exact, but certain of the measurements for the construction of buildings and bridges should be given to the nearest $\frac{1}{16}$ in.

Establishing Points by Intersection.—Where conditions render the use of the tape difficult or impossible, often points are established at the intersection of two transit lines by simultaneous sighting with two transits in known locations. The process is the inverse of that in which the location of a ground point is determined by the method of intersection (Art. 99). By this method, points may be located in elevation as well as in plan. The precision of measurement is made commensurate with the requirements of construction.

180. Survey for Building Site.—In the preparation of his plans for a building, the architect requires a large-scale map of the site to show the information necessary to the proper location of the building both in plan and in elevation. Such maps are usually drawn to the scale of 1 in. = 10 ft. or 1 in. = 20 ft.

The party consists of an instrumentman and one or two chainmen, and the equipment includes that of a transit party. Because of the large number of elevations to be determined, frequently an engineer's level is also used. The notes may be kept in a transit or topography

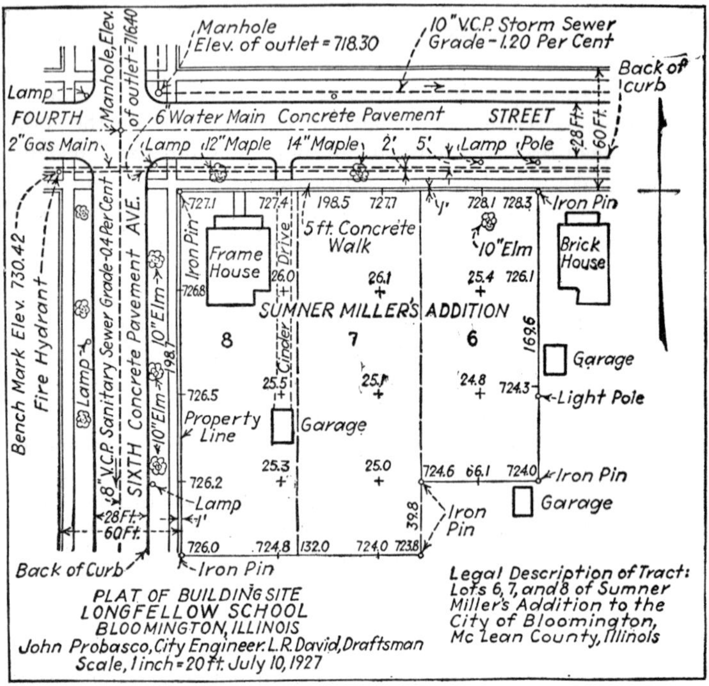

FIG. 180.—Map of building site.

notebook; or a sketch-board may be used to good advantage, as it provides more space for sketching and recording details than does the single page of a notebook.

First the lot corners are located, and permanent markers are set at these points. Then, with the property lines being used as reference lines, all objects are located, usually by tape measurements only, the instrumentman recording the data and drawing such sketches as are necessary. On extensive sites, or where it is not convenient for the chainmen to locate objects by coordinate measurements, transit angles and taped distances may be used.

In addition to the details, the map should give the legal description of the tract and the other information ordinarily shown on the plot of an urban land map. A typical map is shown in Fig. 180.

181. Surveys for Building.—At the beginning of excavation, the corners of the building are marked by stakes, which will, of course, be lost as excavation proceeds. Sighting lines are established and referenced on each outside building line and line of columns. A batter board is set at each end of each outside building line, about 3 ft. outside the excavation. If the ground permits, the tops of all boards are set at the same elevation; in any event the boards at opposite ends of a given line or portion thereof are set at the same elevation so that the cord stretched between them will be level. The elevations are chosen at some whole number of feet above the bottom of the excavation, usually that for the floor rather than that for the footings. A nail is driven in the top edge of the board on the building line, which is given by the transit. Carpenter's lines stretched between opposite batter boards define both the line and the grade, and measurements can be made conveniently by the workmen for excavation, setting forms, and alining masonry and framing. If the space around the building is obstructed so that batter boards cannot be set, other means of marking the line and grade are substituted to meet the requirements of the situation.

When excavation is completed, grades for column and wall footings are given by ground pegs driven to the elevation of either the top of the footing or the top of the floor. Lines for footings are given by batter boards set in the bottom of the excavation. Column bases and wall plates are set to grade directly by the leveler. The position of each column or wall is marked in advance on the footing; and when a concrete form, a steel member, or a first course of masonry has been placed on the footing, its alinement and grade are checked directly.

In setting the form for a concrete wall, the bottom is alined and fixed in place before the top is alined.

Similarly, at each floor level the governing lines and grade are set and checked, except that for prefabricated steel framing the structure as a whole is plumbed by means of the transit at every second or third story level. Notes are kept in a field book used especially for the purpose, principally by means of sketches.

Throughout the construction of large buildings, selected key points are checked by means of stretched wires, plumb lines, transit, or level in order to detect settlement, excessive deflection of forms or members, or mistakes.

182. Photogrammetric Surveying.—*Photogrammetry* is the science of measurement by means of photographs. The scale and position of objects in photographs vary according to the distance and position of the corresponding actual objects relative to the camera. By measurement of these differences in scale and displacements in position in single or overlapping photographs, the location and elevation of ground points and other details can be determined for use in mapping.

Although oblique photographs taken from ground stations or from airplanes are used to some extent, the principal development has been in the field of vertical photographs taken from airplanes. Aerial photogrammetric surveying is used extensively in both military and civil mapping, and its advantages of speed, accuracy, low cost, and wealth of detail make its use likely to become general except for small areas or where small differences in elevation, say. 5 ft. or less, are to be determined. The methods are described in more detailed texts (Refs. 5, 7, 26, 29, and 40, p. 205); herein is given a brief description of the general process.

Aerial photographs of the area to be surveyed are taken by means of specially designed single-lens or multiple-lens automatic cameras, with the axis of the camera vertical (as nearly as may be). The airplane is flown at a constant elevation (as nearly as may be) above sea level, as indicated by corrected barometer readings. The area is covered in a series of parallel strips, with photographs of the same strip overlapping about 60 per cent and photographs of adjacent strips overlapping 30 to 50 per cent. From the photographs an index map is prepared, showing the relative position and the coverage of each photograph.

If the photographs were truly vertical and if the terrain were flat, the photographs would be true maps and all the features would be shown, to some scale, in their proper scale relationship. In general, however, the features are displaced and their size is varied because of longitudinal and transverse inclination (*tip* and *tilt*) of the camera axis and because of the relief of the ground. The effect of tip and tilt can be corrected either by photographic methods or (more laboriously) by computation; usually the combined average tip and tilt is less than 3° and is neglected. The displacement, or *parallax*, of features due to ground relief makes necessary special methods of plotting to secure a true scale map; but it is useful, in fact necessary, as a means of determining the elevations of points by computation or by stereoscopic measurement.

A *mosaic* is an assemblage of separate photographs. It is not a scale map but is a map substitute which is subject to the errors and displacements just mentioned and yet is useful for some purposes such as regional planning. Mosaics are not considered herein.

It is necessary in connection with photogrammetric mapping that *ground control* be established, in the form of a relatively small number of points whose location and elevation are known. Ground-control points either may be visible stations established by other surveys such as the Federal triangulation and land surveys or may be established especially for the photogrammetric survey. With adjacent ground-control points identified on a photograph or series of photographs and with the flight altitude and the focal length of the camera known, the scale of the photograph can be computed. A series of photographs can be brought to the same scale by enlargement or reduction of the individual photographs.

For purposes of planimetric mapping, selected image points on the photographs are used to bring a series of photographs between adjacent ground-control points into the desired relations for plotting to scale. The selected points are called the *secondary control.* One method of plotting the secondary control, called the *radial-line method,* is as follows: Several points (objects) common to two or three overlapping photographs are selected and marked on each of the photographs. A radial line is drawn through each of the control points from the center of the photograph. Beginning at one of the ground-control points, each photograph in turn is laid in its correct relative position under a transparent sheet, and the radial lines are traced. The correct plotted location of each secondary-control point is at the intersection of the corresponding lines traced from successive photographs, regardless of parallactic displacements and of differences in scale as between the individual photographs. The scale for the series of photographs is then determined from the plotted distance between the two ground-control points. The scale can be brought to any desired value, within limits, by graphical means or by photography.

The map details surrounding each control point are traced or sketched from the photograph onto the transparent sheet. Conventional signs are employed as necessary to represent the pictured objects and the types of land culture.

For purposes of representing relief, or *contouring*, use is made of the difference in parallactic displacement between selected ground points of unknown elevation and control points of known elevation. To compute the displacements of all the image points of controlling points of the terrain would be unduly laborious, and in practice it is customary to employ automatic plotting machines based on the principle of stereoscopic observation. A stereoscope is a mechanical device, usually consisting of lenses and mirrors, to facilitate seeing the photographs as in three dimensions; by viewing simultaneously one of a pair of overlapping photographs with the right eye and the other photograph with the left eye, the terrain appears in relief.

A plotting machine consists essentially of a viewing system, a measuring system, and a drawing system. The elevation of a given point shown on the photographs is determined by manipulating an index mark of the machine until it apparently touches that point in the three-dimensional stereoscopic image. With some elaborate types of plotting machine the ground points and map details are located in plan and elevation simultaneously, and no secondary control is required. A less elaborate type, which is increasing in use, is employed primarily for contouring only, after the secondary control has been plotted by other means such as the radial-line method previously described. If desired, contours can be drawn directly on the photographs.

Contouring may be accomplished in the field with the plane table, a copy of the planimetric map being used as the plane-table sheet.

183. State Systems of Plane Coordinates.—In each state, Federal and state agencies have established a system of plane rectangular coordinates for horizontal control; one set of coordinates is made to serve the whole area of a small state or a considerable portion (usually half) of a large state. Many triangulation stations have been established and monumented, and at each station a line of known true azimuth is established. In addition, a system of bench marks has been established for vertical control. The x and y coordinates and the azimuth at each station and the elevation of each bench mark are published for the information of engineers and surveyors.

To refer a local survey to the state system, the surveyor orients the transit on the line of known azimuth at a near-by triangulation station and runs a survey (by traversing or triangulation) to the area under consideration. The coordinates of any point in the local survey can then be computed in terms of the state system, by the ordinary methods of plane surveying. Preferably the survey should be checked by traversing back either to the original triangulation station or to another triangulation station. If elevations are determined, they are referred to the established system of bench marks.

184. Field Problems.—Field problems 18 and 23 provide exercises in the controlling-point method of locating details; and field problem 27 provides an exercise in the cross-profile method. Elementary field problems in topographic surveying are given in Chap. VIII ("Stadia Surveying") and Chap. XIII ("The Plane Table").

Problem 25. Topographic Survey by Checkerboard Method

Object.—To obtain sufficient data for an accurate topographic map of large scale and small contour interval. The area to be mapped is small and possesses few details, and the topography is smooth. The data may be used in office problem 21 (p. 132).

Procedure.—(1) With transit and tape divide the tract into 100-ft. squares, setting stakes at the corners. Letter and number each stake to conform to a coordinate system. (2) With the engineer's level run levels over the area, taking rod readings at summits, depressions, corner stakes, and points of change in slope along the sides of the squares. (3) Locate the details (either definite details or ground points) inside the squares by taking offsets, ties, etc. (Arts. 107 and 107*a*).

Problem 26. Topographic Survey by Trace-contour Method Using Plane Table and Engineer's Level

Object.—To obtain sufficient data for an accurate topographic map of intermediate or large scale and small contour interval. The area to be mapped is small, and the topography is irregular. It is assumed that control points have been established at advantageous locations within the tract and that the control has been plotted (office problem 21, p. 132) on the plane-table sheet.

Procedure.—(1) The plane-table man sets up and orients the table at one or more control stations such that the entire area may be mapped. (2) The levelman sets up the engineer's level, takes a backsight on a bench mark, and computes a rod reading such that the foot of the rod will be at the elevation of a given contour. (3) The rodman moves about as directed by the levelman, locating critical contour points. When a contour point has been located, the plane-table man sights on the rod, determines the distance by stadia, and plots the contour point.

CHAPTER XV

ROUTE SURVEYING

185. General.—Route surveys are made for the purpose of locating and building cross-country utilities such as highways, railways, and canals. Route surveying consists in determining the ground configuration and the location of objects along the proposed route, establishing the line on the ground, and computing volumes of earthwork. The methods described in this chapter apply primarily to highway and railway surveys; some special considerations pertaining to surveys for route structures, sewers, and pipe lines are stated in Arts. 191 and 192.

For the location of a new route, first a general study called a *reconnaissance* is made of the whole area under consideration, and one or more general routes are selected for further investigation. A *preliminary survey* is then run over each selected route, and a topographic map and a profile are prepared. A center line of the proposed roadway is then tentatively established on the topographic map; this is called the *paper location*. The line is then located on the ground by transit surveys; this is called the *field location*. Subsequently *construction surveys* may be necessary to establish lines and grades for the structures or to measure the amount of earthwork. Land boundaries near the route are surveyed and monumented, and right-of-way maps are drawn.

LOCATION SURVEYS

186. Reconnaissance.—Usually a reconnaissance involves field exploration of the area under consideration, supplemented by study of available maps. Particular attention is given to streams and divides, and decision is made as to certain *controlling points* such as bridge sites, passes, and towns. With these controlling points as a guide, one or more general routes are selected for further investigation.

Many highway surveys are made along established routes, and little or no reconnaissance is necessary.

187. Preliminary Survey.—A narrow strip of country along each proposed general route is surveyed and mapped, the strip being of sufficient width to contain the final location.

1. *Transit-tape-level Method.*—The survey corps consists of a transit party, a level party, and a topography party. The chief of the transit party, following the instructions of the locating engineer, runs an open traverse at random approximately along the middle of the strip through which it appears that the final line will lie. Stakes are set at all full stations, and hubs are set at all transit stations. Usually the results of each day's work are plotted at the close of the day.

The level party (levelman and rodman) follows the transit party, takes profile levels along the traverse, and establishes bench marks at intervals of a mile or less. Ground elevations are determined on the traverse line at all stakes set by the transit party, at changes in slope, and at roads and streams. Usually the preliminary profile is brought up to date at the close of each day's work.

The topography party (topographer, rodman, and frequently a tapeman) follows the level party, taking preliminary cross-sections as described in Art. 178*b*. The topography party also locates all pertinent objects and notes the character of cultivation and the quality of the land.

2. *Transit-stadia Method.*—Where the country is fairly open, often the transit-stadia method of running the preliminary survey is the most satisfactory. The usual procedure is to run the traverse, measuring the vertical angles and stadia distances, and to take side shots at the same time, as described in Arts. 115 and 178*a*; thus the horizontal and vertical control are established and the details are observed in one operation. Usually the party consists of a chief of party who may also act as recorder, a transitman, two or more rodmen, and sometimes a recorder. Hubs are set at all transit stations, but no intermediate stakes are set.

3. *Plane-table Method.*—The field procedure employing the plane table is much the same as that just described for the transit-stadia method, except that the map of the strip of country is constructed in the field as the work progresses. Usually the plane table is employed for locating details only, the control being established by a transit-tape traverse and direct leveling.

188. Preliminary Profile and Map.—A profile of the ground along the traverse line is prepared as described in Art. 65. For highway surveys, the usual horizontal scale is 1 in. = 100 ft. and the usual vertical scale is 1 in. = 10 ft. From the notes of the preliminary survey there is also prepared a map showing the topography and details along the selected strip of country. Both the profile and the map are employed by the locating engineer as a guide during the progress of the preliminary survey, and hence each day's work is usually plotted before the next day's work is begun.

189. Location Survey: Paper Location.—Based upon a study of the preliminary map and profile and upon further detailed study of the ground surface, the tentative alinement of a route (including curves) is chosen and is drawn on the map. Usually a profile of this paper location is drawn by use of elevations taken from the contour lines, a grade line is fixed on the profile (Art. 66), and the cost of construction is roughly estimated. The paper location is then used as a guide in locating the line in the field.

In fixing the location and grade of a roadway, some of the primary considerations are (1) to keep changes in alinement at a minimum; (2) to keep grades at a minimum; (3) to make the sum of the volumes in cut and borrow as small as possible consistent with suitable alinement and grades, by making the volume of earthwork in fills as nearly as practicable equal to that in adjacent cuts; (4) to keep at a minimum the amount of *haul* that will be necessary to transport excavated material from the cuts or borrow pits to the adjacent fills; and (5) to provide for drainage.

For an example of the use of a contour map in the location of a route for a highway, see Art. 157.

189a. Location Survey: Field Location and Office Work.—The work of field location consists first in laying off the line in the field so that it bears the same relation to the preliminary traverse on the ground that the paper location bears to the preliminary traverse on the map. This relation may be determined either by intersections between paper location and preliminary traverse or by scaling from the map the offsets from stations on the preliminary traverse to the tangents of the paper location. As the field location is run (see following paragraph), the tangents of the field location are established on the ground either by intersections with the preliminary traverse or by chaining the scaled offsets from the various stations on the preliminary traverse. Where practicable, adjoining tangents are run to an intersection, the intersection angles are measured, and the curve notes are computed as explained in Art. 195.

Beginning at some point for which the chainage is taken as zero, the field location is extended in the manner described in the preceding paragraph. Usually the located line is the center line of the roadway, but for highway surveys sometimes the line is located offset from one edge of the roadway. Stakes are set on tangents and circular curves at all full stations and in some cases at 50-ft. or even 25-ft. stations; and hubs are set at all P.I.'s, P.C.'s, P.T.'s (Art. 193), and intermediate transit stations. Transit notes are kept up the page in the manner illustrated by Table 195.

Profile levels are then run over the located line in the same manner as for the preliminary line, and from the data thus obtained a location profile is prepared showing the ground and grade lines. For railway location usually an alinement diagram (Fig. 65) is drawn on the location profile, whereas for highway location usually the plan (without contours) and profile are drawn on the same sheet of standard form called a *Federal Aid plan-profile sheet.*

In the light of a study of this profile and of the preliminary map, the grades are adjusted so that the line will better fit the ground. If minor modifications in alinement appear desirable, parts of the location are revised in the field. The line as finally located on the ground is plotted both in plan and in profile; it is then termed the *final location.* On the final-location map are shown all features of importance, including the location of bench marks and of the objects to which hubs are referenced.

Cross-sections of the located line are plotted (Art. 202) in order to estimate earthwork quantities and for purposes of letting the construction contract. For approximate estimates the cross-sections may be plotted from the data of the preliminary contour map; but usually the final cross-sections are taken while the slope stakes are being set (Art. 201), for which case the form of notes is shown in Fig. 199. The cross-sections are plotted on cross-ruled paper, which can be obtained either in rolls 20 in. wide or in standard 23 by 36-in. sheets called *Federal Aid cross-section sheets* to match the Federal Aid plan-profile sheets.

When the line is located definitely, a survey is run to determine and monument the boundaries of property that will be needed for the project, in order to secure rights of way. The results of the survey are plotted on a *property-line map,* or *right-of-way map,* in the usual form for a land map (Chap. XVI); and legal descriptions of the property are prepared.

CONSTRUCTION SURVEYS

190. Construction Surveys for Roadways.—The construction surveys for a roadway consist essentially in (1) staking out earthwork and structures preparatory to, and during the process of, grading and construction, and (2) making the measurements necessary to determine the volume of work actually performed up to a given date, as a basis for payment to the contractor. With regard to the final cross-sectioning, setting of slope stakes, and staking out of curves, the dividing line between the location survey and the construction survey is not definite; the practice varies according to the organization.

Some general considerations relating to construction surveys are given in Art. 179.

Highways.—Generally just prior to the beginning of construction of a section of highway, the located line is rerun, missing stakes are replaced, and hubs are referenced. Borrow pits (if necessary) are staked out and cross-sectioned as described in Art. 67. Lines and grades are staked out for bridges, culverts, and other structures. If slope stakes have not already been set during the location survey, they are set except where clearing is necessary; in that case they are set when the right of way has been cleared. For purposes of clearing, only rough measurements from the center-line stakes are necessary.

The method of setting and marking slope stakes is described in Art. 201. Where the depth of cuts and fills does not average more than about 3 ft., the slope stakes may be omitted; in this case the line and grade for earthwork may be indicated by a line of pegs (with guard stakes) along one side (or both sides) of the road and offset a uniform distance such that they will not be disturbed by the grading operations.

When rough grading has been completed, a line of finishing stakes is set to grade on both sides of the roadway at the edge of the shoulder, as a guide in trimming the slopes.

To give line and grade for the pavement, a line of stakes is set along each side, offset a uniform distance (usually 2 ft.) from the edge of the pavement. The grade of the top of the pavement, at the edge, is indicated either by the top of the stake or by a nail or line on the side of the stake. The alinement is indicated on one side of the roadway only, by means of a tack in the top of each stake. The distance between stakes in a given line is usually 100 or 50 ft. on tangents at uniform grade and half the normal distance on horizontal or vertical curves. The dimensions of the finished subgrade and of the finished pavement are checked by the construction inspector, usually by means of a templet.

As construction proceeds, monthly estimates are made of the work completed to date. A quantity survey is made near the close of each month, and the volumes of earthwork, etc., are classified and summarized as a basis for payment.

Streets.—For street construction the procedure of surveying is similar to that just described for highways. Ordinarily the curb is built first. The line and grade for the top of each curb are indicated by pegs driven just outside the curb line, usually at 50-ft. intervals. The grade for the edge of the pavement is then marked on the face of the completed curb; or for a combined curb and gutter it is indicated by the completed gutter. Ground pegs are set on the center line of the pavement, either at the grade of the finished subgrade or with the cut or fill indicated on the peg or on an adjacent stake. Where the street is wide, an intermediate row of pegs may be set between center line and each curb. It is usually necessary to reset the pegs after the street is graded. Where driving stakes is impractical because of hard or paved ground, nails or spikes may be driven or marks may be cut or painted on the surface. The surveys for street location and construction should determine the location of all surface and underground utilities that may affect the project.

Railways.—Surveys for railway earthwork are similar to those previously described for highways. Prior to construction the located line is rerun, missing stakes are replaced, hubs are referenced, borrow pits are staked out, slope stakes are set, and lines and grades for structures are established on the ground. When rough grading is completed, finishing stakes are set at the outer edges of the roadbed.

When the roadbed has been graded, alinement is established accurately by setting tacked stakes along the center line at full stations on tangents and usually at fractional stations on horizontal and vertical curves. Spiral curves are staked out at this time. An additional line of pegs is set on one side of the track and perhaps 3 ft. from the proposed line of the rail, with the top of the peg usually at the elevation of top of rail. Track is usually laid on the subgrade and is lifted into position after the ballast has been dumpod.

Near the end of each month, a survey of completed work is made and the monthly estimate is prepared.

191. Surveys for Bridge. *Site.*—Normally the location survey will provide sufficient information for use in the design of culverts; but for bridges and for grade-separation structures usually a special topographic survey of the site is necessary. The site map should include all the data of the location survey. The usual map scale is 1 in. = 100 ft., and the usual contour interval is 5 ft. on steep slopes and 2 ft. over flat areas. The preliminary report submitted with the site map should give all available information necessary for economic design, such as the character of the watershed and the stream bed, the elevation of the highest water, and the character of the foundation.

Construction.—For a short bridge with no offshore piers, first the center line of the roadway is established accurately, the stationing of some governing line such as the abutment face is established on the located line, and the angle of intersection of the face with the located line is turned off. This governing crossline may be established by two well-referenced transit stations at each end of the crossline beyond the limits of excavation or, if the face of the abutment is in the stream, by a similar transit line offset on the shore. Similarly, governing lines for each of the wing walls are established on shore beyond the limits of excavation. If the faces are battered, usually one line is established for the bottom of the batter and another for the top. Stakes are set as a guide to the excavation and are replaced as necessary. When the foundation concrete is cast, line is given on the footings for the setting of forms and then by sighting with the transit for the top of the forms. As the structure is built up, grades are carried up by leveling, with marks on the forms or on the hardened portions of the concrete. Also, the alinement is established on completed portions of the structure. The data are recorded in field books kept especially for the structure, principally by means of sketches.

For long sights or for work of high precision, as in the case of offshore piers, various transit stations are established on shore by a system of triangulation, such that favorable intersection angles and checks will be obtained for all parts of the work. To establish the offshore piers, simultaneous sights are taken from the ends of a line of known length. Where cofferdams are used, reference points are established on the cofferdams for measurements to the pier.

Culverts.—At the intersection of the center line of the culvert with the located line, the angle of intersection is turned off, and a survey line defining the direction of the culvert is projected for a short distance beyond its ends and is well referenced. At (or offset from) each end of the culvert, a line defining the face is turned off and referenced. If excavation is necessary for the channel to and from the culvert, it is staked out in a manner similar to that for a roadway cut. Bench marks are established near by, and pegs are set for convenient leveling to the culvert. Line and grade are given as required for the particular type of structure.

192. Surveys for Sewers and Pipe Lines.—The center line for a proposed sewer is located on the ground with stakes or other marks set usually at 50-ft. intervals. At one side of this line, just far enough from it to prevent being disturbed by the excavation, a parallel line of ground pegs is set. A guard stake is driven beside each peg, with the side to the line; on the side of the guard stake farthest from the line is marked the station number and offset, and on the side nearest the line is marked the cut (to the nearest ⅛ in.). In paved streets or hard roads where it is impossible to drive stakes and pegs, the line and grade are marked with spikes (driven flush), chisel marks, or paint marks.

When the trench has been excavated, batter boards are set across the trench at the intervals used for stationing. The top of the board is set at a fixed whole number of feet above the sewer invert (inside surface of bottom of sewer); and a measuring stick of the same length is prepared. A nail is driven in the top edge of each batter board to define the line. As the sewer is being laid, a cord is stretched tightly between these nails, and the free end of each tile is set at the proper distance below the cord as determined by measuring with the stick.

If the trench is to be excavated by hand, the side pegs may be omitted and the batter boards set at the beginning of excavation.

For pipe lines the procedure is similar to that for sewers, but the interval between grade pegs or batter boards may be greater, and less care need be taken to lay the pipe at the exact grade.

For both sewers and pipe lines the extent of excavation in earth and in rock is measured in the trench, and the volumes of each class of excavation are computed as a basis of payment to the contractor.

The records of the survey should include the location of underground utilities crossed by, or adjacent to, the trench.

CURVES

193. Circular Curves.—In highway and railway location the horizontal curves employed at points of change in direction are arcs of circles. The straight lines connecting these *circular curves* are tangent to them and are therefore called *tangents.* For the completed line the transition from tangent to circular curve and from circular curve to tangent may be accomplished gradually by means of a segment in the form of a *spiral,* described in more detailed textbooks on surveying.

The stationing of a route progresses around a curve in the same manner as along a tangent. The point where a circular curve begins is commonly called the *point of curve,* written P.C.; that where the curve ends is called the *point of tangent,* written P.T.; and that where two tangents produced intersect is called the *point of intersection* or the *vertex,* written P.I. or V.

In the field, the distances from station to station (usually 100 ft.) on a curve are necessarily measured in straight lines, so that essentially the curve consists of a succession of 100-ft. chords. Where the curve is of long radius, as in railroad practice, the distances along the arc of the curve are considered to be the same as along the chords. Where the curve is of short radius, as in highway practice and along curved property boundaries, either a correction is applied for the difference between arc length and chord length or the chords are made so short as to reduce the error to a negligible amount.

The sharpness of curvature may be expressed in any of three ways:

1. *Radius.*—By stating the length of the radius. This method is often employed in highway work, with the radius for a given curve taken as a multiple of 100 ft.

2. *Degree of Curve, Arc Basis.*—By stating the "degree of curve," or the angle subtended at the center by an *arc* 100 ft. long. This method is generally followed in highway practice. Thus, if the degree of curve is D and the radius R,

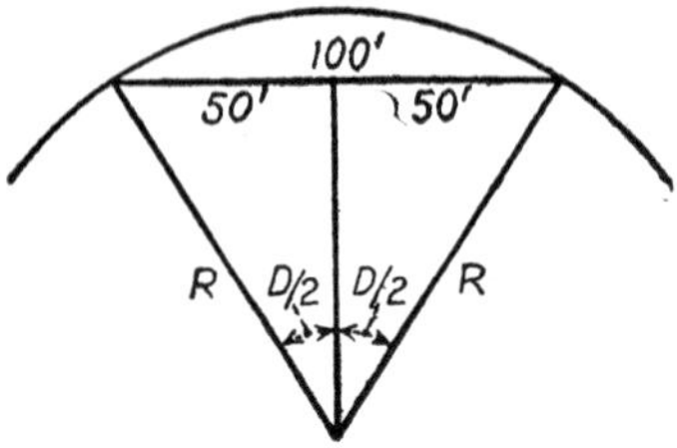

FIG. 193.—Degree of curve.

$$R = \left(\frac{360°}{D°}\right)\left(\frac{100}{2\pi}\right) \quad (1)$$

From this relation the radius may be found if the degree of curve is known, and vice versa.

3. *Degree of Curve, Chord Basis.*—By stating the degree of curve as the angle subtended by a *chord* of 100 ft. This method is followed in railroad practice. Thus in Fig. 193,

$$R = \frac{50}{\sin \frac{1}{2}D} \quad (2)$$

Except as specifically stated, hereinafter the discussions refer to the chord basis for expressing curvature.

194. Curve Formulas.—In the following discussion the geometrical fact is employed that either an inscribed angle or an angle formed by a tangent and a chord is measured by one-half its intercepted arc and is therefore equal to one-half the corresponding central angle.

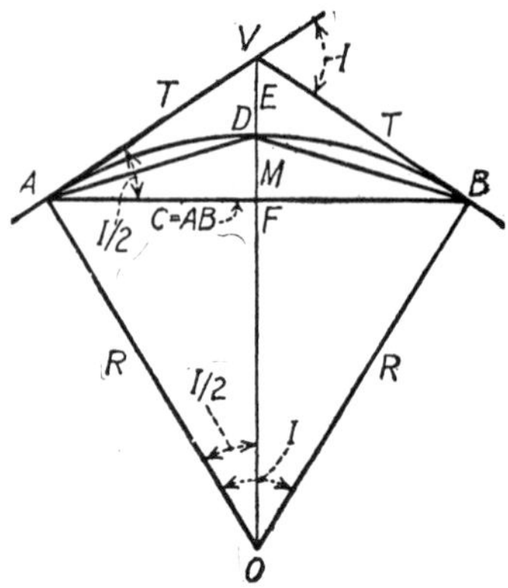

FIG. 194.—Basis for curve formulas.

Figure 194 represents a circular curve joining two tangents. In the field the intersection angle I between the two tangents is measured. The line OV bisects the angles at V and at O, bisects the chord AB and the arc ADB, and is perpendicular to the chord AB at F. From the figure, $\angle AOB = I$ and $\angle AOV = \angle VOB = \frac{1}{2}I$.

The chord $AB = C$ from the beginning to the end of the curve is called the *long chord.* The distance $AV = BV = T$ from vertex to P.C. or P.T. is called the *tangent distance.* The distance $DF = M$ from the mid-point of the arc to the mid-point of the chord is called the *middle ordinate.* The distance $DV = E$ from the mid-point of the arc to the vertex is called the *external distance.*

Given the radius of the curve $OA = OB = R$ and the intersection angle I, then in the triangle OAV

$$T = R \tan \tfrac{1}{2}I = \text{tangent distance} \tag{3}$$

$$E = R \sec \tfrac{1}{2}I - R = R \operatorname{exsec} \tfrac{1}{2}I = \text{external distance} \tag{4}$$

From the triangle AOF, in which $AF = \frac{1}{2}C$,

$$C = 2R \sin \tfrac{1}{2}I = \text{long chord} \tag{5}$$

$$M = R - R \cos \tfrac{1}{2}I = R \operatorname{vers} \tfrac{1}{2}I = \text{middle ordinate} \tag{6}$$

From the triangle AVF, in which $\angle VAF = \frac{1}{2}I$ and $AF = \frac{1}{2}C$,

$$C = 2T \cos \tfrac{1}{2}I \tag{7}$$

From the triangle ADF, in which $\angle DAF = \frac{1}{4}I$,

$$M = \tfrac{1}{2}C \tan \tfrac{1}{4}I \tag{8}$$

On either the arc basis or the chord basis, the length of curve is

$$L = 100 \frac{I}{D} \tag{9}$$

in which D is on the same basis as that for the desired length L.

195. Laying Out Curve by Deflection Angles.—Curves are staked out usually by the use of deflection angles turned at the P.C. from the tangent to stations along the curve together with the use of chords measured from station to station along the curve. The method is illustrated in Fig. 195, in which ABC represents the curve, AX the tangent to the curve at A, and angles XAB and XAC the deflection angles from the tangent to the chords AB and AC.

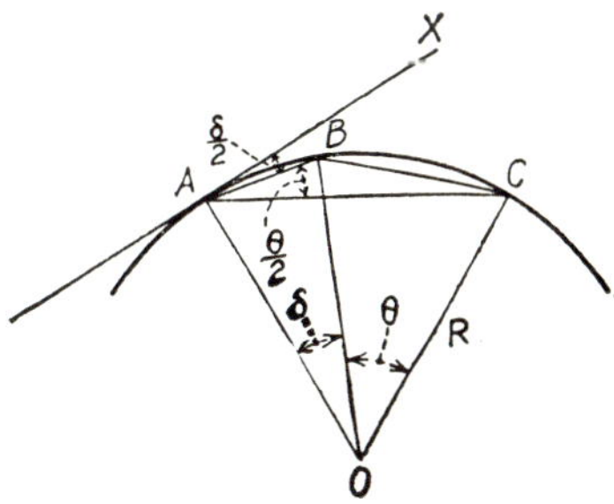

FIG. 195.—Curve by deflection angles.

Assume the transit to be set up at A. Given R, δ, θ. Required to locate B and C. Considering the point B,

$$\angle XAB = \tfrac{1}{2}\delta \tag{10}$$

$$AB = 2R \sin \tfrac{1}{2}\delta \tag{11}$$

In the field, the point B is located as follows: The deflection angle $XAB = \frac{1}{2}\delta$ is set off from the tangent; the distance AB is measured from A; and the forward end of the tape at B is lined in with the transit.

Considering the point C,

$$\angle BAC = \tfrac{1}{2}\theta \tag{12}$$
$$BC = 2R \sin \tfrac{1}{2}\theta \tag{13}$$
$$\angle XAC = \tfrac{1}{2}(\delta + \theta) \tag{14}$$

TABLE 195.—FIELD NOTES FOR CIRCULAR CURVE

Station	Point	Deflection angle	Curve or bearing
100			
99			N73°10′W
98 + 56.7	P.T. ⊙	52°18′	
98		48°54′	
97		42°54′	
96		36°54′	$D = 12°$
95		30°54′	$I = 104°36'$
94	⊙	24°54′	$T = 618.9$
93		18°54′	$R = 478.3$
92		12°54′	$L = 871.7$
91		6°54′	
90		0°54′	
89 + 85.0	P.C. ⊙	0°00′	12°L
89			
88			N31°26′E

In the field, the point C is located as follows: With the transit still at A, the deflection angle XAC is set off from the tangent, the distance BC is measured from B, and the forward end of the tape at C is lined in with the transit sighted along the line AC. Succeeding stations on the curve are located in similar manner.

A curve is located in the field normally as follows: The P.C. and P.T. are marked on the ground. The deflection angle from the P.C. is computed for each full station on the curve and for any intermediate stations that are to be located. The transit is set up at the P.C.; a backsight is taken along the tangent with telescope inverted; the telescope is plunged; and each point on the curve is located by deflection angle and by distance measured from the preceding *full station*. Table 195 illustrates the usual form of field notes.

196. Transit Set-ups on the Curve.—Often it is impossible to run all of a given curve with the transit at the P.C., and one or more set-ups are required along the curve. In any case, all the deflection angles are computed as if for use at the P.C. (see field notes of preceding article).

The transit is set up at any point along the curve, and a backsight (with telescope inverted) is taken to any preceding transit station with the vernier reading the computed deflection angle for the point sighted. If the backsight is to the P.C., the vernier is set at zero. The telescope is plunged, and points ahead of the transit are then set by using the deflection angles previously computed for these points.

CROSS-SECTIONS

197. Form of Cross-sections.—Figures 200 and 201 illustrate typical highway or railroad cross-sections in fill and in cut. The rough subgrade is usually a plane surface, transversely level, and on a given road is of uniform width in cut and of uniform but usually a smaller width in fill. The side slopes are plane surfaces of constant slope for a given material of excavation. The rate of the side slope (as 2:1) is stated in terms of the number of units measured horizontally (as 2 ft.) to one unit measured vertically (as 1 ft.). The finished cross-section may be sloped variously to provide shoulders, drainage, and rounded corners.

If the ground is level in a direction transverse to the center line, the cross-section is called a *level section*. When rod readings are taken at each slope stake in addition to the reading taken at the center, as will normally be done where the ground is sloping, the cross-section is called a *three-level section* (Fig. 203). When rod readings are taken at the center stake, the slope stakes, and points on each side of the center at a distance of half the width of the roadbed, the cross-section is called a *five-level section*. A cross-section for which observations are taken to points between center and slope stakes at irregular intervals is called an *irregular section*. Where the cross-section passes from cut to fill, it is called a *side-hill section*.

198. Preliminary Route Cross-sections.—To furnish data for route location studies and for estimating volumes of earthwork, it is customary to determine the shape of the ground on both sides of the traverse line by running levels over crosslines at right angles to the traverse, usually at each 100-ft. station. Generally the elevations are determined with the hand level in rough country and with the engineer's level in flat country. For each crossline the height of instrument is established by a backsight on the ground at the center stake. The rod is then held on the crossline at breaks in the surface slope, and distances from the traverse line to these points are measured with the metallic tape. The direction of short crosslines is

PRELIMINARY CROSS-SECTIONS C. & R. EXTENSION						O.H. Ellis, ⊼ C.O. Lord, Rod J.A. Crum, Tape		Jan. 20, 1942 Cold, Snow
Sta.	B.S.	H.I.	F.S.	Elev.		Left	℄	Right
405	12.4	633.0		620.6	(Dist.) (Elev.) (Rod)	300 210 123 80 632.1 630.8 627.0 626.7 0.9 2.2 6.0 6.3	620.6	
405	0.6	621.2		620.6				50 160 250 350 617.0 612.3 610.0 609.7 4.2 8.9 11.2 11.5
406	12.1	628.9	0.7	616.8		90 50 628.2 624.3 0.7 4.6	616.8	
406	11.5	639.7		628.2		280 200 120 638.3 635.6 632.2 1.4 4.1 7.5		
406	1.9	618.7		616.8				60 100 200 300 615.5 611.2 610.9 609.3 3.2 4.5 7.8 9.4
407	4.7	615.9		611.2		300 180 100 615.6 614.7 612.6 0.3 1.2 3.3	611.2	75 155 270 606.2 604.5 602.9 9.7 11.4 13.0
408	10.6	615.9		605.3		280 200 100 611.6 609.2 607.7 4.3 6.7 8.2	605.3	100 200 300 604.1 603.2 603.0 11.8 12.7 12.9

Fig. 198.—Preliminary route cross-section notes.

laid off by eye; that of long crosslines by means of a compass, transit, or other suitable instrument.

Notes may be kept in the form shown in Fig. 198. The center line of the right-hand page represents the traverse line, and to the right and left of this line are recorded the observed distances and rod readings and the computed elevations.

199. Final Route Cross-sections.—The preliminary survey is made, and the profile is plotted. With the profile and the data of the preliminary cross-sections as a guide, the grade line of the subgrade is established on the sheet with the plotted profile. The center cut or fill to be made at each station is then equal to the difference between the observed elevation of the ground line and the established elevation of the grade line.

Prior to actual construction, final cross-sections are taken, and slope stakes marking the intersection of the side slopes with the natural ground surface are set opposite each center stake (Art. 201).

Usually level readings for final cross-sections are taken with the engineer's level, and distances to the right or left of the center stake are measured with the metallic tape, all to tenths of feet. At each station a foresight is taken to check the ground elevation against that obtained by profile leveling, and the computed cut or fill is

CROSS-SECTIONS FOR I.N. RY. FINAL LOCATION

Cox Brook to Big Forks — June 4, 1942 — F.F. Smith, ⊼

Roadbed 20 ft. in Cut, 16 ft. in Fill — Cloudy — J. Richie, Rod

Slope 1½:1 — O. Byram, Tape

Sta.	B.S.	H.I.	F.S.	Elev.	Grade	Left	Ctr.	Right	Remarks
B.M.28	2.67	566.98		(564.31)					50 ft. Lt. Sta. 605+10.
605			1.9	565.1	556.00	C8.6 / 22.9	C9.1	C11.2 / 26.8	Gravel in this hill.
606			4.0	563.0	555.60	C5.0 / 17.5	C7.4	C8.4 / 22.6	
607			4.5	562.5	555.20	C4.6 / 16.9	C7.3	C8.0 / 22.0	
608			8.0	559.0	554.80	C1.8 / 12.7	C4.2	C5.1 / 17.7	
+25			9.2	557.8	554.70	0.0 / 10.0	C3.1	C2.6 / 13.9	
T.P.	1.94	557.19	(11.73)	(555.25)					On plug.
+90			2.8	554.4	554.44	F1.8 / 10.7	0.0	C2.4 / 13.6	
609			3.4	553.8	554.40	F3.2 / 12.8	F0.6	0.0 / 4.0 C1.0 / 11.5	
+20			5.6	551.6	554.32	F3.6 / 13.4	F2.7	0.0 / 8.0	
610			11.1	546.1	554.00	F8.2 / 20.3	F7.9	F6.3 / 17.5	
+40			11.2	546.0	553.84	F8.0 / 20.0	F7.8	F7.8 / 19.7	Top of bank Cox Brook
+45			14.6	542.6	553.82	F11.2 / 24.8	F11.2	F11.2 / 24.8	In brook.
+60			14.5	542.7	553.76	F11.1 / 24.7	F11.1	F11.1 / 24.7	" "
+65			11.6	545.6	553.74	F8.2 / 20.3	F8.1	F8.0 / 20.0	Top of bank.
611			10.9	546.3	553.60	F7.8 / 19.7	F7.3	F7.2 / 18.8	
612			9.7	547.5	553.20	F7.6 / 19.4	F5.7	F3.1 / 12.7	
613			10.8	546.4	552.80	F7.8 / 19.7	F6.4	F2.0 / 11.0	
T.P.	11.96	559.69	(9.46)	(547.73)					On stump.
	16.57	564.31	21.19						
		559.69	16.57						
		4.62=4.62 ck.							

FIG. 199.—Cross-section notes for final location of roadway.

marked on the back of the center stake. A crossline is established, and the slope stakes are set as later described.

Figure 199 illustrates a suitable form of final cross-section notes. The left-hand page is essentially the same as for profile leveling except that a column is added for grade elevations. Some engineers prefer to have the notes read up the page. The notes shown are for a portion of the line for which profile-level notes are shown in Fig. 63. The values in columns marked "Left" or "Right" are for points at which the slope stakes are driven; for each such point the upper number is the cut or fill, and the lower number is the distance out from center. The slope is such that three-level sections are adequate. Cross-sections are taken where the left edge, center, and right edge of the roadbed pass from cut to fill. The cross-sections at 608 + 90 and 609 are side-hill sections.

200. Cuts and Fills.—Figure 200 shows at *A* the engineer's level in position above grade, for taking rod readings at a section in fill. The height of instrument has been determined; the elevation of grade at the particular station is known. The leveler computes the difference between the H.I. and the grade elevation, a difference known as the *grade rod;* that is, H.I. − el. of grade = grade rod. The rod is held at any point for which the fill is desired, and a reading called the *ground rod* is taken. The difference between the grade rod and the ground rod is equal to the fill. Similarly, for a section in cut this difference is equal to the cut.

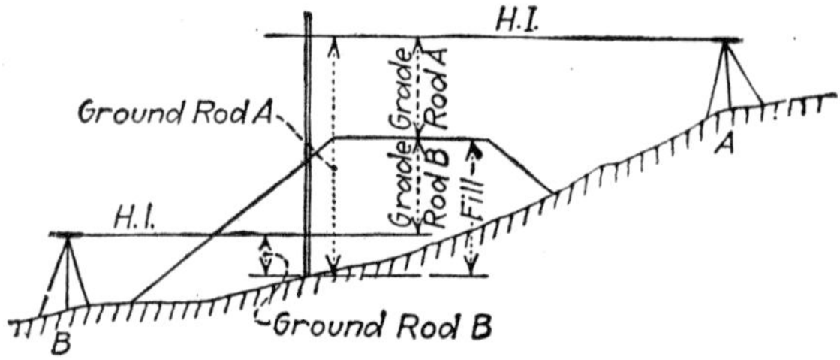

FIG. 200.—Road cross-section in fill.

If the H.I. is *below* grade, as at *B* in the figure, it is clear that the fill is the *sum* of the grade rod and the ground rod.

201. Setting Slope Stakes.—If w is the width of roadbed, d the measured distance from center to slope stake, s the side-slope ratio (ratio of horizontal distance to drop or rise), and h the cut (or fill) at the slope stake, then by Fig. 201 when the slope stake is in the correct position (at C),

$$d = \frac{w}{2} + hs \qquad (15)$$

The following numerical example for a cut illustrates the steps involved in establishing the correct location for a slope stake in the field; the same procedure is followed in fill.

Let w = 20 ft.; side slope = 1½ to 1; grade rod = 15.2 ft. As a first trial the rod is held at A (Fig. 201); ground rod = 6.6 ft.,

$$h_1 = \text{grade rod} - \text{ground rod} = 15.2 - 6.6 = 8.6 \text{ ft.}$$

The computed distance for this value of h_1 is

$$\frac{w}{2} + h_1 s = 10.0 + 8.6 \times 3/2 = 22.9 \text{ ft.}$$

Measurement from the center stake shows d_1 to be 18.2 ft.; hence the rodman should go farther out. A second trial is made perhaps at B, and by similar computation it is found that the rod is too far out. Eventually by trial the rod will be held at C;

$$\text{Ground rod} = 7.8 \text{ ft.}; \qquad h = 15.2 - 7.8 = 7.4 \text{ ft}$$

The computed distance for this value of h is

$$\frac{w}{2} + hs = 10.0 + 7.4 \times 3/2 = 21.1 \text{ ft.}$$

The measured value of d is 21.1 ft.; hence this is the correct location for the slope stake.

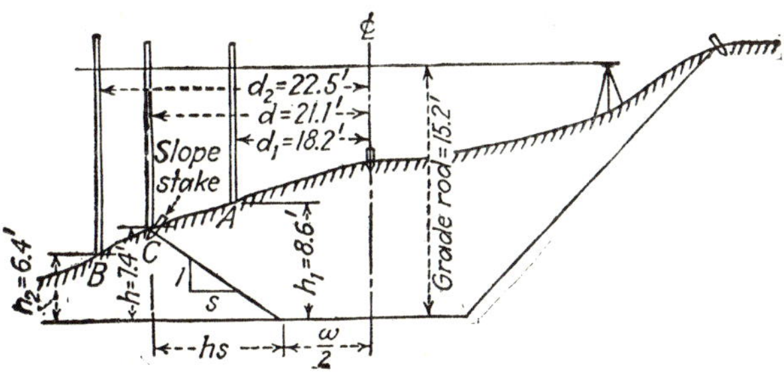

FIG. 201.—Setting slope stakes.

In the notes the coordinates of the slope stake are given by the expression $\frac{c7.4}{21.1}$ in the form of a fraction, but the trial observations are not recorded.

Slope stakes are set side to the line, sloping outward in fill and inward in cut. On the back of the stake is marked the station number. On the front (side nearest the center line) is marked the cut or fill at the stake, and sometimes the distance from center to slope stake. The numbers read down the stake.

202. Plotting Cross-sections.—Irregular cross-sections for earthwork are commonly drawn to scale on cross-section paper which is ruled usually with 10 divisions to the inch in both directions. Governing points of the cross-section are plotted either from the cross-section notes or from the data of a topographic map. The surface may be indicated either by an irregular line or by a series of straight lines connecting the points; the fixed portions of the cross-section either may be plotted directly or may be drawn by means of a templet. For large cross-sections a scale of 1 in. = 10 ft. is common.

If the cross-sections are shallow, sometimes the vertical scale is exaggerated. Usually cross-sections are plotted in pencil only.

The cross-section for the first station of the route is usually placed in the upper left corner of the sheet, and successive cross-sections are placed one below the other (Fig. 202). Below each cross-section is shown its station number. Within each cross-section is shown its calculated area in square feet, and between successive cross-sections is shown the calculated volume of earthwork in cubic yards. The coordinates may or may not be noted on the sheet. Some engineers

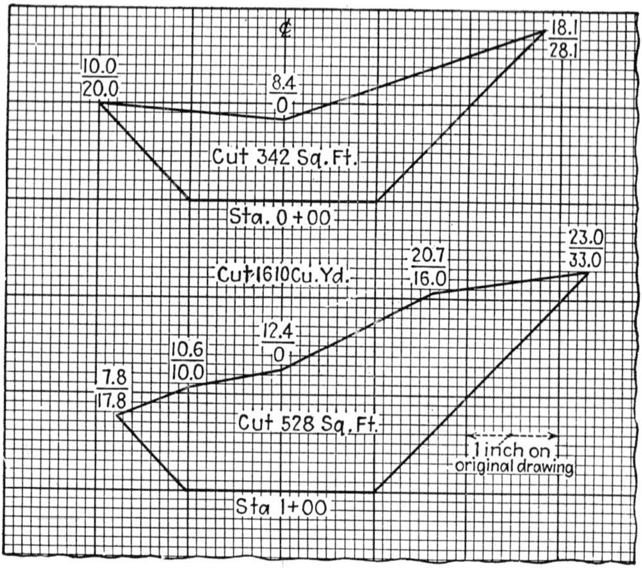

FIG. 202.—Plotted cross-sections.

place the first cross-section in the *lower* left corner of the sheet, show the elevation of the ground line at each station, and place the notations at locations other than those shown in the figure.

203. Areas of Regular Cross-sections.—Areas of regular cross-sections are readily determined by numerical calculations without plotting.

For a trench the cross-sectional area at any point is determined by multiplying the average of the top and bottom widths by the depth.

The same method may be applied to level cross-sections for highways and railroads; if d is the distance to either slope stake from the center, w is the width of the roadbed, and c is the center cut or fill, then the area A of the level section is

$$A = c\left(d + \frac{w}{2}\right) \tag{16}$$

A three-level section may be divided into four triangles, as shown in Fig. 203. Then from the figure the area A is

$$A = \frac{w}{4}(h_l + h_r) + \frac{c}{2}(d_l + d_r) \tag{17}$$

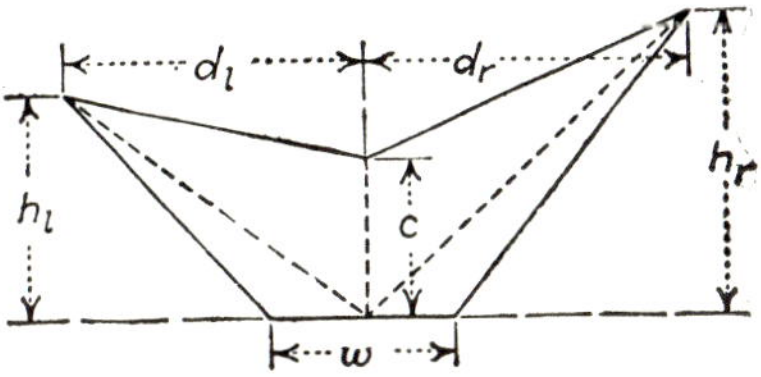

FIG. 203.—Three-level section.

204. Areas of Irregular Road Cross-sections.—Areas of irregular sections may be calculated by plotting the sections and dividing them into trapezoids and triangles, but the computations are tedious except for side-hill sections. The method of coordinates (Art. 142) may be adapted in simplified form for the particular case. If the cross-sections are bounded by curved lines or are very irregular, usually they are plotted, and the area is determined by traversing the perimeter of each plotted cross-section with a polar planimeter.

VOLUMES OF EARTHWORK

205. General.—Volumes of earthwork are calculated by a variety of methods. If cross-sections have been taken along a route, their areas are determined and the volumes of the successive prismoids are determined either by the method of average end areas (Art. 207) or by the prismoidal formula (Art. 208). The same procedure may be followed for borrow pits and similar excavations; or if elevations are observed at the same points before and after excavating, the volume may be calculated by dividing it into vertical truncated prisms. Volumes of earthwork may be determined by the use of contours (see Art. 148).

206. Volume of Borrow Pit.—The common method of determining the volume of a borrow pit is to cross-section the area before and after excavating (Art. 67). The pit is then plotted in plan, and its area is divided into rectangles and triangles, thus dividing the volumes into truncated prisms. Actually the upper and lower surfaces of the prisms are warped, but for earthwork calculations they are assumed to be plane.

The volume V of a triangular truncated prism of horizontal sectional area A is

$$V = \frac{A}{3}(h_1 + h_2 + h_3) \tag{18}$$

in which h_1, h_2, and h_3 are the corner heights.

Any rectangular prism may be divided into two triangular prisms by either of two diagonal planes; and where differences in corner heights are large, the proper division into triangles should be indicated in the field notes. Where differences in corner heights are not large, the volume of earthwork in rectangular prisms is computed with sufficient accuracy by multiplying the average of the corner heights by the horizontal sectional area.

207. Volumes by Average End Areas.—The most common method of determining volumes of excavation along a line is that of *average end areas*. It is assumed that the volume between successive cross-sections is the average of their areas multiplied by the distance between them, or

$$V = \frac{l}{2}(A_1 + A_2) \tag{19}$$

in which V is the volume (cubic feet) of the prismoid of length l (feet) between cross-sections having areas (square feet) A_1 and A_2.

The formula is exact only when $A_1 = A_2$ but is sufficiently accurate for ordinary earthwork. The maximum error (50 per cent) occurs as one of the end areas approaches zero, but in this case the volume is usually computed as a pyramid; that is,

$$\text{Volume} = \tfrac{1}{3} \text{ area of base} \times \text{length}$$

208. Volumes by the Prismoidal Formula.—It can be shown that the volume of a prismoid is

$$V = \frac{l}{6}(A_1 + 4A_m + A_2) \tag{20}$$

in which V, l, A_1, and A_2 are the same as in Art. 207, and A_m is the area halfway between the end sections. A_m is determined by averaging the corresponding linear dimensions of the end sections and not by averaging the end areas.

The use of the prismoidal formula in earthwork calculations is justified only if cross-sections are taken at short intervals, if small surface deviations are observed, and if the areas of successive cross-sections differ widely. Usually it yields smaller values than those computed from average end areas.

209. Numerical Problems.

1. Given $I = 30°40'$, D (chord basis) $= 4°00'$, and P.C. = station 18 + 16.0. Required R, L, T, E, C, and M; also deflection angles arranged in notebook form for staking out this curve, using 100-ft. stations.

2. The width of roadbed of a proposed railroad is 24 ft. in cut, and the side slopes are 1½ to 1. At a given station the elevation of grade is 607.98. For obtaining the cross-section at the station the H.I. of the level is 620.56. The ground rod at the center stake is 5.6. Compute the grade rod and the center cut. The rod reading at the right slope stake is 1.9 and at the left slope stake is 8.2. Compute the cut and the distance out to each slope stake.

3. Following are the notes for an irregular cross-section. The width of the roadbed is 22 ft., and the side slope is 1½:1. Compute the cross-sectional area.

$$\frac{c3.6}{16.4} \quad \frac{c5.7}{11.0} \quad \frac{c7.4}{0} \quad \frac{c13.2}{8.0} \quad \frac{c12.1}{15.0} \quad \frac{c6.8}{21.2}$$

4. From the notes of Fig. 199 compute by the method of average end areas the volume in cut and in fill between station 607 and station 612. Tabulate the data in columns headed "Station," "Cross-section," "Area," and "Volume."

5. From the data of the preceding problem, compute volumes by the prismoidal formula.

210. Field and Office Problems.

PROBLEM 27. PRELIMINARY SURVEY FOR ROAD (TOPOGRAPHY BY CROSS-PROFILE METHOD)

Object.—To obtain data for a topographic map along the route of a proposed highway or railroad, locating the contours directly by means of the hand level and metallic tape. The data may be used in office problem 21 (p. 132). Steps 1 and 2 of the following procedure may have been accomplished in field problems 13 (p. 85) and 5 (p. 53). For a preliminary survey run by an alternative method, employing transit and stadia, see field problem 17 (p. 96).

Procedure.—(1) Over the assigned route run an open deflection-angle traverse with transit and tape. (2) Establish vertical control for the route by direct profile leveling. (3) At each full station and at any necessary plus stations locate the 5-ft. contours on a crossline extending 300 to 800 ft. on either side of the line; also locate the points where the 5-ft. contours cross the traverse line. Employ the hand level, leveling rod, and metallic tape (see Art. 178*b*). (4) Measure distances from the traverse line to other topographic features such as land lines, streams, roads, and buildings. (5) Note the quality of the land, as to whether it is clay, rock, or sand. Note the condition of the land, as to whether it is cleared land, pasture land, or woods; note the kind of trees and density of growth in woods. (6) Keep notes as shown in Fig. 101.

PROBLEM 28. LAYING OUT A CIRCULAR CURVE

Object.—To lay out a circular curve, as for the curb line of a driveway, by the use of deflection angles.

Procedure.—(1) From the assigned central angle and degree of curve compute the tangent distance and the length of curve. (2) Assume that the P.C. is station 9 + 83.2, and compute deflections for each full station and +50. Prepare notes in the form shown in Table 195. (3) In the field, locate two tangents making the assumed angle. Locate the P.C. and the P.T. (4) Set up the transit at the P.C., orient the instrument, and stake out the full and +50 stations by the method described in Art. 195. Report the error observed at the P.T. (5) Set the transit at the P.T., and check the angle. (6) If transit set-ups on the curve are required, follow the method of Art. 196.

Problem 29. Setting Slope Stakes; Cross-sections

Object.—To prepare a proposed highway or railway for grading and to obtain data for calculating earthwork (see office problem 30, below).

Procedure.—(1) From the level notes of field problem 5 plot a profile and fix a grade such that the amount of cut will approximately balance the amount of fill. (2) Keep field notes in the form of Fig. 199. (3) Drive short pegs flush with the ground against the center stakes, on the side farthest from the beginning of the line. Run profile levels over the line of pegs, checking the elevations obtained with those of field problem 5; and mark on the back of each center stake the cut or fill at that point, as *C* 3.9 or *F* 4.7. (4) Opposite each center stake, at right angles to and on both sides of the line, set and mark slope stakes as described in Art. 201. Assume a roadway 20 ft. wide with side slopes 1½ to 1. (5) Drive ground pegs at "grade points" where the center line and each edge of roadbed pass from cut to fill; mark the location of these grade pegs by stakes marked "grade."

Problem 30. Plotting Cross-sections; Quantities of Earthwork

Object.—To plot cross-sections of a roadway from field notes and to calculate quantities of earthwork. It is assumed that the cross-section notes give cut and fill (see field problem 29).

Procedure.—(1) Beginning near the upper left corner of a sheet of cross-section paper, choose convenient heavy horizontal and vertical lines as grade and center lines. With these as coordinates plot the cross-section notes of the first station (see Fig. 202). Mark the plotted points with dimensions identical with those of corresponding points in the notes. Draw straight lines showing roadbed and side slopes of cut or fill and the original ground, thus enclosing the section. Below the cross-section, mark its station number. (2) At a convenient distance below and on the same center line, plot the next section in similar manner. When the bottom of the sheet is reached, plot the next section a little farther to the right and at the top of the sheet; and in this way continue until all plotting is done. (3) Compute the area of each section, and show its value within the section (as 123 sq. ft.). Irregular sections may be planimetered. (4) Calculate volumes by the average-end-area method and by the prismoidal formula. Show the volume of each prismoid (as 97 cu. yd.) between its end sections. (5) By each method find the total yardage in each cut and fill, and mark these totals conspicuously. (6) Make an appropriate title.

CHAPTER XVI

LAND SURVEYING

211. General.—A land survey may be (1) an *original survey,* made for the purposes of writing a legal description and of finding the area of a designated tract; (2) a *resurvey,* made to reestablish the boundaries of a tract for which the description as defined by a previous survey is known; or (3) a *subdivision survey,* made for the purpose of subdividing a tract into parcels such as blocks and lots of a city addition. The land surveyor makes the field surveys, calculates dimensions and areas, prepares maps, and writes legal descriptions for deeds.

212. Instruments and Methods.—Nearly all land surveys are run by traversing with the transit and tape, as described in Chap. VII. Usually the directions of lines are expressed as true bearings. Ordinarily distances are measured to feet and decimals, but on United States public-land surveys the unit of length prescribed by law is the Gunter's (66-ft.) chain.

Formerly, the surveyor's compass and 66-ft. link chain were used extensively, and the descriptions in many old deeds are given in terms of magnetic bearings and Gunter's chains. In retracing old surveys of this character, allowance must be made for change in magnetic bearing since the time of the original survey. Also, it must be kept in mind that the compass and link chain used on old surveys were relatively inaccurate instruments and that great precision was not regarded as necessary since generally the land values were low. Many of the lines and corners established by old surveys are not where they theoretically should be; nevertheless these boundaries legally remain fixed as they were originally established.

213. Corners and Monuments.—It is customary to mark the corners of landed property by visible monuments. The term *corner* is applied to a point established by a survey or by an agreement; the term *monument* is applied to an object placed to mark the corner point upon the surface of the earth. If there is a possibility that a corner monument will become displaced, the corner should be *referenced* (Art. 106) to near-by objects of more or less permanent character. Usually the recorded measurement is called a *connection,* and the object is called a *reference mark* or a *corner accessory.*

If the location of a corner within very narrow limits can be determined beyond all reasonable doubt, the corner is said to *exist;* otherwise it is said to be *lost.* If the monument marking an existing corner cannot be found, the corner is said to be *obliterated,* but it is not necessarily lost. Where a corner falls in such location as to make it impossible or impracticable to establish a monument at that point, it is customary to establish a *witness corner* on one or more of the boundary lines leading to the corner, as near to the true corner as possible.

214. Meander Lines.—In surveying along the shore of a lake or the bank of a stream a traverse called a *meander line* is run, roughly following the shore or bank line. The process of establishing such a line is called *meandering.* Meander lines are for surveying and mapping purposes only and are not property lines. If more accurate location of the shore line is desired, offsets are taken from the meander line.

215. Boundary Records.—Descriptions of the boundaries of real property may be found from deeds, official plats or maps, or notes of original surveys. Records of the transfer of land from one owner to another are kept in the country registry of deeds, exact copies of all deeds of transfer being filed in deed books. These files are open to the public and are a frequent source of information for the land surveyor.

In most cases the deeds of transfer of city lots give only the lot or block number and the name of the addition or the subdivision. The official plat or map showing the dimensions of all lots and the character and location of permanent monuments is on file either in the office of the city clerk or in the county registry of deeds; copies are also on file in the offices of city and county assessors.

As the United States public lands are subdivided, official plats are prepared showing the dimensions of subdivisions and the character of monuments marking the corners. When the surveys within a state have been completed, all records are given to the state. Usually information concerning these records can be obtained from the state secretary of state.

216. Legal Interpretation of Deed Description.—The descriptions of the boundaries of a tract include the objects that fix the corners, the lengths and directions of lines between the corners, and the area of the tract. A deed description may contain errors or mistakes of measurement or mistakes of calculation or record, thus introducing inconsistencies which cannot be reconciled completely when retracement becomes necessary. In such cases, where uncertainty has arisen as to the location of property lines, it is a universal principle of law that the endeavor is to make the deed effectual rather than void and to execute the intention of the contracting parties. The following general rules are pursuant to this principle.

1. *Monuments.*—A corner established by an existing material object or by conclusive evidence as to the previous location of the object will prevail against all other conflicting evidence.

2. *Distance and Direction vs. Area.*—In case of discord, the deed-description requirements for distances or directions of courses will prevail against the requirements for area.

3. *Mistakes.*—A deed description which taken as a whole plainly indicates the intentions of the parties concerned will not be invalidated by evident mistakes or omissions.

4. *Purchaser Favored.*—In a description that is capable of two or more interpretations, that one will prevail which favors the purchaser.

217. Legal Authority and Liability of Surveyor.—A resurvey may be run to settle a controversy between owners of adjoining property. The surveyor should understand that, although he may act as an arbiter in such cases, it is not within his power to fix boundaries without the mutual consent and authority of all interested parties. If he cannot bring his clients to an agreement, the boundaries in dispute become valid and defined only by a decision of the court. In boundary disputes the surveyor is an expert witness, not a judge.

The right to enter upon property for the purpose of making public surveys is generally provided by law, but there is no similar provision regarding private surveys. The surveyor (or his employer, whether public or private) is liable for damage caused by cutting trees, destroying crops or fences, etc.

It has been held in court decisions that county surveyors and surveyors in private practice are members of a learned profession and may be held liable for incompetent services rendered. The general principle is that the surveyor is bound to exhibit that degree of prudence, judgment, and skill which may reasonably be expected of a member of his profession. Ruling Case Law says, " . . . a person undertaking to make a survey does not insure the correctness of his work, nor is absolute correctness the test of the amount of skill the law requires. Reasonable care, honesty, and a reasonable amount of skill are all he is bound to bring to the discharge of his duties."

RURAL-LAND SURVEYS

218. Description of Rural Land.—In describing a tract not within the system of United States public-land surveys, the lengths and bearings of the several courses are given in order, and the objects marking the corners are described, if any boundary follows some prominent feature of the terrain, the fact is stated; and the calculated area of the tract is given. When the bearings and lengths of the sides are thus given, the tract is said to be described by *metes and bounds.* Lands that have been divided in accordance with the rectangular system of the General Land Office are described as illustrated in Arts. 228, 229, and 230. In some of the states the locations

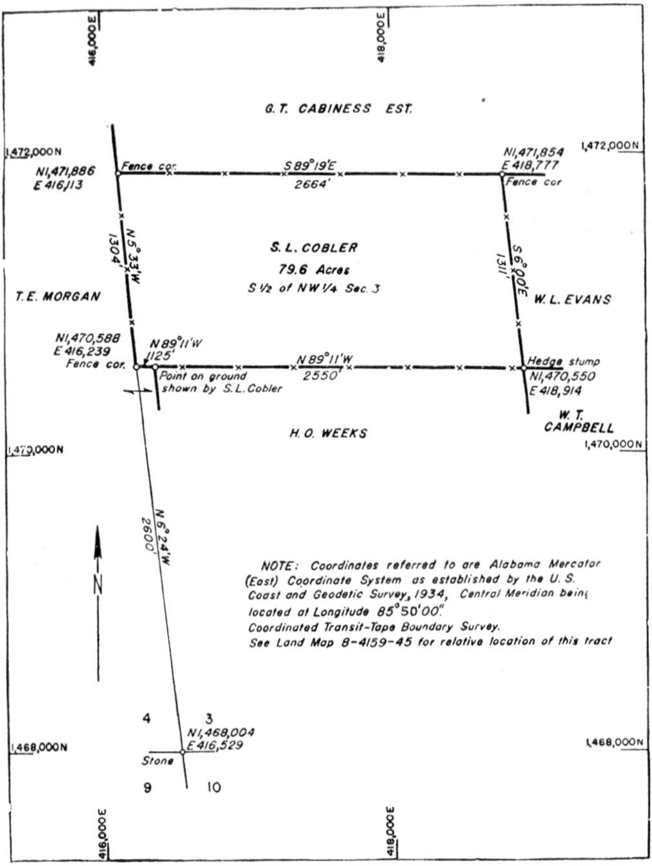

Fig. 218.—Land map.

of land corners are legally described by their coordinates with respect to the state-wide plane coordinate system. The practice of the Tennessee Valley Authority is to describe land both by metes and bounds and by coordinates, with further reference to the subdivisions of the United States public-land survey; a typical map is shown in Fig. 218.

219. Original Survey.—With the desired boundaries of the land given, the surveyor establishes monuments at the corners and runs a closed transit traverse around the property, measuring the lengths of lines and the angles between intersecting lines. Curved boundaries are located by offsets from traverse lines; and obstructed boundary lines are located by linear and angular measurements from near-by traverse lines. The direction of the true meridian is determined, usually by a solar observation (Art. 238). From the field data the true bearings of the boundaries and the area of the tract are calculated.

A description of the tract, usually by metes and bounds, is prepared. Usually a plat is drawn. A copy of the description and a tracing of the plat are submitted to the person for whom the survey is made.

220. Resurvey.—The purpose of a resurvey is to reestablish boundaries in their original positions. To guide him the surveyor has available the description contained in the deed or obtained from old records and descriptions of adjacent property.

First the surveyor critically examines the descriptions for mistakes; he then calculates the latitudes and departures of the several courses as given in the description, determines the error of closure, and plots the boundaries of the tract to scale.

If original bearings are magnetic, the magnetic declination at the time of the original survey is found, and true bearings are calculated. If true bearings cannot be found in this manner (as when the date of the original survey is unknown), and if one or more boundaries can be positively identified, observations are made to determine the true bearings of these known lines; from these the true bearings of the other lines are computed. If no other evidence exists, the magnetic declination at the time of the original survey is estimated as closely as possible.

One or More Boundaries Evident.—If one or more boundary lines can be identified, a comparison is obtained between the length of the chain or tape used on the original survey and that to be used on the resurvey; the proportionate lengths of the other sides of the tract are then computed.

With the computed directions and proportionate lengths the surveyor starts from a known corner and reruns the courses; at each estimated location of a corner he seeks physical evidence (such as monuments, the remains of monuments, or reference marks) of the location of the original corner. If such evidence is found, the survey is modified as necessary to conform to the existing corner.

One Corner Evident.—Where only a single corner can be found, about the only course open to the surveyor is to establish the true meridian and to rerun the survey in accordance with the old description, distributing the error of closure proportionately among the several courses.

No Corner Evident.—If a description is available but all evidence of the location of original corners is lost, the surveyor will find it expedient to search the records for descriptions of adjoining property and by means of these descriptions to reestablish by measurement as many corners of the tract in question as seems feasible. Having thus tentatively fixed the location of one or more corners, the surveyor attempts to reconcile these locations with the description of the given tract. Readjustments are made to conform to the judgment of the surveyor in the light of the information that he obtains as the survey progresses.

220a. When a resurvey has been completed, it is the duty of the surveyor to render a report to his client stating exactly what he found and what course of procedure he employed in attempting to reestablish missing corners. The report should be accompanied by a plat similar to that of an original survey. In addition, it should indicate which are original monuments and which are monuments established at the time of the resurvey. Mistakes in the original description should be pointed out, but the surveyor should clearly understand that it is his function to reestablish boundaries of a given tract in as nearly as possible their original location.

221. Subdivision Survey of Rural Land.—The subdivision of the United States public lands is described in Art. 230, and the subdivision of urban lands into lots in Art. 223. Surveys of irregular subdivisions for rural lands may be made for a variety of purposes, such as the establishing on the ground of a right of way for a highway or railway or the division of a tract of land into parts stipulated by a will. In such cases a resurvey of the tract is run, new monuments are established on the new boundary lines, and a new plat and description are prepared as in the case of an original survey.

URBAN-LAND SURVEYS

222. Description of Urban Land.—If the boundaries of a tract coincide exactly with a lot that is part of a municipal subdivision or addition for which there is recorded an official map, the tract may be legally described by a statement giving the lot and block numbers and the name and date of filing of the official map.

If the boundaries of a given tract do not conform exactly to boundaries shown on the official map, the tract is described by metes and bounds with the point of beginning referred either to a corner shown on the official map or to a reference monument of the city or state coordinate system. If the tract is within an area not so monumented, the point of beginning of the boundary description may be referred by direction and distance to the intersection of the center lines of streets.

223. Subdivision Survey of Urban Land.—As a city or town develops, unimproved lands are subdivided into lots which are placed on sale as residential or business property. A tract thus divided according to an acceptable plan is known as an *addition* or *subdivision*. For the ordinary real-estate development the owner usually calls for the services of an engineer or surveyor who has had experience in such work. The surveyor makes a resurvey of the entire property; in some cases a complete topographic survey may be made. With the general plan fixed, the surveyor works out a detailed plan on paper and prepares a report. This detailed plan, when approved by the owner, is submitted to the governing body in the municipality. If it meets with the requirements of this body, it is approved. Upon the authority of the owner, the surveyor then proceeds to execute the necessary subdivision surveys, including the laying out of roads, walks, blocks, and lots.

224. City Surveying.—The term *city surveying* is sometimes applied to the surveying operations within a municipality with regard to mapping its area, laying out new streets and lots, and constructing streets, sewers and other public utilities, and buildings. Although the principles of city surveying are not different from those of ordinary surveying, there are some differences in the details of the methods employed. Some features pertinent to city surveying are as follows:

1. Measurements are made with a greater degree of refinement than for land of less value.
2. Some cities maintain a standard of length with which tapes may be compared.
3. Usually the horizontal control of the survey for the map of a city is by triangulation rather than by traversing, which would be employed for an equal area outside the city.
4. A system of reference points and bench marks is established at points a few blocks apart (usually at street intersections), and preferably this system is tied in with the United States precise surveys. Points are located in the street, at the curb, or on the sidewalk.
5. Maps showing the location of proposed sewers, street extensions, and other improvements usually show, to scale and in figures, the exact dimensions of adjacent lots and of all other lots that will be benefited by, or assessed for, the proposed improvement.
6. Sometimes separate maps are made of surface and underground utilities such as car lines, sewers, water lines, gas lines, electric power

and telephone lines and conduits, tunnels, etc., both for convenient reference and in order to avoid interference in the location of new projects.

224a. Recently the term *city survey* has come to mean a comprehensive survey of the area in and near a city for the purposes of fixing reference monuments, locating property lines and improvements, and determining the configuration and physical features of the land. Such a survey is of value for a wide variety of purposes, particularly for planning city improvements. The technical procedure for a city survey of this type is described in detail in Ref. 39, p. 205. Briefly, the work consists in

1. Establishing horizontal and vertical control, as described for topographic surveying.
2. Making a topographic survey and topographic map.
3. Monumenting a system of selected points at suitable locations such as street corners, for reference in subsequent surveys.
4. Making a property map, which shows the length and bearing of all street lines and boundaries of public property, coordinates of governing points, control, monuments, important structures, natural features of the terrain, etc., all with appropriate legends and notes.
5. Making a wall map which shows essentially the same information as the topographic map but which is drawn to a smaller scale.
6. Making a map, or maps, to show underground utilities.

225. Cadastral Surveying.—Cadastral surveying is a general term referring to extensive surveys relating to land boundaries and subdivisions, whether they are city surveys as described in the preceding article or surveys of rural land. The term is sometimes applied to the surveys of the public lands of the United States. A cadastral map shows individual tracts of land with corners, length and bearing of boundaries, acreage, ownership, and sometimes the cultural and drainage features. The surveying methods are the same as those described for land and topographic surveying.

UNITED STATES PUBLIC-LAND SURVEYS

226. General.—The public lands of the United States are subdivided into townships, sections, and quarter sections by United States land surveyors under the direction of the General Land Office, in a manner prescribed by law. Further subdivision of such lands is made after the lands have passed into the hands of private owners, the work being carried out by surveyors in private practice.

The laws regulating the subdivision of public lands and the surveying methods employed are fully described in the "Manual of Instructions for the Survey of the Public Lands of the United States," published by the General Land Office (Ref. 34, p. 205). From the manual is drawn much of the material for the succeeding articles of this chapter.

The early surveys were made with crude instruments and often under unfavorable field conditions; hence, often the lines and corners will be found in other than their theoretical locations. However, the original corners as established legally stand as the true corners, and the surveyor must be guided by them in making resurveys or subdivisions, regardless of irregularities in the original survey.

The unit of length in public-land surveys is the Gunter's (66-ft.) chain divided into 100 links each 7.92 in. long.

The primary unit of subdivision is the *township*, bounded by meridional and latitudinal lines and as nearly as may be 6 miles square. The township is divided into 36 secondary units called *sections*, each as nearly as may be 1 mile square. Because the meridians converge, it is impossible to lay out a square township by such lines; and because the township is not square, not all of the 36 sections can be square.

227. Standard Lines.—The townships and sections are located with respect to principal axes passing through an origin called an *initial point;* the north-south axis is a true meridian called the *principal meridian*, and the east-west axis is a true parallel of latitude called the *base line*.

The principal meridian is given a name to which all subdivisions are referred. Thus the principal meridian which governs the rectangular surveys (wholly or in part) of the states of Ohio and Indiana is called the First Principal Meridian. The extent of the surveys which are referred to a given initial point may be found by consulting a map, published by the General Land Office, entitled "United States, Showing Principal Meridians, Base Lines, and Areas Governed Thereby," or from Ref. 34, p. 205.

Secondary axes are established at intervals of 24 miles east or west of the principal meridian and at intervals of 24 miles north or south of the base line,[1] thus dividing the tract being surveyed into quadrangles bounded by true meridians 24 miles long and by true parallels, the south boundary of each quadrangle being 24 miles long and the north boundary being 24 miles long less the convergency of the meridians in that distance. The secondary parallels are called *standard parallels* or *correction lines*, and each is continuous throughout its length. The secondary meridians are called *guide meridians*, and each is broken at the base line and at each standard parallel.

The principal meridian, base line, standard parallels, and guide meridians are called *standard lines*.

[1] In some early surveys, these intervals were 30 or 36 miles.

A typical system of principal and secondary axes is shown in Fig. 227. The base line and standard parallels, being everywhere perpendicular to the direction of the meridian, are laid out on the ground as curved latitudinal lines, the rate of curvature depending upon the latitude. The principal meridian and guide meridians, being true north-and-south lines, are laid out as straight lines but converge toward the north, the rate of convergency depending upon the latitude.

Ordinarily a curved latitudinal line is laid off on the ground by offsets from a straight line. The amounts of the offsets may be obtained from published tables (Ref. 37, p. 205).

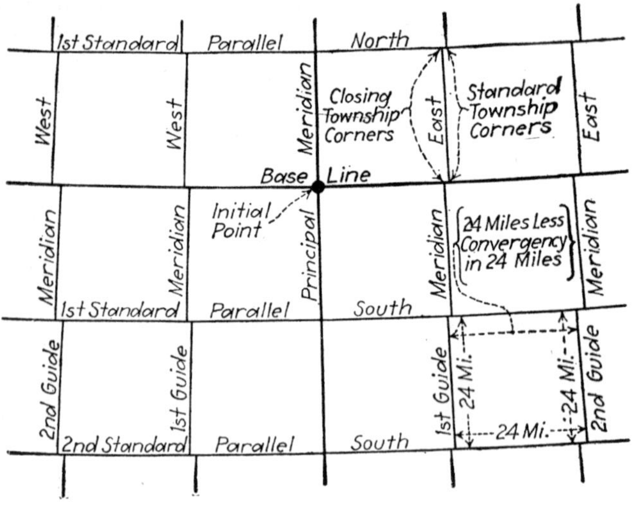

FIG. 227.—Standard lines.

228. Townships.—The division of the 24-mile quadrangles into townships is accomplished by laying off true meridional lines called *range lines* at intervals of 6 miles along each standard parallel, the range line extending north 24 miles to the next standard parallel, and by joining the township corners established at intervals of 6 miles on the range lines, guide meridians, and principal meridian with curved latitudinal lines called *township lines*.

The plan of subdivision is illustrated by Fig. 228. A row of townships extending north and south is called a *range;* and a row extending east and west is called a *tier*. Ranges are counted east or west of the principal meridian, and tiers are counted north or south of the base line. Usually for purposes of description the word "township" is substituted for "tier." A township is designated by the number of its tier and range and the name of the principal meridian. For example, the location of a township is defined by the description T7S, R7W (read *township seven south, range seven west*), of the Third Principal Meridian.

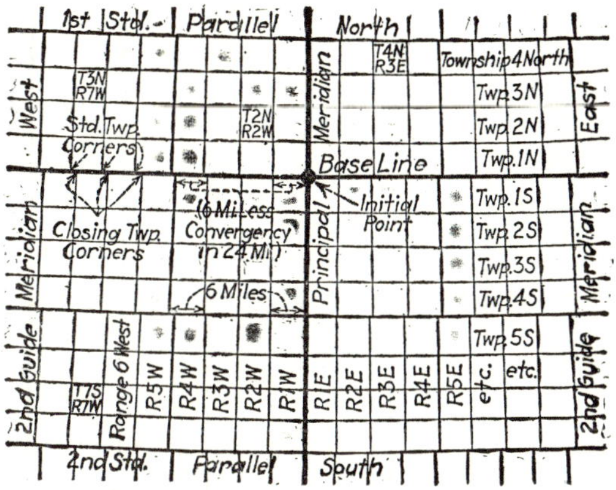

FIG. 228.—Township and range lines.

229. Sections.—The division of townships into sections is performed by establishing, at intervals of 1 mile, meridional lines[1] parallel to the east boundary of the township and by joining the section corners established at intervals of 1 mile with straight latitudinal lines. These lines, called *section lines*, divide each township into 36 sections, as shown in Fig. 229. The sections are numbered consecutively from east to west and from west to east, beginning with No. 1 in the northeast corner of the township and ending with No. 36 in the southeast corner.

A section is legally described by giving its number, the tier and range of the township, and the name of the principal meridian; for example, Section 16, T7S, R7W, of the Third Principal Meridian.

On account of the convergency of the range lines, which are true meridians, the latitudinal lines forming the north and south boundaries of townships are less than 6 miles in length, except for the south boundary of townships that lie just north of a standard parallel. As the meridional[1] section lines are run parallel to the east boundary of the township, it follows that all sections except those adjacent to the west boundary will be 1 mile square but those adjacent to the west

[1] These lines are not exactly meridional but are parallel to the east boundary of the township, which is a meridional line.

boundary will have a latitudinal dimension less than 1 mile by an amount equal to the convergency of the range lines.

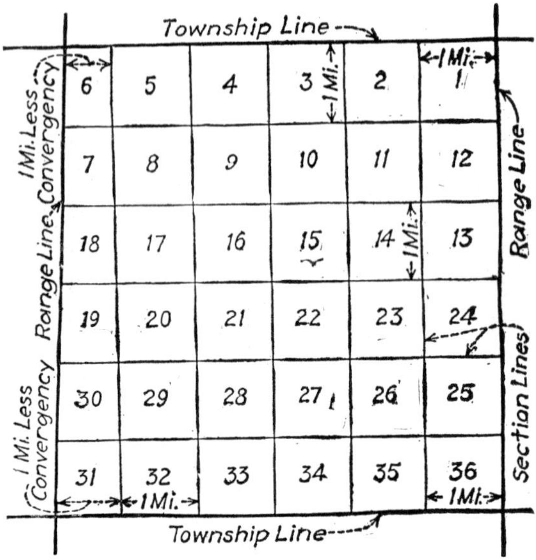

FIG. 229.—Numbering of sections.

230. Subdivision of Townships and Sections.—In the normal subdivision of townships with regular boundaries, the south and east boundaries are the governing lines. The method of establishing section corners is such that all sections except the 11 that lie next to the north and west boundaries of the township are as nearly as may be 1 mile square, containing only the errors of measurement. The maximum permissible error of closure in either latitude or departure is $\frac{1}{640}$; this error is determined for each section established during the survey, and if necessary the survey is rerun to bring the error within the permissible limit. Any excess or deficiency in the measured distance between the south and north boundaries of the township is placed in the most northerly half mile. Any deficiency due to the convergency of meridians and any excess or deficiency in the measured distance between the east and west boundaries of the township are placed in the most westerly half mile. (If the section is further subdivided, all of the excess or deficiency is placed in the most northerly quarter mile and the most westerly quarter mile.) Typical dimensions, in chains, are shown in Fig. 230 for four sections which lie in the northwest corner of a township.

The *regular* subdivisions of a section are the quarter section (½ mile square), the half-quarter section (¼ by ½ mile), and the quarter-quarter section (¼ mile square); the last contains 40 acres and is the legal minimum for purposes of disposal under the general land laws. An example of a complete legal description of a normal quarter-quarter section is as follows: "The north-east quarter of the south-west quarter of section ten (10), Township four (4) south, Range six (6) east, of the Mount Diablo Meridian, containing forty (40) acres, more or less, according to the United States Survey."

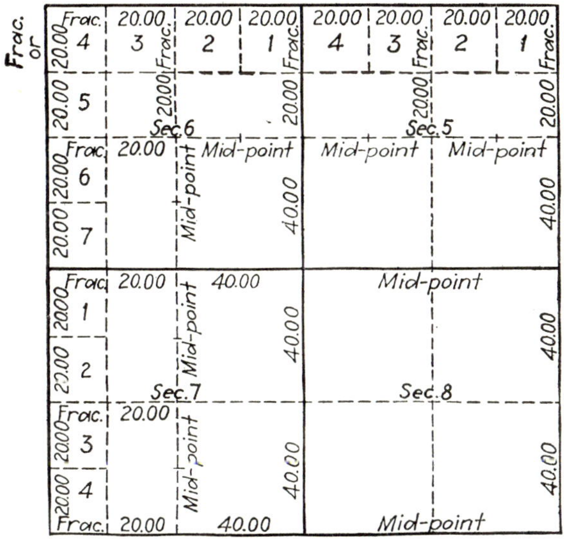

FIG. 230.—Subdivision of sections.

Of the 36 sections in each normal township (Fig. 229), 25 are returned as containing 640 acres each; 10 sections adjacent to the north and west boundaries (comprising sections 1–5, 7, 18, 19, 30, and 31) each contain regular subdivisions totaling 480 acres and in addition 4 *fractional lots* each containing 40 acres plus or minus definite differences to be determined in the survey (see Fig. 230); and 1 section (section 6) in the northwest corner contains regular subdivisions totaling 360 acres and in addition 7 fractional lots (Fig. 230). The system of numbering fractional lots is shown in the figure. Sections or lots may also be made fractional on account of meanderable bodies of water (Art. 214), mining claims, and other segregated areas within their limits.

On all section lines, monuments are placed at intervals of 40 chains, thus marking all section corners and all exterior quarter-section corners. Further subdivision of sections must be done by surveyors in private practice, but such subdivision must conform to the established dimensions of the official plat, including certain lines of subdivision which are shown on the plat but are not laid off on the ground. For example, in Fig. 230 the dash lines are established on the plat by protraction from definite points on the opposite boundaries (represented by solid lines) which have been established in the field.

230a. A section cannot be legally subdivided until the section and exterior quarter-section corners have been either found or restored and until the resulting courses and distances have been determined in the field. When the opposite quarter-section corners have been located, the interior quarter-section corner may be established. If the boundaries of quarter-quarter sections or of fractional lots are to be established on the ground, it is necessary to measure the boundaries of the quarter section and to fix thereon the quarter-quarter-section corners at distances in proportion to those given on the official plat; then the interior quarter-quarter-section corner may be placed. The subdivisional lines of fractional quarter sections are run from properly established quarter-quarter-section corners, with courses governed by the conditions represented on the official plat.

231. Field and Office Problem.

PROBLEM 31. SURVEY OF TRACT FOR DEED DESCRIPTION

Object.—To obtain sufficient data for a proper legal description of a tract and to prepare such a description.

Procedure.—(1) Around the assigned field run an azimuth traverse with the transit, measuring the sides with a steel tape and setting hubs at the corners. The angular error of closure in minutes should not exceed $\frac{1}{2} \times \sqrt{\text{number of sides}}$. Distribute the error of closure among the angles of the traverse. Refer azimuths to the true meridian. (2) Calculate the latitudes and departures and the linear error of closure; the linear error of closure should not exceed 1/5,000. (3) Balance the survey, and calculate the area of the tract by the coordinate method. (4) Determine the location of one corner of the traverse from an established reference point, and beginning at this corner write a description of the tract by metes and bounds.

CHAPTER XVII

DETERMINATION OF AZIMUTH

232. General.—In plane surveying, often it is desired to determine the absolute direction of a line on the surface of the earth, as defined by the azimuth of the line from the true meridian. The azimuth of a line is determined by angular observations on some celestial body, most commonly (1) by observations on Polaris, the north star, when it is at elongation; or (2) by observations on the sun, preferably near midmorning or midafternoon. In general, more accurate results are obtained from Polaris than from the sun.

For the purpose of computing the azimuth from an astronomical observation, it is necessary that the latitude of the place be known. Ordinarily the latitude can be determined with sufficient accuracy by scaling from a map. Methods of observing to determine latitude are indicated in Arts. 237 and 244.

The requirements of plane surveying are met if the true azimuth of the line is established with a degree of precision at least equal to that with which the angles between survey lines are measured. For plane surveying of ordinary precision, the use of the engineer's transit and methods described herein will yield sufficiently accurate results.

The discussions herein are intended to apply in the Northern Hemisphere and for longitudes west of Greenwich.

Although more elaborate astronomical tables are available (Refs. 30 and 31, p. 205), the work described herein can be accomplished by use of the "Ephemeris of the Sun and Polaris" (12 pp.), published by the U. S. General Land Office and sold by the Superintendent of Documents, Washington, D. C. Useful condensed tables of data regarding the sun and Polaris are furnished to surveyors free of charge by various manufacturers of surveying instruments.

For astronomical computations, the hours of the day are reckoned consecutively from 0 to 24, beginning at midnight.

POLARIS OBSERVATIONS

233. Polaris.—Polaris is a second-magnitude star the position of which is readily identified by the neighboring constellations of Ursa Major and Cassiopeia (Fig. 233). Its *polar distance*, or angular distance from the pole, is approximately 1°, and its annual change in polar distance is less than 01′ (Table III).

Observations to determine azimuth are usually made when Polaris is at eastern or western *elongation*, when the star is farthest east or farthest west of the pole, and when it appears to be traveling vertically for some time (Figs. 233 and 235). When Polaris is at its highest altitude, above the pole, and is moving almost horizontally westward, it is said to be at *upper culmination*.

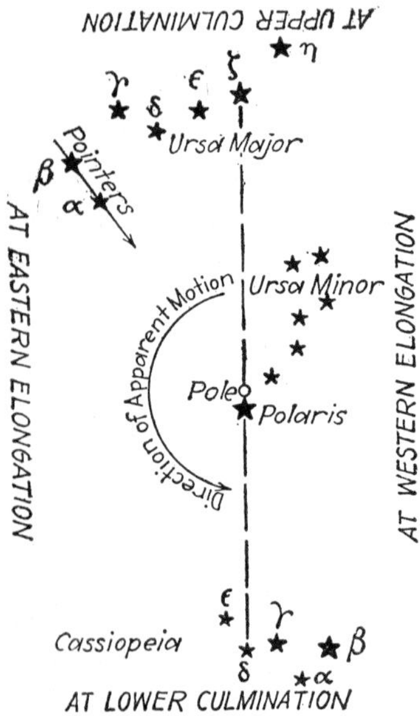

FIG. 233.—Positions of constellations near the North Pole when Polaris is at culmination and elongation.

234. Time of Elongation.—Although it is not necessary that the time of elongation be known, this information is of convenience in scheduling the observations. In Tables IV and IV*a* are given values by means of which the approximate standard time of elongation of Polaris can be determined for any place and date. (It must be kept in mind that "war time" is 1 hr. faster than standard time on which published tables are based.) Accompanying the tables is an explanation of their use for any year, any day of the year, any longitude, and standard time.

It is not essential that the observation be made at the exact instant of elongation. Within 10 min. before to 10 min. after elongation, the maximum change in azimuth of Polaris is only 0.1′.

235. Azimuth of Polaris at Elongation.[1]—In Fig. 235, O represents both the earth and the location of the observer, and Z, P, and S are points in the celestial sphere, an imaginary sphere of infinite radius with the earth at its center. The point Z directly above the observer is called the *zenith*. The point S represents a celestial body, in this case a circumpolar star such as Polaris. The point P represents the north celestial pole, or intersection of the earth's axis with the celestial sphere. The latitude of the observer is the angular distance ϕ north of the equator. The star's declination, or angular distance δ north

[1] For rough determination of meridian by ranging plumb lines on Polaris, see Art. 73.

of the equator, is equal to 90° minus the polar distance p. The azimuth of the star is the horizontal angle (at Z or O) between vertical planes passing through the pole and the star respectively.

At the instant of elongation, as a circumpolar star appears to be traveling vertically, its apparent path in the celestial sphere is tangent to the vertical plane through the observer's zenith, as illustrated by Fig. 235; therefore the angle at S is a right angle. In the spherical triangle PZS, known as the *astronomical triangle* or *PZS triangle*, the angle $S = 90°$, the side $PS = p = 90° - \delta$ is given in published tables of polar distance p or declination δ, and the side

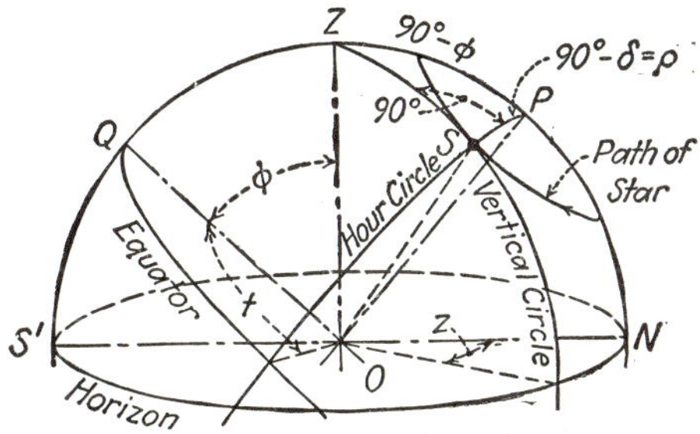

FIG. 235.—Star at elongation.

$ZP = 90° - \phi$ is known provided the latitude of the observer is known. By spherical trigonometry it can be shown that

$$\sin Z = \frac{\sin p}{\cos \phi} \tag{1}$$

in which the azimuth Z from north is reckoned east or west of north according as the star is at eastern or western elongation.

Table V gives values of the azimuth of Polaris at elongation, for various latitudes.

236. Field Procedure.—The direction of Polaris from the observer's station at the time of elongation is established by projecting a vertical plane from the star to the earth. The terrestrial line thus established has the same azimuth as the star. The angle between this line and an established reference line of the survey is measured, and the azimuth of the reference line is computed.

The usual procedure is as follows: The standard time of elongation at the given station is determined within perhaps 5 min. by Table IV herein or by use of an ephemeris. A few minutes before the estimated time of elongation the transit is set up over the station and is carefully leveled. The telescope is focused for a star; this can be done approximately by focusing on a distant object in the landscape. The latitude of the place is laid off on the vertical circle to facilitate finding Polaris. When Polaris is sighted, he horizontal and vertical motions are clamped, and the star is continuously bisected with the vertical cross-hair. If necessary, the cross-hairs are illuminated by holding a flashlight a few inches in front of the objective and a little to one side of the telescope barrel, thus causing the rays to enter the telescope diagonally. When during a period of 2 or 3 min. Polaris no longer appears to move away from the hair but moves vertically along it, the star is practically at elongation. The telescope is depressed, and a point on the line of sight is marked on a stake or other reference monument 300 ft. or more away; the location of the mark is indicated by a light, usually a slit in the side of a box containing a lamp. The telescope is then plunged, and another sight is taken on Polaris. The line of sight is again depressed, and a second point is set on the stake beside the first. Usually the transit is releveled, and a second pair of observations is made.

Later the mean of the points is found and marked on the stake, thus defining the direction from the observer's station to Polaris at elongation. A true meridian can be accurately established by a perpendicular offset from the established point on the stake.

By the methods just described, the azimuth of any line may be determined within perhaps 0.5′, or less if the mean of several observations is taken. If an error of 01′ is allowable, points need not be set beneath the star, and the angle may be measured directly.

237. Latitude by Observation on Polaris at Culmination.—If it is desired to determine the azimuth of Polaris with more than ordinary precision, the latitude of the observer's station may be determined within 01′ or less by observing Polaris at culmination. It can be shown that if h is the true altitude of Polaris as it crosses the meridian and p its polar distance, then the latitude ϕ is

$$\phi = h \pm p \tag{2}$$

in which the sign preceding p is positive or negative according as the star is at lower or upper culmination. The observed altitude must be corrected for refraction (Table II) and index error to determine the true altitude. The field procedure of observing is similar to that described for azimuth. Within 12 min. before to 12 min. after culmination, the maximum change in the altitude of Polaris is only 0.1′.

SOLAR OBSERVATIONS

238. General.—The azimuth of the sun at any instant of observation can be determined by observing its altitude and employing the following relation derived by spherical trigonometry:

$$\cos A = \tan h \tan \phi - \frac{\sin \delta}{\cos h \cos \phi} \tag{3}$$

in which A is the azimuth of the sun from south, reckoned counter-clockwise in the morning or clockwise in the afternoon; δ is the declination of the sun, obtained from published tables (Art. 239); h is the true altitude of the sun's center, determined by observation (Art. 243); and ϕ is the latitude of the place of observation, either scaled from a map or determined by separate observation as described in Art. 244.

If the sign of cos A is found to be $\left\{\begin{matrix}\text{positive}\\ \text{negative}\end{matrix}\right\}$, then A is $\left\{\begin{matrix}\text{less}\\ \text{greater}\end{matrix}\right\}$ than 90°.

Simultaneously with the observation of altitude to determine the sun's azimuth, the horizontal angle between the sun and a reference line on the earth's surface is measured, and thus the azimuth of the reference line is determined.

239. Declination of the Sun.—The declination of the sun is the angular distance from the equatorial plane to the sun, considered as $\left\{\begin{matrix}\text{positive}\\ \text{negative}\end{matrix}\right\}$ when the sun is $\left\{\begin{matrix}\text{north}\\ \text{south}\end{matrix}\right\}$ of the equator. Each year the sun crosses the equator northward about Mar. 21, reaches a maximum positive declination about June 21, crosses the equator southward about Sept. 22, and reaches a maximum negative declination about Dec. 21. The declination at a given instant is obtained by interpolating between values given in a solar ephemeris for the current year.

One type of ephemeris, especially adapted for use when the standard time is known, gives the declination for each day of the year at the instant of 0^h (midnight) Greenwich civil time. The Greenwich civil time at the instant of observation is computed by adding 5 hr. to Eastern standard time, 6 hr. to Central standard time, 7 hr. to Mountain standard time, or 8 hr. to Pacific standard time.

Example: An observation is taken on the sun at 10^h00^m a.m. Eastern standard time, on Dec. 15, 1940. It is desired to determine the declination at the instant of observation.

The Greenwich civil time at the instant of observation is $10^h00^m + 5^h = 15^h00^m = 15.00^h$. By ephemeris for 0^h Greenwich civil time the declination for 0^h on Dec. 15 is $-23°15'10''$, and the change per day is $-182.7''$. The change in declination since 0^h Greenwich civil time is $-182.7'' \times (15.00/24) = -01'54''$. The declination at the instant of observation is $-23°15'10'' - 01'54'' = -23°17'04''$.

Another type of ephemeris in general use gives the declination for each day of the year at the instant of Greenwich apparent noon

(12^h). (Apparent time is based upon the actual position of the sun, whereas civil time is based upon the assumption that the sun moves at a uniform rate throughout the year.) For the purpose of determining azimuth as described in this chapter where the time is used only for the purpose of determining declination, Greenwich apparent noon may be considered to be the same as Greenwich civil noon, since the difference between the two never exceeds about 16 min.

240. Observations on the Sun.—Whenever the sun is to be observed, the transit should be leveled with great care preferably by means of the telescope level. Also, the index error of the vertical circle should be determined. It is important that the horizontal axis be in adjustment with respect to the vertical axis.

To observe the sun directly through the telescopic eyepiece may result in serious injury to the eye. A piece of colored or smoked glass may be held between the eye and the eyepiece. Good observations can be made by bringing the sun's image to a focus on a white card held several inches in the rear of the telescopic eyepiece. The eyepiece should be drawn back and the objective focused until the sun's image and the cross-hairs are clearly seen on the card. At altitudes too high to be sighted directly, the sun may be sighted by means of a white card as previously described, a *solar screen* of ground glass utilizing the principle of the eyepiece, or a *prismatic eyepiece* that reflects the rays through an angle of 90° with the axis of the telescope.

240a. Semidiameter.—Because the sun is large, its center cannot be sighted accurately with the ordinary transit, and it is customary to bring the cross-hairs tangent to the sun's image. When a vertical angle is measured to the sun's $\left\{\begin{matrix}\text{upper}\\ \text{lower}\end{matrix}\right\}$ edge, or limb, it is necessary to $\left\{\begin{matrix}\text{subtract}\\ \text{add}\end{matrix}\right\}$ the sun's semidiameter in order to obtain the observed altitude of the sun's center. For the work described in this chapter, the sun's semidiameter may be taken as 16′; more exact values are given for each day in the year in solar ephemerides.

When a horizontal angle is measured to the sun's right or left limb, a correction equal to the sun's semidiameter times the secant of the altitude is applied. Thus if the altitude h is 60° and the semidiameter is 16′, the correction to a horizontal angle is $16' \sec h = 32'$.

For most solar observations, an equal number of sights are taken to opposite limbs of the sun; the mean of the horizontal angles and the mean of the vertical angles at the mean of the times of observation are taken, and no corrections for semidiameter are necessary.

240b. Method of Sighting.—The process of bringing both cross-hairs tangent to the sun is illustrated by Fig. 240, which shows the position of the sun in the field of view just prior to tangency, in the morning. For the first of a pair of observations, the horizontal cross-hair is sighted a short distance above the sun's lower limb as illustrated. As the altitude of the sun is increasing, the horizontal cross-hair approaches tangency owing to the sun's apparent movement. At the same time, the vertical cross-hair is kept continuously on the sun's western limb by means of the upper-motion tangent-screw. At the instant when the vertical and horizontal cross-hairs are simultaneously tangent to the sun's disk, the motion of the telescope is stopped, the time is observed, and the horizontal and vertical circles are read. The telescope is then plunged, and the second observation is taken with the sun in the lower right-hand quadrant as shown in Fig. 240, as follows: The vertical cross-hair is

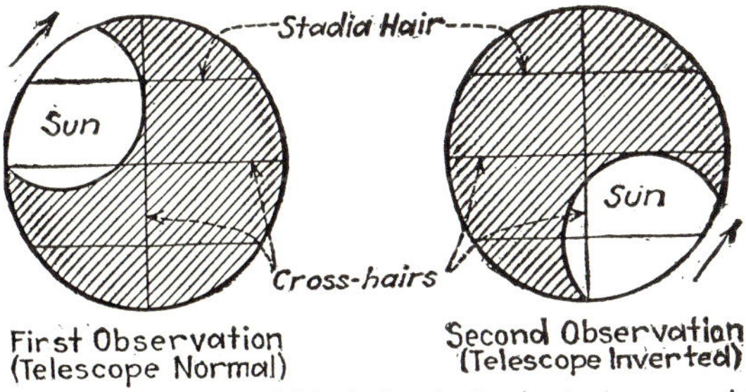

Fig. 240.—Position of sun in field of view just prior to tangency in morning.

set a short distance to the right of the sun's eastern limb. As the sun is traveling westward, the vertical cross-hair approaches tangency owing to the sun's apparent movement. At the same time, the horizontal cross-hair is kept continuously on the sun's upper limb by means of the telescope tangent-screw. As before, observations are taken for the instant when both horizontal and vertical cross hairs are simultaneously tangent to the sun's disk.

For afternoon observations the procedure is the same as that just described, except that the sun is sighted first in the upper right-hand quadrant and then in the lower left-hand quadrant.

If the transit is not equipped with a full vertical circle, it cannot be plunged between sights, but otherwise the procedure may be as just described, and the means for two observations taken as the observed angles.

241. Parallax Correction.—In the previous discussions it has been assumed that the celestial sphere is of infinite radius and that a vertical angle measured from a station on the surface of the earth is the same as it would be if measured from a station at the center of the earth. For the fixed stars this assumption yields results that are sufficiently accurate. However, the distance between the sun and the earth is relatively small, and for solar observations a correction is added to the observed altitude to obtain the altitude of the sun from the center of the earth. This *parallax correction* is small; it is always less than 09″. Corrections for parallax and refraction (see Art. 242) are usually made together; values of the combined correction are given in Table I.

242. Refraction Correction.—From an observed altitude it is necessary to subtract a *refraction correction* to determine the actual altitude. Table II gives values of refraction correction for a barometric pressure of 29.5 in. (which may be assumed with sufficient accuracy for practical purposes), for various temperatures, and for altitudes between 10° and 90°. Owing to the uncertainties of the refraction correction for low altitudes, observations should never be taken near the horizon.

For solar observations, corrections for refraction and parallax (see Art. 241) are usually made together; values of the combined correction are given in Table I.

243. Field Procedure.—The usual procedure is as follows: The transit is set up and carefully leveled over one end of the reference line. The *A* vernier is set at zero on the horizontal circle, and the reading of the *B* vernier is noted. A sight is taken along the given line with the telescope in, say, the normal position, and the lower motion is clamped. The upper motion is loosened, and sights are taken on the sun as described in Art. 240*b*. The field work is completed by again sighting along the line and reading the horizontal circle, this time with the telescope still inverted. The watch is compared with a timepiece keeping correct standard time, and the error is noted. The mean of each pair of observed angles is taken as the angle to the sun's center at the mean of the observed times. The observed values are corrected, and the azimuth of the sun is computed as previously described, five-place logarithms being used. The azimuth of the line is computed by subtracting algebraically the mean of the horizontal angles from the azimuth of the sun, angles taken in a clockwise direction from line to sun being considered positive. The field notes and computations are made in a form similar to that shown in Fig. 243.

The accuracy with which azimuths may be determined depends not only upon the precision of field observations and the exactitude of the corrections in altitude but also upon the shape of the astronomical triangle. In order to avoid either large uncertainties in the refraction correction (when the sun is low) or weak *PZS* triangles (when the sun is high), observations are usually taken between the hours of 8 and 10 a.m. or 2 and 4 p.m. For observations taken at these times, roughly an error of 01′ in latitude, declination, or altitude produces an error of 01′ to 02′ in the azimuth. Under ordinary conditions the azimuth of a line may be determined within about 01′.

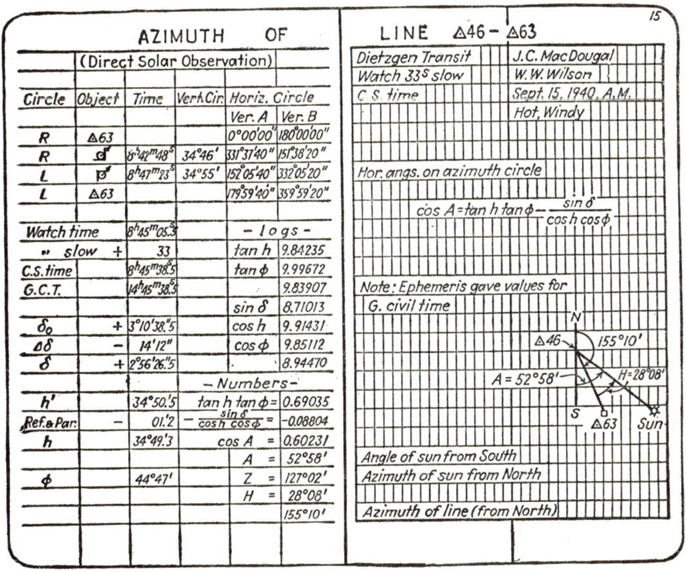

Fig. 243.—Azimuth by observation on sun.

244. Latitude by Observation on Sun at Noon.—The latitude of a given station may be determined with a fair degree of precision, ordinarily within 01′, by observing with the engineer's transit the altitude of the sun when the sun crosses the meridian. It can be shown that the latitude ϕ is equal to the zenith distance $(90° - h)$ plus the declination δ (north declinations considered as positive and south declinations as negative), or

$$\phi = 90° - h + \delta \tag{4}$$

As before, the true altitude h of the sun's center is determined by applying to the observed altitude the appropriate corrections for index error, semidiameter, and parallax and refraction. The sun's declination δ at the instant of sighting is computed from values given in an ephemeris, the standard time of the observation being roughly known. It is not necessary that the direction of the meridian be known.

The horizontal cross-hair is sighted continuously on the upper limb of the sun until the sun reaches its maximum altitude and begins its apparent descent. The standard time is observed. The maximum altitude and the time are recorded. The telescope may be plunged, and a second sight taken on the sun's lower limb; if the time between sights does not exceed 3 or 4 min., the mean altitude may be considered as the altitude at apparent noon. For the purpose of determining the declination, differences between civil time and apparent time may be neglected.

245. Numerical Problems.

1. From Table IV find the Pacific standard time of eastern elongation of Polaris on Aug. 15 of the current year for latitude 37°53′15″N and longitude 122°08′46″W.
2. From Table V find the azimuth of Polaris when at elongation on Aug. 15 of the current year at a place the latitude of which is 37°53′15″N.
3. By Eq. (1), (p. 195), and using Table III to determine the polar distance, compute the azimuth of Polaris when at elongation on Aug. 15 of the current year at a place the latitude of which is 37°53′15″N. Compare with the result obtained in the preceding problem.
4. From Table IV find the Mountain standard time of upper culmination of Polaris on Nov. 10 of the current year at a place the longitude of which is 102°14′30″W.
5. At a given place, on Nov. 10 of the current year the observed altitude of Polaris at upper culmination is 39°57′25″. The temperature is 35°F. What is the latitude of the place?
6. From an ephemeris find the declination of the sun for the instant of $2^h44^m10^s$ p.m. Pacific standard time on Mar. 26 of the current year.
7. In connection with a solar observation, sights to determine index error are taken on a mark in the general direction of the sun. The vertical angles to the mark are −3°16′40″ with telescope normal and −3°18′20″ with telescope inverted. The sun is sighted with telescope normal, and the vertical angle is observed to be 38°22′00″. Correct the observed angle for index error of the vertical circle.
8. The observed altitude of the sun's center is 18°46′30″. The temperature is 38°F. By Table I find the correction for refraction and parallax, and compute the true altitude of the sun.
9. At a place the latitude of which is 41°58′30″N, the observed altitude of the sun's center (taken as the mean obtained by double-sighting) at 3^h06^m p.m. Central standard time on Oct. 21 of the current year is 20°04′20″. The horizontal angle measured clockwise from a reference line to the sun is 73°39′40″. The temperature is 45°F. Compute the azimuth of the sun at the given instant and the azimuth of the reference line.

246. Field Problems.

PROBLEM 32. AZIMUTH BY OBSERVATION ON POLARIS AT ELONGATION

Object.—To determine the azimuth of a line by observation on Polaris at eastern or western elongation.

Procedure.—(1) Follow the procedure outlined in Art. 236. If the transit is not equipped with a full vertical circle, make two observations with the telescope normal, releveling the instrument between observations. (2) As a check, observe the magnetic bearing of the established line.

Hints and Precautions.—(1) Pay particular attention to algebraic signs. (2) As a check, the mean of the times of observation should agree (within a few minutes) with the computed time of elongation.

Problem 33. Azimuth by Direct Solar Observation

Object.—To determine the true azimuth of a line by an observation on the sun with the engineer's transit.

Procedure.—(1) Follow the procedure outlined in Art. 243. If the transit is not equipped with a full vertical circle, the correction for semidiameter (taken as 16′) must be applied to the altitude; further, either sights must be taken to both right and left limbs of the sun, or the correction for semidiameter (taken as $16' \times \sec h$) must be applied to the observed horizontal angle. (2) As a check, observe the magnetic bearing of the line.

Hints and Precautions.—(1) See Art. 240. (2) Pay particular attention to algebraic signs.

REFERENCES

1. ALLEN, C. F., "Railroad Curves and Earthwork; Field and Office Tables," 7th ed., McGraw-Hill Book Company, Inc., New York, 1931.
2. BIRDSEYE, C. H., "Topographic Instructions of the United States Geological Survey," *Bull.* 788, U. S. Geological Survey, Government Printing Office, Washington, D. C., 1928.
3. BRUCE, A. G., "Highway Design and Construction," 2d ed., International Textbook Company, Scranton, Pa., 1937.
4. CARY, E. F., "Geodetic Surveying," John Wiley & Sons, Inc., New York, 1916.
5. CHURCH, EARL, "Analytical Computations in Aerial Photogrammetry," Syracuse University, 1936.
6. CLARK, F. E., "Law of Surveying and Boundaries," 2d ed., Bobbs-Merrill Company, Indianapolis, 1939.
7. DAVIS, R. E., and F. S. FOOTE, "Surveying: Theory and Practice," 3d ed., McGraw-Hill Book Company, Inc., New York, 1940.
8. DEETZ, C. H., and O. S. ADAMS, "Elements of Map Projection," *Special Pub.* 68, U. S. Coast and Geodetic Survey, Government Printing Office, Washington, D. C., 1935.
9. DURHAM, E. B., "Mine Surveying," McGraw-Hill Book Company, Inc., New York, 1927.
10. ETCHEVERRY, B. A., and S. T. HARDING, "Irrigation Practice and Engineering," 2d ed., McGraw-Hill Book Company, Inc., New York, 1933.
11. FRANKLIN, W. S., "An Elementary Treatise on Precision of Measurement," Franklin and Charles, Lancaster, Pa., 1925.
12. GRIEVES, CAPT. L. C., "Military Sketching and Map Reading," 4th ed., U. S. Infantry Association, Washington, D. C., 1921.
13. HAWLEY, J. H., "Hydrographic Manual," *Special Pub.* 143, U. S. Coast and Geodetic Survey, Government Printing Office, Washington, D. C., 1928.
14. HAZARD, DANIEL L., and W. N. MCFARLAND, "Magnetic Declination in the United States—1935," *Serial* 592, U. S. Coast and Geodetic Survey, Government Printing Office, Washington, D. C., 1937. (New bulletin published every five years.)
15. HODGMAN, F., "Land Surveying," The F. Hodgman Company, Climax, Mich., 1907.
16. HODGSON, C. V., "Manual of Second- and Third-order Triangulation and Traverse," *Special Pub.* 145, U. S. Coast and Geodetic Survey, Government Printing Office, Washington, D. C.
17. HOSMER, G. L., "Geodesy," 2d ed., John Wiley & Sons, Inc., New York, 1930.
18. JORDAN, H. H., and R. P. HOELSCHER, "Engineering Drawing, with Problems," 2d ed., John Wiley & Sons, Inc., New York, 1928.
19. LEWIS, N. P., "The Planning of the Modern City," 2d ed., John Wiley & Sons, Inc., New York, 1923.

20. Nassau, J. J., "Textbook of Practical Astronomy," McGraw-Hill Book Company, Inc., New York, 1932.
21. Pickels, G. W., and C. C. Wiley, "Route Surveying," 2d ed., John Wiley & Sons, Inc., New York, 1939.
22. Rubey, Harry, "Route Surveys," The Macmillan Company, New York, 1938.
23. Searles, W. H., and H. C. Ives, "Field Engineering," 21st ed., John Wiley & Sons, Inc., New York, 1936.
24. Sloane, R. C., and J. M. Montz, "Elements of Topographic Drawing," McGraw-Hill Book Company, Inc., New York, 1930.
25. Swainson, O. W., "Topographic Manual," *Special Pub.* 144, U. S. Coast and Geodetic Survey, Government Printing Office, Washington, D. C., 1928.
26. Talley, Major B. B., "Engineering Applications of Aerial and Terrestrial Photogrammetry," Pitman Publishing Corp., New York, 1938.
27. Webb, W. L., "Railroad Construction," 9th ed., John Wiley & Sons, Inc., New York, 1932.
28. Wright, T. W., and J. F. Hayford, "The Adjustment of Observations," D. Van Nostrand Company, Inc., New York, 1906.
29. "Advanced Map and Aerial Photograph Reading," *Basic Field Manual* 21–26, U. S. War Department, Government Printing Office, Washington, D. C., 1941.
30. "American Ephemeris and Nautical Almanac," U. S. Naval Observatory, Government Printing Office, Washington, D. C., annual (in advance).
31. "American Nautical Almanac," U. S. Naval Observatory, Government Printing Office, Washington, D. C., annual (in advance).
32. "Ephemeris of the Sun and Polaris," U. S. General Land Office, Government Printing Office, Washington, D. C., annual (in advance).
33. "Manual of the American Railway Engineering Association," American Railway Engineering Association, Chicago, 1929.
34. "Manual of Instructions for the Survey of the Public Lands of the United States," U. S. General Land Office, Government Printing Office, Washington, D. C., 1930.
35. Manuals of instruction for surveys and for construction of roads and bridges, various state highway departments.
36. "Restoration of Lost or Obliterated Corners and Subdivision of Sections," U. S. General Land Office, Government Printing Office, Washington, D. C., 1939.
37. "Standard Field Tables," U. S. General Land Office, Government Printing Office, Washington, D. C.
38. "State Plane Coordinates," *Geodetic Letter* 1, Vol. 4, U. S. Coast and Geodetic Survey, Government Printing Office, Washington, D. C., January, 1937.
39. "Technical Procedure for City Surveys," "Location of Underground Utilities," and "Land Subdivision," *Manuals of Engineering Practice* 10, 14, and 16, American Society of Civil Engineers, New York, 1934, 1937, and 1939.
40. "Aerial Phototopography," *Technical Manual* 5–240, U. S. War Department, Government Printing Office, Washington, D. C., 1941.

TABLES

TABLE I.—CORRECTION FOR REFRACTION AND PARALLAX, TO BE SUBTRACTED FROM THE OBSERVED ALTITUDE OF THE SUN

App't alt.	Temperature										App't alt.
	−10° C. +14° F.	−5° C. +23° F.	0° C. +32° F.	+5° C. +41° F.	+10° C. +50° F.	+15° C. +59° F.	+20° C. +68° F.	+25° C. +77° F.	+30° C. +86° F.	+35° C. +95° F.	
°	′	′	′	′	′	′	′	′	′	′	°
10	5.52	5.42	5.30	5.20	5.10	5.00	4.92	4.83	4.75	4.67	10
11	5.02	4.92	4.82	4.73	4.63	4.55	4.47	4.38	4.32	4.23	11
12	4.60	4.50	4.42	4.33	4.25	4.17	4.10	4.03	3.97	3.88	12
13	4.23	4.15	4.07	4.00	3.92	3.85	3.78	3.72	3.65	3.58	13
14	3.92	3.83	3.77	3.70	3.62	3.55	3.50	3.45	3.37	3.32	14
15	3.65	3.58	3.50	3.43	3.37	3.32	3.25	3.20	3.13	3.08	15
16	3.43	3.35	3.30	3.23	3.17	3.12	3.07	3.00	2.95	2.90	16
17	3.22	3.15	3.10	3.03	2.98	2.92	2.88	2.82	2.77	2.72	17
18	3.02	2.95	2.90	2.85	2.80	2.75	2.70	2.65	2.60	2.55	18
19	2.83	2.78	2.73	2.68	2.63	2.58	2.53	2.48	2.43	2.40	19
20	2.68	2.63	2.58	2.53	2.48	2.43	2.38	2.33	2.30	2.27	20
21	2.53	2.48	2.43	2.38	2.35	2.30	2.27	2.22	2.17	2.13	21
22	2.38	2.35	2.30	2.25	2.22	2.18	2.13	2.08	2.05	2.02	22
23	2.28	2.25	2.20	2.15	2.12	2.08	2.03	1.98	1.95	1.93	23
24	2.17	2.13	2.08	2.05	2.02	1.98	1.93	1.88	1.87	1.83	24
25	2.07	2.03	1.98	1.95	1.92	1.88	1.83	1.80	1.77	1.75	25
26	1.99	1.95	1.90	1.87	1.83	1.80	1.75	1.72	1.70	1.67	26
27	1.88	1.85	1.82	1.78	1.75	1.72	1.68	1.63	1.62	1.60	27
28	1.80	1.77	1.72	1.70	1.67	1.63	1.60	1.57	1.53	1.52	28
29	1.72	1.68	1.65	1.63	1.60	1.57	1.53	1.50	1.47	1.46	29
30	1.65	1.62	1.58	1.57	1.53	1.50	1.47	1.45	1.42	1.40	30
32	1.53	1.50	1.47	1.45	1.42	1.38	1.35	1.33	1.30	1.28	32
34	1.41	1.37	1.35	1.32	1.30	1.27	1.25	1.23	1.20	1.18	34
36	1.30	1.27	1.25	1.22	1.20	1.18	1.15	1.13	1.10	1.08	36
38	1.20	1.18	1.15	1.13	1.12	1.10	1.07	1.05	1.02	1.02	38
40	1.11	1.10	1.07	1.05	1.03	1.02	0.98	0.97	0.95	0.93	40
42	1.03	1.00	0.98	0.97	0.95	0.93	0.90	0.88	0.87	0.87	42
44	0.96	0.93	0.92	0.90	0.88	0.87	0.85	0.83	0.82	0.80	44
46	0.89	0.88	0.87	0.85	0.83	0.82	0.80	0.78	0.77	0.75	46
48	0.83	0.82	0.80	0.78	0.77	0.75	0.73	0.72	0.70	0.68	48
50	0.77	0.75	0.73	0.72	0.70	0.68	0.67	0.67	0.65	0.63	50
55	0.63	0.62	0.60	0.60	0.58	0.57	0.57	0.55	0.53	0.52	55
60	0.52	0.52	0.50	0.50	0.48	0.47	0.47	0.45	0.45	0.43	60
65	0.42	0.40	0.40	0.40	0.38	0.38	0.37	0.37	0.35	0.33	65
70	0.32	0.32	0.32	0.30	0.30	0.30	0.28	0.28	0.28	0.27	70
75	0.23	0.23	0.23	0.22	0.22	0.22	0.20	0.20	0.20	0.18	75
80	0.15	0.15	0.13	0.13	0.13	0.13	0.13	0.12	0.12	0.12	80
85	0.07	0.07	0.07	0.07	0.07	0.07	0.07	0.05	0.05	0.05	85
90	0.00	0.00	0.00	0.00	0.00	0.00	0.00	0.00	0.00	0.00	90

Table II.—Correction for Refraction, to Be Subtracted from the Observed Altitude of a Star

App't alt.	Temperature −10° C. +14° F.	−5° C. +23° F.	0° C. +32° F.	+5° C. +41° F.	+10° C. +50° F.	+15° C. +59° F.	+20° C. +68° F.	+25° C. +77° F.	+30° C. +86° F.	+35° C. +95° F.	App't alt.
°	′	′	′	′	′	′	′	′	′	′	°
10	5.67	5.57	5.45	5.35	5.25	5.15	5.07	4.98	4.90	4.82	10
11	5.17	5.07	4.97	4.88	4.78	4.70	4.62	4.53	4.47	4.38	11
12	4.75	4.65	4.57	4.48	4.40	4.32	4.25	4.18	4.12	4.03	12
13	4.38	4.30	4.22	4.15	4.07	4.00	3.93	3.87	3.80	3.73	13
14	4.06	3.97	3.91	3.84	3.76	3.69	3.64	3.59	3.51	3.46	14
15	3.79	3.72	3.64	3.57	3.51	3.46	3.39	3.34	3.27	3.22	15
16	3.57	3.49	3.44	3.37	3.31	3.26	3.21	3.14	3.09	3.04	16
17	3.36	3.29	3.24	3.17	3.12	3.06	3.02	2.96	2.91	2.86	17
18	3.16	3.09	3.04	2.99	2.94	2.89	2.84	2.79	2.74	2.69	18
19	2.97	2.92	2.87	2.82	2.77	2.72	2.67	2.62	2.57	2.54	19
20	2.82	2.77	2.72	2.67	2.62	2.57	2.52	2.47	2.44	2.41	20
21	2.67	2.62	2.57	2.52	2.49	2.44	2.41	2.36	2.31	2.27	21
22	2.52	2.49	2.44	2.39	2.36	2.32	2.27	2.22	2.19	2.16	22
23	2.42	2.39	2.34	2.29	2.26	2.22	2.17	2.12	2.09	2.07	23
24	2.31	2.27	2.22	2.19	2.16	2.12	2.07	2.02	2.01	1.97	24
25	2.21	2.17	2.12	2.09	2.06	2.02	1.97	1.94	1.91	1.89	25
26	2.12	2.08	2.03	2.00	1.96	1.93	1.88	1.85	1.83	1.80	26
27	2.01	1.98	1.95	1.91	1.88	1.85	1.81	1.76	1.75	1.73	27
28	1.93	1.90	1.85	1.83	1.80	1.76	1.73	1.70	1.66	1.65	28
29	1.85	1.81	1.78	1.76	1.73	1.70	1.66	1.63	1.60	1.59	29
30	1.78	1.75	1.71	1.70	1.66	1.63	1.60	1.58	1.55	1.53	30
32	1.65	1.62	1.59	1.57	1.54	1.50	1.47	1.45	1.42	1.40	32
34	1.53	1.49	1.47	1.44	1.42	1.39	1.35	1.35	1.32	1.30	34
36	1.42	1.39	1.37	1.34	1.32	1.30	1.27	1.25	1.22	1.20	36
38	1.32	1.30	1.27	1.25	1.24	1.22	1.19	1.17	1.14	1.14	38
40	1.22	1.21	1.18	1.16	1.14	1.13	1.09	1.08	1.06	1.04	40
42	1.14	1.11	1.09	1.08	1.06	1.04	1.01	0.99	0.98	0.98	42
44	1.07	1.04	1.03	1.01	0.99	0.98	0.96	0.94	0.93	0.91	44
46	0.99	0.98	0.97	0.95	0.93	0.92	0.90	0.88	0.87	0.85	46
48	0.93	0.92	0.90	0.88	0.87	0.85	0.83	0.82	0.80	0.78	48
50	0.86	0.84	0.82	0.81	0.79	0.77	0.76	0.76	0.74	0.72	50
55	0.72	0.71	0.69	0.69	0.67	0.66	0.66	0.64	0.62	0.61	55
60	0.59	0.59	0.57	0.57	0.55	0.54	0.54	0.52	0.52	0.50	60
65	0.48	0.46	0.46	0.46	0.44	0.44	0.43	0.43	0.41	0.39	65
70	0.37	0.37	0.37	0.35	0.35	0.35	0.33	0.33	0.33	0.32	70
75	0.27	0.27	0.27	0.26	0.26	0.26	0.24	0.24	0.24	0.22	75
80	0.18	0.18	0.16	0.16	0.16	0.16	0.16	0.15	0.15	0.15	80
85	0.08	0.08	0.08	0.08	0.08	0.08	0.08	0.06	0.06	0.06	85
90	0.00	0.00	0.00	0.00	0.00	0.00	0.00	0.00	0.00	0.00	90

TABLE III.—POLAR DISTANCE OF POLARIS FOR EACH YEAR, 1937–1945

Date	1937	1938	1939	1940	1941	1942	1943	1944	1945
	° ′ ″	° ′ ″	° ′ ″	° ′ ″	° ′ ″	° ′ ″	° ′ ″	° ′ ″	° ′ ″
Jan. 15........	1 1 43	1 1 27	1 1 11	1 0 56	1 0 40	1 0 24	1 0 07	0 59 49	0 59 30
Feb. 15........	1 1 45	1 1 29	1 1 13	1 0 58	1 0 42	1 0 26	1 0 09	0 59 51	0 59 32
Mar. 15........	1 1 51	1 1 35	1 1 19	1 1 04	1 0 48	1 0 32	1 0 15	0 59 57	0 59 38
Apr. 15........	1 2 00	1 1 44	1 1 28	1 1 13	1 0 57	1 0 41	1 0 24	1 0 06	0 59 47
May 15........	1 2 09	1 1 53	1 1 37	1 1 22	1 1 06	1 0 50	1 0 33	1 0 15	0 59 56
June 15........	1 2 14	1 1 58	1 1 42	1 1 27	1 1 11	1 0 55	1 0 38	1 0 20	1 0 01
July 15........	1 2 15	1 1 59	1 1 43	1 1 28	1 1 12	1 0 56	1 0 39	1 0 21	1 0 02
Aug. 15........	1 2 11	1 1 55	1 1 39	1 1 24	1 1 08	1 0 52	1 0 35	1 0 17	0 59 58
Sept. 15........	1 2 02	1 1 46	1 1 30	1 1 15	1 0 59	1 0 43	1 0 26	1 0 08	0 59 49
Oct. 15........	1 1 52	1 1 36	1 1 20	1 1 05	1 0 49	1 0 33	1 0 16	0 59 58	0 59 39
Nov. 15........	1 1 40	1 1 24	1 1 08	1 0 53	1 0 37	1 0 21	1 0 04	0 59 46	0 59 27
Dec. 15........	1 1 31	1 1 15	1 0 59	1 0 44	1 0 28	1 0 12	0 59 55	0 59 37	0 59 18

TABLE IV.—LOCAL CIVIL TIME OF UPPER CULMINATION OF POLARIS IN THE YEAR 1937*

(*Computed for* 90°, *or* 6 *hours west of Greenwich*)

Date, 1937	Civil time of upper culmination			Variation per day		Date, 1937	Civil time of upper culmination			Variation per day	
	h	m	s	m	s		h	m	s	m	s
Jan. 1	18	56	26	−3	57	July 10	6	29	15	−3	55
Jan. 11	18	16	56	3	57	July 20	5	50	9	−3	55
Jan. 21	17	37	25	−3	57	July 30	5	11	1	−3	55
Jan. 31	16	57	54	−3	57	Aug. 9	4	31	54	−3	55
Feb. 10	16	18	24	−3	57	Aug. 19	3	52	46	−3	55
Feb. 20	15	38	55	−3	57	Aug. 29	3	13	37	−3	55
Mar. 2	14	59	26	−3	57	Sept. 8	2	34	27	−3	55
Mar. 12	14	19	59	−3	57	Sept. 18	1	55	17	−3	55
Mar. 22	13	40	36	−3	56	Sept. 28	1	16	5	−3	55
Apr. 1	13	1	14	−3	56	Oct. 8	0	36	50	−3	56
Apr. 11	12	21	53	−3	56	Oct. 17	23	57	33	−3	56
Apr. 21	11	42	35	−3	56	Oct. 27	23	18	16	−3	56
May 1	11	3	21	−3	56	Nov. 6	22	38	57	−3	56
May 11	10	24	7	−3	55	Nov. 16	21	59	34	−3	56
May 21	9	44	55	−3	55	Nov. 26	21	20	10	−3	56
May 31	9	5	45	−3	55	Dec. 6	20	40	45	−3	57
June 10	8	26	36	−3	55	Dec. 16	20	1	18	−3	57
June 20	7	47	28	−3	55	Dec. 26	19	21	49	−3	57
June 30	7	8	21	−3	55	Jan. 5, 1938	18	42	20	−3	57

TABLE IV(a).—MEAN TIME INTERVAL BETWEEN UPPER CULMINATION AND ELONGATION

Latitude	Time interval		Latitude	Time interval		Latitude	Time interval		Latitude	Time interval	
°	h	m	°	h	m	°	h	m	°	h	m
10	5	58.3	35	5	56.1	48	5	54.4	58	5	52.5
15	5	57.9	40	5	55.6	50	5	54.1	60	5	51.9
20	5	57.5	42	5	55.3	52	5	53.8	62	5	51.3
25	5	57.1	44	5	55.0	54	5	53.4	64	5	50.6
30	5	56.6	46	5	54.7	56	5	52.9	66	5	49.8

Eastern elongation precedes and western elongation follows upper culmination by the time interval given in Table IV(a). Lower culmination precedes or follows upper culmination by $11^h58.0^m$. It should be noted that there are two upper culminations (when culmination occurs near midnight) on one day in October (16th in 1937) and two lower culminations in April (17th in 1937). There are also two western elongations on one day in January and two eastern elongations on one day in July.

* To refer the times to other years, other days, other longitudes, or standard time, see next page.

A. To refer the times in Table IV to other years:

For Year	m
1936	subtract 1.7 on and after Mar. 1
1938	add 1.7
1939	add 3.3
1940	add 4.8 up to Mar. 1
1940	add 0.9 on and after Mar. 1
1941	add 2.4
1942	add 3.9
1943	add 5.3
1944	add 6.6 up to Mar. 1
1944	add 2.7 on and after Mar. 1
1945	add 4.1

B. To refer to other than the tabular days: SUBTRACT from the time for the preceding tabular day the product of the variation per day and the days elapsed, as given below:

Days elapsed	Variation per day			Days elapsed	Variation per day		
	$3^m\ 57^s$	$3^m\ 56^s$	$3^m\ 55^s$		$3^m\ 57^s$	$3^m\ 56^s$	$3^m\ 55^s$
	m s	m s	m s		m s	m s	m s
1	3 57	3 56	3 55	6	23 42	23 36	23 30
2	7 54	7 52	7 50	7	27 39	27 32	27 25
3	11 51	11 48	11 45	8	31 36	31 28	31 20
4	15 48	15 44	15 40	9	35 33	35 24	35 15
5	19 45	19 40	19 35				

C. To refer to any other than the tabular longitude (90°)*:* ADD 0.1^m for each 10° east of the ninetieth meridian or SUBTRACT 0.1^m for each 10° west of the ninetieth meridian.

D. To refer to standard time: ADD to the quantities in Table IV four minutes for every degree of longitude the place of observation is west of the standard meridian (60°, 75°, 90°, etc.). SUBTRACT when the place is east of the standard meridian.

TABLE V.—AZIMUTH OF POLARIS AT ELONGATION, 1936–1945*

Latitude	1936	1937	1938	1939	1940	1941	1942	1943	1944	1945
°	° ′	° ′	° ′	° ′	° ′	° ′	° ′	° ′	° ′	° ′
10	1 3.3	1 3.0	1 2.8	1 2.5	1 2.2	1 2.0	1 1.7	1 1.4	1 1.1	1 0.8
12	1 3.8	1 3.5	1 3.2	1 2.9	1 2.7	1 2.4	1 2.1	1 1.8	1 1.5	1 1.2
14	1 4.3	1 4.0	1 3.7	1 3.4	1 3.2	1 2.9	1 2.6	1 2.3	1 2.0	1 1.7
16	1 4.9	1 4.6	1 4.3	1 4.0	1 3.8	1 3.5	1 3.2	1 2.9	1 2.6	1 2.3
18	1 5.6	1 5.3	1 5.0	1 4.7	1 4.5	1 4.2	1 3.9	1 3.6	1 3.3	1 3.0
20	1 6.4	1 6.1	1 5.8	1 5.5	1 5.2	1 5.0	1 4.7	1 4.4	1 4.0	1 3.7
22	1 7.4	1 7.0	1 6.7	1 6.4	1 6.1	1 5.8	1 5.5	1 5.2	1 4.9	1 4.6
24	1 8.3	1 8.0	1 7.7	1 7.4	1 7.1	1 6.8	1 6.5	1 6.2	1 5.9	1 5.5
26	1 9.4	1 9.1	1 8.8	1 8.5	1 8.2	1 7.9	1 7.6	1 7.3	1 7.0	1 6.6
28	1 10.7	1 10.3	1 10.0	1 9.7	1 9.4	1 9.1	1 8.8	1 8.5	1 8.2	1 7.8
30	1 12.0	1 11.7	1 11.4	1 11.1	1 10.8	1 10.5	1 10.2	1 9.8	1 9.5	1 9.1
32	1 13.6	1 13.2	1 12.9	1 12.6	1 12.3	1 12.0	1 11.6	1 11.3	1 11.0	1 10.6
34	1 15.3	1 14.9	1 14.6	1 14.3	1 13.9	1 13.6	1 13.3	1 13.0	1 12.6	1 12.2
36	1 17.1	1 16.7	1 16.4	1 16.1	1 15.8	1 15.4	1 15.1	1 14.8	1 14.4	1 14.0
38	1 19.2	1 18.8	1 18.4	1 18.1	1 17.8	1 17.5	1 17.1	1 16.8	1 16.4	1 16.0
40	1 21.4	1 21.0	1 20.7	1 20.4	1 20.0	1 19.7	1 19.3	1 19.0	1 18.6	1 18.2
42	1 24.0	1 23.5	1 23.2	1 22.8	1 22.5	1 22.1	1 21.8	1 21.4	1 21.0	1 20.6
44	1 26.7	1 26.3	1 25.9	1 25.6	1 25.2	1 24.8	1 24.5	1 24.1	1 23.7	1 23.2
46	1 29.8	1 29.4	1 29.0	1 28.6	1 28.2	1 27.9	1 27.5	1 27.1	1 26.6	1 26.2
48	1 33.3	1 32.8	1 32.4	1 32.0	1 31.6	1 31.2	1 30.8	1 30.4	1 29.9	1 29.5
50	1 37.1	1 36.6	1 36.2	1 35.8	1 35.4	1 35.0	1 34.5	1 34.1	1 33.6	1 33.1

* See Table V(a) for correction for day of the year.

TABLE V(a).—CORRECTION TO AZIMUTH OF POLARIS, FOR DAY OF THE YEAR

Date	Latitude			Date	Latitude		
	10°	40°	50°		10°	40°	50°
	′	′	′		′	′	′
Jan. 1	−0.3	−0.4	−0.5	July 9	+0.2	+0.2	+0.3
Jan. 10	−0.3	−0.4	−0.5	July 19	+0.2	+0.2	+0.3
Jan. 20	−0.4	−0.5	−0.6	July 29	+0.2	+0.2	+0.2
Jan. 30	−0.4	−0.5	−0.5	Aug. 8	+0.1	+0.2	+0.2
Feb. 9	−0.3	−0.4	−0.5	Aug. 18	+0.1	+0.1	+0.1
Feb. 19	−0.3	−0.4	−0.5	Aug. 28	+0.1	+0.1	+0.1
Mar. 1	−0.3	−0.4	−0.4	Sept. 7	0.0	0.0	0.0
Mar. 11	−0.3	−0.3	−0.4	Sept. 17	0.0	−0.1	−0.1
Mar. 21	−0.2	−0.3	−0.3	Sept. 27	−0.1	−0.1	−0.2
Mar. 31	−0.2	−0.2	−0.2	Oct. 7	−0.2	−0.2	−0.3
Apr. 10	−0.1	−0.1	−0.2	Oct. 17	−0.2	−0.3	−0.4
Apr. 20	−0.1	−0.1	−0.1	Oct. 26	−0.3	−0.4	−0.5
Apr. 30	0.0	0.0	0.0	Nov. 5	−0.4	−0.5	−0.5
May 10	0.0	+0.1	+0.1	Nov. 15	−0.4	−0.5	−0.6
May 20	+0.1	+0.1	+0.1	Nov. 25	−0.5	−0.6	−0.7
May 30	+0.1	+0.2	+0.2	Dec. 5	−0.5	−0.7	−0.8
June 9	+0.1	+0.2	+0.2	Dec. 15	−0.6	−0.7	−0.9
June 19	+0.2	+0.2	+0.3	Dec. 25	−0.6	−0.8	−0.9
June 29	+0.2	+0.2	+0.3	Dec. 31	−0.6	−0.8	−0.9

TABLE VI.*—HORIZONTAL DISTANCES AND ELEVATIONS FROM STADIA READINGS

	0°		1°		2°		3°	
Minutes	Hor. Dist.	Diff. Elev.	Hor. Dist.	Diff. Elev.	Hor. Dist.	Diff. Elev.	Hor. Dist.	Diff. Elev.
0	100.00	0.00	99.97	1.74	99.88	3.49	99.73	5.23
2	100.00	0.06	99.97	1.80	99.87	3.55	99.72	5.28
4	100.00	0.12	99.97	1.86	99.87	3.60	99.71	5.34
6	100.00	0.17	99.96	1.92	99.87	3.66	99.71	5.40
8	100.00	0.23	99.96	1.98	99.86	3.72	99.70	5.46
10	100.00	0.29	99.96	2.04	99.86	3.78	99.69	5.52
12	100.00	0.35	99.96	2.09	99.85	3.84	99.69	5.57
14	100.00	0.41	99.95	2.15	99.85	3.90	99.68	5.63
16	100.00	0.47	99.95	2.21	99.84	3.95	99.68	5.69
18	100.00	0.52	99.95	2.27	99.84	4.01	99.67	5.75
20	100.00	0.58	99.95	2.33	99.83	4.07	99.66	5.80
22	100.00	0.64	99.94	2.38	99.83	4.13	99.66	5.86
24	100.00	0.70	99.94	2.44	99.82	4.18	99.65	5.92
26	99.99	0.76	99.94	2.50	99.82	4.24	99.64	5.98
28	99.99	0.81	99.93	2.56	99.81	4.30	99.63	6.04
30	99.99	0.87	99.93	2.62	99.81	4.36	99.63	6.09
32	99.99	0.93	99.93	2.67	99.80	4.42	99.62	6.15
34	99.99	0.99	99.93	2.73	99.80	4.48	99.62	6.21
36	99.99	1.05	99.92	2.79	99.79	4.53	99.61	6.27
38	99.99	1.11	99.92	2.85	99.79	4.59	99.60	6.33
40	99.99	1.16	99.92	2.91	99.78	4.65	99.59	6.38
42	99.99	1.22	99.91	2.97	99.78	4.71	99.59	6.44
44	99.98	1.28	99.91	3.02	99.77	4.76	99.58	6.50
46	99.98	1.34	99.90	3.08	99.77	4.82	99.57	6.56
48	99.98	1.40	99.90	3.14	99.76	4.88	99.56	6.61
50	99.98	1.45	99.90	3.20	99.76	4.94	99.56	6.67
52	99.98	1.51	99.89	3.26	99.75	4.99	99.55	6.73
54	99.98	1.57	99.89	3.31	99.74	5.05	99.54	6.78
56	99.97	1.63	99.89	3.37	99.74	5.11	99.53	6.84
58	99.97	1.69	99.88	3.43	99.73	5.17	99.52	6.90
60	99.97	1.74	99.88	3.49	99.73	5.23	99.51	6.96
C = 0.75	0.75	0.01	0.75	0.02	0.75	0.03	0.75	0.05
C = 1.00	1.00	0.01	1.00	0.03	1.00	0.04	1.00	0.06
C = 1.25	1.25	0.02	1.25	0.03	1.25	0.05	1.25	0.08

* From "Theory and Practice of Surveying" by Prof. J. B. Johnson. By permission of the publishers, John Wiley & Sons, Inc., New York.

TABLE VI.—HORIZONTAL DISTANCES AND ELEVATIONS FROM STADIA READINGS.—*Continued*

	4°		5°		6°		7°	
Minutes	Hor. Dist.	Diff. Elev.	Hor. Dist.	Diff. Elev.	Hor. Dist.	Diff. Elev.	Hor. Dist.	Diff. Elev.
0	99.51	6.96	99.24	8.68	98.91	10.40	98.51	12.10
2	99.51	7.02	99.23	8.74	98.90	10.45	98.50	12.15
4	99.50	7.07	99.22	8.80	98.88	10.51	98.48	12.21
6	99.49	7.13	99.21	8.85	98.87	10.57	98.47	12.26
8	99.48	7.19	99.20	8.91	98.86	10.62	98.46	12.32
10	99.47	7.25	99.19	8.97	98.85	10.68	98.44	12.38
12	99.46	7.30	99.18	9.03	98.83	10.74	98.43	12.43
14	99.46	7.36	99.17	9.08	98.82	10.79	98.41	12.49
16	99.45	7.42	99.16	9.14	98.81	10.85	98.40	12.55
18	99.44	7.48	99.15	9.20	98.80	10.91	98.39	12.60
20	99.43	7.53	99.14	9.25	98.78	10.96	98.37	12.66
22	99.42	7.59	99.13	9.31	98.77	11.02	98.36	12.72
24	99.41	7.65	99.11	9.37	98.76	11.08	98.34	12.77
26	99.40	7.71	99.10	9.43	98.74	11.13	98.33	12.83
28	99.39	7.76	99.09	9.48	98.73	11.19	98.31	12.88
30	99.38	7.82	99.08	9.54	98.72	11.25	98.29	12.94
32	99.38	7.88	99.07	9.60	98.71	11.30	98.28	13.00
34	99.37	7.94	99.06	9.65	98.69	11.36	98.27	13.05
36	99.36	7.99	99.05	9.71	98.68	11.42	98.25	13.11
38	99.35	8.05	99.04	9.77	98.67	11.47	98.24	13.17
40	99.34	8.11	99.03	9.83	98.65	11.53	98.22	13.22
42	99.33	8.17	99.01	9.88	98.64	11.59	98.20	13.28
44	99.32	8.22	99.00	9.94	98.63	11.64	98.19	13.33
46	99.31	8.28	98.99	10.00	98.61	11.70	98.17	13.39
48	99.30	8.34	98.98	10.05	98.60	11.76	98.16	13.45
50	99.29	8.40	98.97	10.11	98.58	11.81	98.14	13.50
52	99.28	8.45	98.96	10.17	98.57	11.87	98.13	13.56
54	99.27	8.51	98.94	10.22	98.56	11.93	98.11	13.61
56	99.26	8.57	98.93	10.28	98.54	11.98	98.10	13.67
58	99.25	8.63	98.92	10.34	98.53	12.04	98.08	13.73
60	99.24	8.68	98.91	10.40	98.51	12.10	98.06	13.78
C = 0.75	0.75	0.06	0.75	0.07	0.75	0.08	0.74	0.10
C = 1.00	1.00	0.08	0.99	0.09	0.99	0.11	0.99	0.13
C = 1.25	1.25	0.10	1.24	0.11	1.24	0.14	1.24	0.16

TABLE VI.—HORIZONTAL DISTANCES AND ELEVATIONS FROM STADIA READINGS.—*Continued*

	8°		9°		10°		11°	
Minutes	Hor. Dist.	Diff. Elev.	Hor. Dist.	Diff. Elev.	Hor. Dist.	Diff. Elev.	Hor. Dist.	Diff. Elev.
0	98.06	13.78	97.55	15.45	96.98	17.10	96.36	18.73
2	98.05	13.84	97.53	15.51	96.96	17.16	96.34	18.78
4	98.03	13.89	97.52	15.56	96.94	17.21	96.32	18.84
6	98.01	13.95	97.50	15.62	96.92	17.26	96.29	18.89
8	98.00	14.01	97.48	15.67	96.90	17.32	96.27	18.95
10	97.98	14.06	97.46	15.73	96.88	17.37	96.25	19.00
12	97.97	14.12	97.44	15.78	96.86	17.43	96.23	19.05
14	97.95	14.17	97.43	15.84	96.84	17.48	96.21	19.11
16	97.93	14.23	97.41	15.89	96.82	17.54	96.18	19.16
18	97.92	14.28	97.39	15.95	96.80	17.59	96.16	19.21
20	97.90	14.34	97.37	16.00	96.78	17.65	96.14	19.27
22	97.88	14.40	97.35	16.06	96.76	17.70	96.12	19.32
24	97.87	14.45	97.33	16.11	96.74	17.76	96.09	19.38
26	97.85	14.51	97.31	16.17	96.72	17.81	96.07	19.43
28	97.83	14.56	97.29	16.22	96.70	17.86	96.05	19.48
30	97.82	14.62	97.28	16.28	96.68	17.92	96.03	19.54
32	97.80	14.67	97.26	16.33	96.66	17.97	96.00	19.59
34	97.78	14.73	97.24	16.39	96.64	18.03	95.98	19.64
36	97.76	14.79	97.22	16.44	96.62	18.08	95.96	19.70
38	97.75	14.84	97.20	16.50	96.60	18.14	95.93	19.75
40	97.73	14.90	97.18	16.55	96.57	18.19	95.91	19.80
42	97.71	14.95	97.16	16.61	96.55	18.24	95.89	19.86
44	97.69	15.01	97.14	16.66	96.53	18.30	95.86	19.91
46	97.68	15.06	97.12	16.72	96.51	18.35	95.84	19.96
48	97.66	15.12	97.10	16.77	96.49	18.41	95.82	20.02
50	97.64	15.17	97.08	16.83	96.47	18.46	95.79	20.07
52	97.62	15.23	97.06	16.88	96.45	18.51	95.77	20.12
54	97.61	15.28	97.04	16.94	96.42	18.57	95.75	20.18
56	97.59	15.34	97.02	16.99	96.40	18.62	95.72	20.23
58	97.57	15.40	97.00	17.05	96.38	18.68	95.70	20.28
60	97.55	15.45	96.98	17.10	96.36	18.73	95.68	20.34
C = 0.75	0.74	0.11	0.74	0.12	0.74	0.14	0.73	0.15
C = 1.00	0.99	0.15	0.99	0.16	0.98	0.18	0.98	0.20
C = 1.25	1.23	0.18	1.23	0.21	1.23	0.23	1.22	0.25

TABLE VI.—HORIZONTAL DISTANCES AND ELEVATIONS FROM STADIA READINGS.—*Continued*

	12°		13°		14°		15°	
Minutes	Hor. Dist.	Diff. Elev.	Hor. Dist.	Diff. Elev.	Hor. Dist.	Diff. Elev.	Hor. Dist.	Diff. Elev.
0	95.68	20.34	94.94	21.92	94.15	23.47	93.30	25.00
2	95.65	20.39	94.91	21.97	94.12	23.52	93.27	25.05
4	95.63	20.44	94.89	22.02	94.09	23.58	93.24	25.10
6	95.61	20.50	94.86	22.08	94.07	23.63	93.21	25.15
8	95.58	20.55	94.84	22.13	94.04	23.68	93.18	25.20
10	95.56	20.60	94.81	22.18	94.01	23.73	93.16	25.25
12	95.53	20.66	94.79	22.23	93.98	23.78	93.13	25.30
14	95.51	20.71	94.76	22.28	93.95	23.83	93.10	25.35
16	95.49	20.76	94.73	22.34	93.93	23.88	93.07	25.40
18	95.46	20.81	94.71	22.39	93.90	23.93	93.04	25.45
20	95.44	20.87	94.68	22.44	93.87	23.99	93.01	25.50
22	95.41	20.92	94.66	22.49	93.84	24.04	92.98	25.55
24	95.39	20.97	94.63	22.54	93.81	24.09	92.95	25.60
26	95.36	21.03	94.60	22.60	93.79	24.14	92.92	25.65
28	95.34	21.08	94.58	22.65	93.76	24.19	92.89	25.70
30	95.32	21.13	94.55	22.70	93.73	24.24	92.86	25.75
32	95.29	21.18	94.52	22.75	93.70	24.29	92.83	25.80
34	95.27	21.24	94.50	22.80	93.67	24.34	92.80	25.85
36	95.24	21.29	94.47	22.85	93.65	24.39	92.77	25.90
38	95.22	21.34	94.44	22.91	93.62	24.44	92.74	25.95
40	95.19	21.39	94.42	22.96	93.59	24.49	92.71	26.00
42	95.17	21.45	94.39	23.01	93.56	24.55	92.68	26.05
44	95.14	21.50	94.36	23.06	93.53	24.60	92.65	26.10
46	95.12	21.55	94.34	23.11	93.50	24.65	92.62	26.15
48	95.09	21.60	94.31	23.16	93.47	24.70	92.59	26.20
50	95.07	21.66	94.28	23.22	93.45	24.75	92.56	26.25
52	95.04	21.71	94.26	23.27	93.42	24.80	92.53	26.30
54	95.02	21.76	94.23	23.32	93.39	24.85	92.49	26.35
56	94.99	21.81	94.20	23.37	93.36	24.90	92.46	26.40
58	94.97	21.87	94.17	23.42	93.33	24.95	92.43	26.45
60	94.94	21.92	94.15	23.47	93.30	25.00	92.40	26.50
C = 0.75	0.73	0.16	0.73	0.17	0.73	0.19	0.72	0.20
C = 1.00	0.98	0.22	0.97	0.23	0.97	0.25	0.96	0.27
C = 1.25	1.22	0.27	1.21	0.29	1.21	0.31	1.20	0.34

TABLE VI.—HORIZONTAL DISTANCES AND ELEVATIONS FROM STADIA READINGS.—*Continued*

	16°		17°		18°		19°	
Minutes	Hor. Dist.	Diff. Elev.	Hor. Dist.	Diff. Elev.	Hor. Dist.	Diff. Elev.	Hor. Dist.	Diff. Elev.
0	92.40	26.50	91.45	27.96	90.45	29.39	89.40	30.78
2	92.37	26.55	91.42	28.01	90.42	29.44	89.36	30.83
4	92.34	26.59	91.39	28.06	90.38	29.48	89.33	30.87
6	92.31	26.64	91.35	28.10	90.35	29.53	89.29	30.92
8	92.28	26.69	91.32	28.15	90.31	29.58	89.26	30.97
10	92.25	26.74	91.29	28.20	90.28	29.62	89.22	31.01
12	92.22	26.79	91.26	28.25	90.24	29.67	89.18	31.06
14	92.19	26.84	91.22	28.30	90.21	29.72	89.15	31.10
16	92.15	26.89	91.19	28.34	90.18	29.76	89.11	31.15
18	92.12	26.94	91.16	28.39	90.14	29.81	89.08	31.19
20	92.09	26.99	91.12	28.44	90.11	29.86	89.04	31.24
22	92.06	27.04	91.09	28.49	90.07	29.90	89.00	31.28
24	92.03	27.09	91.06	28.54	90.04	29.95	88.96	31.33
26	92.00	27.13	91.02	28.58	90.00	30.00	88.93	31.38
28	91.97	27.18	90.99	28.63	89.97	30.04	88.89	31.42
30	91.93	27.23	90.96	28.68	89.93	30.09	88.86	31.47
32	91.90	27.28	90.92	28.73	89.90	30.14	88.82	31.51
34	91.87	27.33	90.89	28.77	89.86	30.19	88.78	31.56
36	91.84	27.38	90.86	28.82	89.83	30.23	88.75	31.60
38	91.81	27.43	90.82	28.87	89.79	30.28	88.71	31.65
40	91.77	27.48	90.79	28.92	89.76	30.32	88.67	31.69
42	91.74	27.52	90.76	28.96	89.72	30.37	88.64	31.74
44	91.71	27.57	90.72	29.01	89.69	30.41	88.60	31.78
46	91.68	27.62	90.69	29.06	89.65	30.46	88.56	31.83
48	91.65	27.67	90.66	29.11	89.61	30.51	88.53	31.87
50	91.61	27.72	90.62	29.15	89.58	30.55	88.49	31.92
52	91.58	27.77	90.59	29.20	89.54	30.60	88.45	31.96
54	91.55	27.81	90.55	29.25	89.51	30.65	88.41	32.01
56	91.52	27.86	90.52	29.30	89.47	30.69	88.38	32.05
58	91.48	27.91	90.48	29.34	89.44	30.74	88.34	32.09
60	91.45	27.96	90.45	29.39	89.40	30.78	88.30	32.14
C = 0.75	0.72	0.21	0.72	0.23	0.71	0.24	0.71	0.25
C = 1.00	0.96	0.28	0.95	0.30	0.95	0.32	0.94	0.33
C = 1.25	1.20	0.35	1.19	0.38	1.19	0.40	1.18	0.42

TABLE VI.—HORIZONTAL DISTANCES AND ELEVATIONS FROM STADIA READINGS.—*Continued*

	20°		21°		22°		23°	
Minutes.	Hor. Dist.	Diff. Elev.	Hor. Dist.	Diff. Elev.	Hor. Dist.	Diff. Elev.	Hor. Dist.	Diff. Elev.
0	88.30	32.14	87.16	33.46	85.97	34.73	84.73	35.97
2	88.26	32.18	87.12	33.50	85.93	34.77	84.69	36.01
4	88.23	32.23	87.08	33.54	85.89	34.82	84.65	36.05
6	88.19	32.27	87.04	33.59	85.85	34.86	84.61	36.09
8	88.15	32.32	87.00	33.63	85.80	34.90	84.57	36.13
10	88.11	32.36	86.96	33.67	85.76	34.94	84.52	36.17
12	88.08	32.41	86.92	33.72	85.72	34.98	84.48	36.21
14	88.04	32.45	86.88	33.76	85.68	35.02	84.44	36.25
16	88.00	32.49	86.84	33.80	85.64	35.07	84.40	36.29
18	87.96	32.54	86.80	33.84	85.60	35.11	84.35	36.33
20	87.93	32.58	86.77	33.89	85.56	35.15	84.31	36.37
22	87.89	32.63	86.73	33.93	85.52	35.19	84.27	36.41
24	87.85	32.67	86.69	33.97	85.48	35.23	84.23	36.45
26	87.81	32.72	86.65	34.01	85.44	35.27	84.18	36.49
28	87.77	32.76	86.61	34.06	85.40	35.31	84.14	36.53
30	87.74	32.80	86.57	34.10	85.36	35.36	84.10	36.57
32	87.70	32.85	86.53	34.14	85.31	35.40	84.06	36.61
34	87.66	32.89	86.49	34.18	85.27	35.44	84.01	36.65
36	87.62	32.93	86.45	34.23	85.23	35.48	83.97	36.69
38	87.58	32.98	86.41	34.27	85.19	35.52	83.93	36.73
40	87.54	33.02	86.37	34.31	85.15	35.56	83.89	36.77
42	87.51	33.07	86.33	34.35	85.11	35.60	83.84	36.80
44	87.47	33.11	86.29	34.40	85.07	35.64	83.80	36.84
46	87.43	33.15	86.25	34.44	85.02	35.68	83.76	36.88
48	87.39	33.20	86.21	34.48	84.98	35.72	83.72	36.92
50	87.35	33.24	86.17	34.52	84.94	35.76	83.67	36.96
52	87.31	33.28	86.13	34.57	84.90	35.80	83.63	37.00
54	87.27	33.33	86.09	34.61	84.86	35.85	83.59	37.04
56	87.24	33.37	86.05	34.65	84.82	35.89	83.54	37.08
58	87.20	33.41	86.01	34.69	84.77	35.93	83.50	37.12
60	87.16	33.46	85.97	34.73	84.73	35.97	83.46	37.16
C = 0.75	0.70	0.26	0.70	0.27	0.69	0.29	0.69	0.30
C = 1.00	0.94	0.35	0.93	0.37	0.92	0.38	0.92	0.40
C = 1.25	1.17	0.44	1.16	0.46	1.15	0.48	1.15	0.50

Table VI.—Horizontal Distances and Elevations from Stadia Readings.—*Continued*

	24°		25°		26°		27°	
Minutes	Hor. Dist.	Diff. Elev.	Hor. Dist.	Diff. Elev.	Hor. Dist.	Diff. Elev.	Hor. Dist.	Diff. Elev.
0	83.46	37.16	82.14	38.30	80.78	39.40	79.39	40.45
2	83.41	37.20	82.09	38.34	80.74	39.44	79.34	40.49
4	83.37	37.23	82.05	38.38	80.69	39.47	79.30	40.52
6	83.33	37.27	82.01	38.41	80.65	39.51	79.25	40.55
8	83.28	37.31	81.96	38.45	80.60	39.54	79.20	40.59
10	83.24	37.35	81.92	38.49	80.55	39.58	79.15	40.62
12	83.20	37.39	81.87	38.53	80.51	39.61	79.11	40.66
14	83.15	37.43	81.83	38.56	80.46	39.65	79.06	40.69
16	83.11	37.47	81.78	38.60	80.41	39.69	79.01	40.72
18	83.07	37.51	81.74	38.64	80.37	39.72	78.96	40.76
20	83.02	37.54	81.69	38.67	80.32	39.76	78.92	40.79
22	82.98	37.58	81.65	38.71	80.28	39.79	78.87	40.82
24	82.93	37.62	81.60	38.75	80.23	39.83	78.82	40.86
26	82.89	37.66	81.56	38.78	80.18	39.86	78.77	40.89
28	82.85	37.70	81.51	38.82	80.14	39.90	78.73	40.92
30	82.80	37.74	81.47	38.86	80.09	39.93	78.68	40.96
32	82.76	37.77	81.42	38.89	80.04	39.97	78.63	40.99
34	82.72	37.81	81.38	38.93	80.00	40.00	78.58	41.02
36	82.67	37.85	81.33	38.97	79.95	40.04	78.54	41.06
38	82.63	37.89	81.28	39.00	79.90	40.07	78.49	41.09
40	82.58	37.93	81.24	39.04	79.86	40.11	78.44	41.12
42	82.54	37.96	81.19	39.08	79.81	40.14	78.39	41.16
44	82.49	38.00	81.15	39.11	79.76	40.18	78.34	41.19
46	82.45	38.04	81.10	39.15	79.72	40.21	78.30	41.22
48	82.41	38.08	81.06	39.18	79.67	40.24	78.25	41.26
50	82.36	38.11	81.01	39.22	79.62	40.28	78.20	41.29
52	82.32	38.15	80.97	39.26	79.58	40.31	78.15	41.32
54	82.27	38.19	80.92	39.29	79.53	40.35	78.10	41.35
56	82.23	38.23	80.87	39.33	79.48	40.38	78.06	41.39
58	82.18	38.26	80.83	39.36	79.44	40.42	78.01	41.42
60	82.14	38.30	80.78	39.40	79.39	40.45	77.96	41.45
C = 0.75	0.68	0.31	0.68	0.32	0.67	0.33	0.66	0.35
C = 1.00	0.91	0.41	0.90	0.43	0.89	0.45	0.89	0.46
C = 1.25	1.14	0.52	1.13	0.54	1.12	0.56	1.11	0.58

TABLE VI.—HORIZONTAL DISTANCES AND ELEVATIONS FROM STADIA READINGS.—*Concluded*

	28°		29°		30°	
Minutes	Hor. Dist.	Diff. Elev.	Hor. Dist.	Diff. Elev.	Hor. Dist.	Diff. Elev.
0	77.96	41.45	76.50	42.40	75.00	43.30
2	77.91	41.48	76.45	42.43	74.95	43.33
4	77.86	41.52	76.40	42.46	74.90	43.36
6	77.81	41.55	76.35	42.49	74.85	43.39
8	77.77	41.58	76.30	42.53	74.80	43.42
10	77.72	41.61	76.25	42.56	74.75	43.45
12	77.67	41.65	76.20	42.59	74.70	43.47
14	77.62	41.68	76.15	42.62	74.65	43.50
16	77.57	41.71	76.10	42.65	74.60	43.53
18	77.52	41.74	76.05	42.68	74.55	43.56
20	77.48	41.77	76.00	42.71	74.49	43.59
22	77.42	41.81	75.95	42.74	74.44	43.62
24	77.38	41.84	75.90	42.77	74.39	43.65
26	77.33	41.87	75.85	42.80	74.34	43.67
28	77.28	41.90	75.80	42.83	74.29	43.70
30	77.23	41.93	75.75	42.86	74.24	43.73
32	77.18	41.97	75.70	42.89	74.19	43.76
34	77.13	42.00	75.65	42.92	74.14	43.79
36	77.09	42.03	75.60	42.95	74.09	43.82
38	77.04	42.06	75.55	42.98	74.04	43.84
40	76.99	42.09	75.50	43.01	73.99	43.87
42	76.94	42.12	75.45	43.04	73.93	43.90
44	76.89	42.15	75.40	43.07	73.88	43.93
46	76.84	42.19	75.35	43.10	73.83	43.95
48	76.79	42.22	75.30	43.13	73.78	43.98
50	76.74	42.25	75.25	43.16	73.73	44.01
52	76.69	42.28	75.20	43.18	73.68	44.04
54	76.64	42.31	75.15	43.21	73.63	44.07
56	76.59	42.34	75.10	43.24	73.58	44.09
58	76.55	42.37	75.05	43.27	73.52	44.12
60	76.50	42.40	75.00	43.30	73.47	44.15
C = 0.75	0.66	0.36	0.65	0.37	0.65	0.38
C = 1.00	0.88	0.48	0.87	0.49	0.86	0.51
C = 1.25	1.10	0.60	1.09	0.62	1.08	0.64

No. 100 Log. 000 — TABLE VII.—LOGARITHMS OF NUMBERS — No. 109 Log. 040

N.	0	1	2	3	4	5	6	7	8	9	Diff.
100	00 0000	0434	0868	1301	1734	2166	2598	3029	3461	3891	432
1	4321	4751	5181	5609	6038	6466	6894	7321	7748	8174	428
2	8600	9026	9451	9876	0300	0724	1147	1570	1993	2415	424
3	01 2837	3259	3680	4100	4521	4940	5360	5779	6197	6616	420
4	7033	7451	7868	8284	8700	9116	9532	9947	0361	0775	416
105	02 1189	1603	2016	2428	2841	3252	3664	4075	4486	4896	412
6	5306	5715	6125	6533	6942	7350	7757	8164	8571	8978	408
7	9384	9789	0195	0600	1004	1408	1812	2216	2619	3021	404
8	03 3424	3826	4227	4628	5029	5430	5830	6230	6629	7028	400
9	7426	7825	8223	8620	9017	9414	9811	0207	0602	0998	397
	04										

PROPORTIONAL PARTS.

Diff.	1	2	3	4	5	6	7	8	9
434	43.4	86.8	130.2	173.6	217.0	260.4	303.8	347.2	390.6
433	43.3	86.6	129.9	173.2	216.5	259.8	303.1	346.4	389.7
432	43.2	86.4	129.6	172.8	216.0	259.2	302.4	345.6	388.8
431	43.1	86.2	129.3	172.4	215.5	258.6	301.7	344.8	387.9
430	43.0	86.0	129.0	172.0	215.0	258.0	301.0	344.0	387.0
429	42.9	85.8	128.7	171.6	214.5	257.4	300.3	343.2	386.1
428	42.8	85.6	128.4	171.2	214.0	256.8	299.6	342.4	385.2
427	42.7	85.4	128.1	170.8	213.5	256.2	298.9	341.6	384.3
426	42.6	85.2	127.8	170.4	213.0	255.6	298.2	340.8	383.4
425	42.5	85.0	127.5	170.0	212.5	255.0	297.5	340.0	382.5
424	42.4	84.8	127.2	169.6	212.0	254.4	296.8	339.2	381.6
423	42.3	84.6	126.9	169.2	211.5	253.8	296.1	338.4	380.7
422	42.2	84.4	126.6	168.8	211.0	253.2	295.4	337.6	379.8
421	42.1	84.2	126.3	168.4	210.5	252.6	294.7	336.8	378.9
420	42.0	84.0	126.0	168.0	210.0	252.0	294.0	336.0	378.0
419	41.9	83.8	125.7	167.6	209.5	251.4	293.3	335.2	377.1
418	41.8	83.6	125.4	167.2	209.0	250.8	292.6	334.4	376.2
417	41.7	83.4	125.1	166.8	208.5	250.2	291.9	333.6	375.3
416	41.6	83.2	124.8	166.4	208.0	249.6	291.2	332.8	374.4
415	41.5	83.0	124.5	166.0	207.5	249.0	290.5	332.0	373.5
414	41.4	82.8	124.2	165.6	207.0	248.4	289.8	331.2	372.6
413	41.3	82.6	123.9	165.2	206.5	247.8	289.1	330.4	371.7
412	41.2	82.4	123.6	164.8	206.0	247.2	288.4	329.6	370.8
411	41.1	82.2	123.3	164.4	205.5	246.6	287.7	328.8	369.9
410	41.0	82.0	123.0	164.0	205.0	246.0	287.0	328.0	369.0
409	40.9	81.8	122.7	163.6	204.5	245.4	286.3	327.2	368.1
408	40.8	81.6	122.4	163.2	204.0	244.8	285.6	326.4	367.2
407	40.7	81.4	122.1	162.8	203.5	244.2	284.9	325.6	366.3
406	40.6	81.2	121.8	162.4	203.0	243.6	284.2	324.8	365.4
405	40.5	81.0	121.5	162.0	202.5	243.0	283.5	324.0	364.5
404	40.4	80.8	121.2	161.6	202.0	242.4	282.8	323.2	363.6
403	40.3	80.6	120.9	161.2	201.5	241.8	282.1	322.4	362.7
402	40.2	80.4	120.6	160.8	201.0	241.2	281.4	321.6	361.8
401	40.1	80.2	120.3	160.4	200.5	240.6	280.7	320.8	360.9
400	40.0	80.0	120.0	160.0	200.0	240.0	280.0	320.0	360.0
399	39.9	79.8	119.7	159.6	199.5	239.4	279.3	319.2	359.1
398	39.8	79.6	119.4	159.2	199.0	238.8	278.6	318.4	358.2
397	39.7	79.4	119.1	158.8	198.5	238.2	277.9	317.6	357.3
396	39.6	79.2	118.8	158.4	198.0	237.6	277.2	316.8	356.4
395	39.5	79.0	118.5	158.0	197.5	237.0	276.5	316.0	351.5

No. 110 Log. 041 — Table VII.—*Continued* — No. 119 Log. 078

N.	0	1	2	3	4	5	6	7	8	9	Diff.
110	04 1393	1787	2182	2576	2969	3362	3755	4148	4540	4932	393
1	5323	5714	6105	6495	6885	7275	7664	8053	8442	8830	390
2	9218	9606	9993	0380	0766	1153	1538	1924	2309	2694	386
3	05 3078	3463	3846	4230	4613	4996	5378	5760	6142	6524	383
4	6905	7286	7666	8046	8426	8805	9185	9563	9942	0320	379
115	06 0698	1075	1452	1829	2206	2582	2958	3333	3709	4083	376
6	4458	4832	5206	5580	5953	6326	6699	7071	7443	7815	373
7	8186	8557	8928	9298	9668	0038	0407	0776	1145	1514	370
8	07 1882	2250	2617	2985	3352	3718	4085	4451	4816	5182	366
9	5547	5912	6276	6640	7004	7368	7731	8094	8457	8819	363

PROPORTIONAL PARTS.

Diff.	1	2	3	4	5	6	7	8	9
395	39.5	79.0	118.5	158.0	197.5	237.0	276.5	316.0	355.5
394	39.4	78.8	118.2	157.6	197.0	236.4	275.8	315.2	354.6
393	39.3	78.6	117.9	157.2	196.5	235.8	275.1	314.4	353.7
392	39.2	78.4	117.6	156.8	196.0	235.2	274.4	313.6	352.8
391	39.1	78.2	117.3	156.4	195.5	234.6	273.7	312.8	351.9
390	39.0	78.0	117.0	156.0	195.0	234.0	273.0	312.0	351.0
389	38.9	77.8	116.7	155.6	194.5	233.4	272.3	311.2	350.1
388	38.8	77.6	116.4	155.2	194.0	232.8	271.6	310.4	349.2
387	38.7	77.4	116.1	154.8	193.5	232.2	270.9	309.6	348.3
386	38.6	77.2	115.8	154.4	193.0	231.6	270.2	308.8	347.4
385	38.5	77.0	115.5	154.0	192.5	231.0	269.5	308.0	346.5
384	38.4	76.8	115.2	153.6	192.0	230.4	268.8	307.2	345.6
383	38.3	76.6	114.9	153.2	191.5	229.8	268.1	306.4	344.7
382	38.2	76.4	114.6	152.8	191.0	229.2	267.4	305.6	343.8
381	38.1	76.2	114.3	152.4	190.5	228.6	266.7	304.8	342.9
380	38.0	76.0	114.0	152.0	190.0	228.0	266.0	304.0	342.0
379	37.9	75.8	113.7	151.6	189.5	227.4	265.3	303.2	341.1
378	37.8	75.6	113.4	151.2	189.0	226.8	264.6	302.4	340.2
377	37.7	75.4	113.1	150.8	188.5	226.2	263.9	301.6	339.3
376	37.6	75.2	112.8	150.4	188.0	225.6	263.2	300.8	338.4
375	37.5	75.0	112.5	150.0	187.5	225.0	262.5	300.0	337.5
374	37.4	74.8	112.2	149.6	187.0	224.4	261.8	299.2	336.6
373	37.3	74.6	111.9	149.2	186.5	223.8	261.1	298.4	335.7
372	37.2	74.4	111.6	148.8	186.0	223.2	260.4	297.6	334.8
371	37.1	74.2	111.3	148.4	185.5	222.6	259.7	296.8	333.9
370	37.0	74.0	111.0	148.0	185.0	222.0	259.0	296.0	333.0
369	36.9	73.8	110.7	147.6	184.5	221.4	258.3	295.2	332.1
368	36.8	73.6	110.4	147.2	184.0	220.8	257.6	294.4	331.2
367	36.7	73.4	110.1	146.8	183.5	220.2	256.9	293.6	330.3
366	36.6	73.2	109.8	146.4	183.0	219.6	256.2	292.8	329.4
365	36.5	73.0	109.5	146.0	182.5	219.0	255.5	292.0	328.5
364	36.4	72.8	109.2	145.6	182.0	218.4	254.8	291.2	327.6
363	36.3	72.6	108.9	145.2	181.5	217.8	254.1	290.4	326.7
362	36.2	72.4	108.6	144.8	181.0	217.2	253.4	289.6	325.8
361	36.1	72.2	108.3	144.4	180.5	216.6	252.7	288.8	324.9
360	36.0	72.0	108.0	144.0	180.0	216.0	252.0	288.0	324.0
359	35.9	71.8	107.7	143.6	179.5	215.4	251.3	287.2	323.1
358	35.8	71.6	107.4	143.2	179.0	214.8	250.6	286.4	322.2
357	35.7	71.4	107.1	142.8	178.5	214.2	249.9	285.6	321.3
356	35.6	71.2	106.8	142.4	178.0	213.6	249.2	284.8	320.4

No. 120 Log. 079 Table VII.—*Continued* No. 134 Log. 130

N.	0	1	2	3	4	5	6	7	8	9	Diff.
120	07 9181	9543	9904	0266	0626	0987	1347	1707	2067	2426	360
1	08 2785	3144	3503	3861	4219	4576	4934	5291	5647	6004	357
2	6360	6716	7071	7426	7781	8136	8490	8845	9198	9552	355
3	9905	0258	0611	0963	1315	1667	2018	2370	2721	3071	352
4	09 3422	3772	4122	4471	4820	5169	5518	5866	6215	6562	349
125	6910	7257	7604	7951	8298	8644	8990	9335	9681	0026	346
6	10 0371	0715	1059	1403	1747	2091	2434	2777	3119	3462	343
7	3804	4146	4487	4828	5169	5510	5851	6191	6531	6871	341
8	7210	7549	7888	8227	8565	8903	9241	9579	9916	0253	338
9	11 0590	0926	1263	1599	1934	2270	2605	2940	3275	3609	335
130	3943	4277	4611	4944	5278	5611	5943	6276	6608	6940	333
1	7271	7603	7934	8265	8595	8926	9256	9586	9915	0245	330
2	12 0574	0903	1231	1560	1888	2216	2544	2871	3198	3525	328
3	3852	4178	4504	4830	5156	5481	5806	6131	6456	6781	325
4	7105	7429	7753	8076	8399	8722	9045	9368	9690	0012	323
	13										

PROPORTIONAL PARTS.

Diff.	1	2	3	4	5	6	7	8	9
355	35.5	71.0	106.5	142.0	177.5	213.0	248.5	284.0	319.5
354	35.4	70.8	106.2	141.6	177.0	212.4	247.8	283.2	318.6
353	35.3	70.6	105.9	141.2	176.5	211.8	247.1	282.4	317.7
352	35.2	70.4	105.6	140.8	176.0	211.2	246.4	281.6	316.8
351	35.1	70.2	105.3	140.4	175.5	210.6	245.7	280.8	315.9
350	35.0	70.0	105.0	140.0	175.0	210.0	245.0	280.0	315.0
349	34.9	69.8	104.7	139.6	174.5	209.4	244.3	279.2	314.1
348	34.8	69.6	104.4	139.2	174.0	208.8	243.6	278.4	313.2
347	34.7	69.4	104.1	138.8	173.5	208.2	242.9	277.6	312.3
346	34.6	69.2	103.8	138.4	173.0	207.6	242.2	276.8	311.4
345	34.5	69.0	103.5	138.0	172.5	207.0	241.5	276.0	310.5
344	34.4	68.8	103.2	137.6	172.0	206.4	240.8	275.2	309.6
343	34.3	68.6	102.9	137.2	171.5	205.8	240.1	274.4	308.7
342	34.2	68.4	102.6	136.8	171.0	205.2	239.4	273.6	307.8
341	34.1	68.2	102.3	136.4	170.5	204.6	238.7	272.8	306.9
340	34.0	68.0	102.0	136.0	170.0	204.0	238.0	272.0	306.0
339	33.9	67.8	101.7	135.6	169.5	203.4	237.3	271.2	305.1
338	33.8	67.6	101.4	135.2	169.0	202.8	236.6	270.4	304.2
337	33.7	67.4	101.1	134.8	168.5	202.2	235.9	269.6	303.3
336	33.6	67.2	100.8	134.4	168.0	201.6	235.2	268.8	302.4
335	33.5	67.0	100.5	134.0	167.5	201.0	234.5	268.0	301.5
334	33.4	66.8	100.2	133.6	167.0	200.4	233.8	267.2	300.6
333	33.3	66.6	99.9	133.2	166.5	199.8	233.1	266.4	299.7
332	33.2	66.4	99.6	132.8	166.0	199.2	232.4	265.6	298.8
331	33.1	66.2	99.3	132.4	165.5	198.6	231.7	264.8	297.9
330	33.0	66.0	99.0	132.0	165.0	198.0	231.0	264.0	297.0
329	32.9	65.8	98.7	131.6	164.5	197.4	230.3	263.2	296.1
328	32.8	65.6	98.4	131.2	164.0	196.8	229.6	262.4	295.2
327	32.7	65.4	98.1	130.8	163.5	196.2	228.9	261.6	294.3
326	32.6	65.2	97.8	130.4	163.0	195.6	228.2	260.8	293.4
325	32.5	65.0	97.5	130.0	162.5	195.0	227.5	260.0	292.5
324	32.4	64.8	97.2	129.6	162.0	194.4	226.8	259.2	291.6
323	32.3	64.6	96.9	129.2	161.5	193.8	226.1	258.4	290.7
322	32.2	64.4	96.6	128.8	161.0	193.2	225.4	257.6	289.8

No. 135 | Log. 130 — Table VII.—*Continued* — No. 149 | Log. 175

N.	0	1	2	3	4	5	6	7	8	9	Diff.
135	13 0334	0655	0977	1298	1619	1939	2260	2580	2900	3219	**321**
6	3539	3858	4177	4496	4814	5133	5451	5769	6086	6403	**318**
7	6721	7037	7354	7671	7987	8303	8618	8934	9249	9564	**316**
8	9879	0194	0508	0822	1136	1450	1763	2076	2389	2702	**314**
9	14 3015	3327	3639	3951	4263	4574	4885	5196	5507	5818	**311**
140	6128	6438	6748	7058	7367	7676	7985	8294	8603	8911	**309**
1	9219	9527	9835	0142	0449	0756	1063	1370	1676	1982	**307**
2	15 2288	2594	2900	3205	3510	3815	4120	4424	4728	5032	**305**
3	5336	5640	5943	6246	6549	6852	7154	7457	7759	8061	**303**
4	8362	8664	8965	9266	9567	9868	0168	0469	0769	1068	**301**
145	16 1368	1667	1967	2266	2564	2863	3161	3460	3758	4055	**299**
6	4353	4650	4947	5244	5541	5838	6134	6430	6726	7022	**297**
7	7317	7613	7908	8203	8497	8792	9086	9380	9674	9968	**295**
8	17 0262	0555	0848	1141	1434	1726	2019	2311	2603	2895	**293**
9	3186	3478	3769	4060	4351	4641	4932	5222	5512	5802	**291**

PROPORTIONAL PARTS.

Diff.	1	2	3	4	5	6	7	8	9
321	32.1	64.2	96.3	128.4	160.5	192.6	224.7	256.8	288.9
320	32.0	64.0	96.0	128.0	160.0	192.0	224.0	256.0	288.0
319	31.9	63.8	95.7	127.6	159.5	191.4	223.3	255.2	287.1
318	31.8	63.6	95.4	127.2	159.0	190.8	222.6	254.4	286.2
317	31.7	63.4	95.1	126.8	158.5	190.2	221.9	253.6	285.3
316	31.6	63.2	94.8	126.4	158.0	189.6	221.2	252.8	284.4
315	31.5	63.0	94.5	126.0	157.5	189.0	220.5	252.0	283.5
314	31.4	62.8	94.2	125.6	157.0	188.4	219.8	251.2	282.6
313	31.3	62.6	93.9	125.2	156.5	187.8	219.1	250.4	281.7
312	31.2	62.4	93.6	124.8	156.0	187.2	218.4	249.6	280.8
311	31.1	62.2	93.3	124.4	155.5	186.6	217.7	248.8	279.9
310	31.0	62.0	93.0	124.0	155.0	186.0	217.0	248.0	279.0
309	30.9	61.8	92.7	123.6	154.5	185.4	216.3	247.2	278.1
308	30.8	61.6	92.4	123.2	154.0	184.8	215.6	246.4	277.2
307	30.7	61.4	92.1	122.8	153.5	184.2	214.9	245.6	276.3
306	30.6	61.2	91.8	122.4	153.0	183.6	214.2	244.8	275.4
305	30.5	61.0	91.5	122.0	152.5	183.0	213.5	244.0	274.5
304	30.4	60.8	91.2	121.6	152.0	182.4	212.8	243.2	273.6
303	30.3	60.6	90.9	121.2	151.5	181.8	212.1	242.4	272.7
302	30.2	60.4	90.6	120.8	151.0	181.2	211.4	241.6	271.8
301	30.1	60.2	90.3	120.4	150.5	180.6	210.7	240.8	270.9
300	30.0	60.0	90.0	120.0	150.0	180.0	210.0	240.0	270.0
299	29.9	59.8	89.7	119.6	149.5	179.4	209.3	239.2	269.1
298	29.8	59.6	89.4	119.2	149.0	178.8	208.6	238.4	268.2
297	29.7	59.4	89.1	118.8	148.5	178.2	207.9	237.6	267.3
296	29.6	59.2	88.8	118.4	148.0	177.6	207.2	236.8	266.4
295	29.5	59.0	88.5	118.0	147.5	177.0	206.5	236.0	265.5
294	29.4	58.8	88.2	117.6	147.0	176.4	205.8	235.2	264.6
293	29.3	58.6	87.9	117.2	146.5	175.8	205.1	234.4	263.7
292	29.2	58.4	87.6	116.8	146.0	175.2	204.4	233.6	262.8
291	29.1	58.2	87.3	116.4	145.5	174.6	203.7	232.8	261.9
290	29.0	58.0	87.0	116.0	145.0	174.0	203.0	232.0	261.0
289	28.9	57.8	86.7	115.6	144.5	173.4	202.3	231.2	260.1
288	28.8	57.6	86.4	115.2	144.0	172.8	201.6	230.4	259.2
287	28.7	57.4	86.1	114.8	143.5	172.2	200.9	229.6	258.3
286	28.6	57.2	85.8	114.4	143.0	171.6	200.2	228.8	257.4

Table VII.—*Continued*

N.	0	1	2	3	4	5	6	7	8	9	Diff.
150	17 6091	6381	6670	6959	7248	7536	7825	8113	8401	8689	289
1	8977	9264	9552	9839	0126	0413	0699	0986	1272	1558	287
2	18 1844	2129	2415	2700	2985	3270	3555	3839	4123	4407	285
3	4691	4975	5259	5542	5825	6108	6391	6674	6956	7239	283
4	7521	7803	8084	8366	8647	8928	9209	9490	9771	0051	281
155	19 0332	0612	0892	1171	1451	1730	2010	2289	2567	2846	279
6	3125	3403	3681	3959	4237	4514	4792	5069	5346	5623	278
7	5900	6176	6453	6729	7005	7281	7556	7832	8107	8382	276
8	8657	8932	9206	9481	9755	0029	0303	0577	0850	1124	274
9	20 1397	1670	1943	2216	2488	2761	3033	3305	3577	3848	272
160	4120	4391	4663	4934	5204	5475	5746	6016	6286	6556	271
1	6826	7096	7365	7634	7904	8173	8441	8710	8979	9247	269
2	9515	9783	0051	0319	0586	0853	1121	1388	1654	1921	267
3	21 2188	2454	2720	2986	3252	3518	3783	4049	4314	4579	266
4	4844	5109	5373	5638	5902	6166	6430	6694	6957	7221	264
165	7484	7747	8010	8273	8536	8798	9060	9323	9585	9846	262
6	22 0108	0370	0631	0892	1153	1414	1675	1936	2196	2456	261
7	2716	2976	3236	3496	3755	4015	4274	4533	4792	5051	259
8	5309	5568	5826	6084	6342	6600	6858	7115	7372	7630	258
9	7887	8144	8400	8657	8913	9170	9426	9682	9938	0193	256
	23										

PROPORTIONAL PARTS.

Diff.	1	2	3	4	5	6	7	8	9
285	28.5	57.0	85.5	114.0	142.5	171.0	199.5	228.0	256.5
284	28.4	56.8	85.2	113.6	142.0	170.4	198.8	227.2	255.6
283	28.3	56.6	84.9	113.2	141.5	169.8	198.1	226.4	254.7
282	28.2	56.4	84.6	112.8	141.0	169.2	197.4	225.6	253.8
281	28.1	56.2	84.3	112.4	140.5	168.6	196.7	224.8	252.9
280	28.0	56.0	84.0	112.0	140.0	168.0	196.0	224.0	252.0
279	27.9	55.8	83.7	111.6	139.5	167.4	195.3	223.2	251.1
278	27.8	55.6	83.4	111.2	139.0	166.8	194.6	222.4	250.2
277	27.7	55.4	83.1	110.8	138.5	166.2	193.9	221.6	249.3
276	27.6	55.2	82.8	110.4	138.0	165.6	193.2	220.8	248.4
275	27.5	55.0	82.5	110.0	137.5	165.0	192.5	220.0	247.5
274	27.4	54.8	82.2	109.6	137.0	164.4	191.8	219.2	246.6
273	27.3	54.6	81.9	109.2	136.5	163.8	191.1	218.4	245.7
272	27.2	54.4	81.6	108.8	136.0	163.2	190.4	217.6	244.8
271	27.1	54.2	81.3	108.4	135.5	162.6	189.7	216.8	243.9
270	27.0	54.0	81.0	108.0	135.0	162.0	189.0	216.0	243.0
269	26.9	53.8	80.7	107.6	134.5	161.4	188.3	215.2	242.1
268	26.8	53.6	80.4	107.2	134.0	160.8	187.6	214.4	241.2
267	26.7	53.4	80.1	106.8	133.5	160.2	186.9	213.6	240.3
266	26.6	53.2	79.8	106.4	133.0	159.6	186.2	212.8	239.4
265	26.5	53.0	79.5	106.0	132.5	159.0	185.5	212.0	238.5
264	26.4	52.8	79.2	105.6	132.0	158.4	184.8	211.2	237.6
263	26.3	52.6	78.9	105.2	131.5	157.8	184.1	210.4	236.7
262	26.2	52.4	78.6	104.8	131.0	157.2	183.4	209.6	235.8
261	26.1	52.2	78.3	104.4	130.5	156.6	182.7	208.8	234.9
260	26.0	52.0	78.0	104.0	130.0	156.0	182.0	208.0	234.0
259	25.9	51.8	77.7	103.6	129.5	155.4	181.3	207.2	233.1
258	25.8	51.6	77.4	103.2	129.0	154.8	180.6	206.4	232.2
257	25.7	51.4	77.1	102.8	128.5	154.2	179.9	205.6	231.3
256	25.6	51.2	76.8	102.4	128.0	153.6	179.2	204.8	230.4
255	25.5	51.0	76.5	102.0	127.5	153.0	178.5	204.0	229.5

No. 170 Log. 230 TABLE VII.—*Continued* No. 189 Log. 278

N.	0	1	2	3	4	5	6	7	8	9	Diff.
170	23 0449	0704	0960	1215	1470	1724	1979	2234	2488	2742	255
1	2996	3250	3504	3757	4011	4264	4517	4770	5023	5276	253
2	5528	5781	6033	6285	6537	6789	7041	7292	7544	7795	252
3	8046	8297	8548	8799	9049	9299	9550	9800	0050	0300	250
4	24 0549	0799	1048	1297	1546	1795	2044	2293	2541	2790	249
175	3038	3286	3534	3782	4030	4277	4525	4772	5019	5266	248
6	5513	5759	6006	6252	6499	6745	6991	7237	7482	7728	246
7	7073	8219	8464	8709	8954	9198	9443	9687	9932	0176	245
8	25 0420	0664	0908	1151	1395	1638	1881	2125	2368	2610	243
9	2853	3096	3338	3580	3822	4064	4306	4548	4790	5031	242
180	5273	5514	5755	5996	6237	6477	6718	6958	7198	7439	241
1	7679	7918	8153	8398	8637	8877	9116	9355	9594	9833	239
2	26 0071	0310	0548	0787	1025	1263	1501	1739	1976	2214	238
3	2451	2688	2925	3162	3399	3636	3873	4109	4346	4582	237
4	4818	5054	5290	5525	5761	5996	6232	6467	6702	6937	235
185	7172	7406	7641	7875	8110	8344	8578	8812	9046	9279	234
6	9513	9746	9980	0213	0446	0679	0912	1144	1377	1609	233
7	27 1842	2074	2306	2538	2770	3001	3233	3464	3696	3927	232
8	4158	4389	4620	4850	5081	5311	5542	5772	6002	6232	230
9	6462	6692	6921	7151	7380	7609	7838	8067	8296	8525	229

PROPORTIONAL PARTS.

Diff.	1	2	3	4	5	6	7	8	9
255	25.5	51.0	76.5	102.0	127.5	153.0	178.5	204.0	229.5
254	25.4	50.8	76.2	101.6	127.0	152.4	177.8	203.2	228.6
253	25.3	50.6	75.9	101.2	126.5	151.8	177.1	202.4	227.7
252	25.2	50.4	75.6	100.8	126.0	151.2	176.4	201.6	226.8
251	25.1	50.2	75.3	100.4	125.5	150.6	175.7	200.8	225.9
250	25.0	50.0	75.0	100.0	125.0	150.0	175.0	200.0	225.0
249	24.9	49.8	74.7	99.6	124.5	149.4	174.3	199.2	224.1
248	24.8	49.6	74.4	99.2	124.0	148.8	173.6	198.4	223.2
247	24.7	49.4	74.1	98.8	123.5	148.2	172.9	197.6	222.3
246	24.6	49.2	73.8	98.4	123.0	147.6	172.2	196.8	221.4
245	24.5	49.0	73.5	98.0	122.5	147.0	171.5	196.0	220.5
244	24.4	48.8	73.2	97.6	122.0	146.4	170.8	195.2	219.6
243	24.3	48.6	72.9	97.2	121.5	145.8	170.1	194.4	218.7
242	24.2	48.4	72.6	96.8	121.0	145.2	169.4	193.6	217.8
241	24.1	48.2	72.3	96.4	120.5	144.6	168.7	192.8	216.9
240	24.0	48.0	72.0	96.0	120.0	144.0	168.0	192.0	216.0
239	23.9	47.8	71.7	95.6	119.5	143.4	167.3	191.2	215.1
238	23.8	47.6	71.4	95.2	119.0	142.8	166.6	190.4	214.2
237	23.7	47.4	71.1	94.8	118.5	142.2	165.9	189.6	213.3
236	23.6	47.2	70.8	94.4	118.0	141.6	165.2	188.8	212.4
235	23.5	47.0	70.5	94.0	117.5	141.0	164.5	188.0	211.5
234	23.4	46.8	70.2	93.6	117.0	140.4	163.8	187.2	210.6
233	23.3	46.6	69.9	93.2	116.5	139.8	163.1	186.4	209.7
232	23.2	46.4	69.6	92.8	116.0	139.2	162.4	185.6	208.8
231	23.1	46.2	69.3	92.4	115.5	138.6	161.7	184.8	207.9
230	23.0	46.0	69.0	92.0	115.0	138.0	161.0	184.0	207.0
229	22.9	45.8	68.7	91.6	114.5	137.4	160.3	183.2	206.1
228	22.8	45.6	68.4	91.2	114.0	136.8	159.6	182.4	205.2
227	22.7	45.4	68.1	90.8	113.5	136.2	158.9	181.6	204.3
226	22.6	45.2	67.8	90.4	113.0	135.6	158.2	180.8	203.4

No. 190
Log. 278

TABLE VII.—*Continued*

No. 214
Log. 332

N.	0	1	2	3	4	5	6	7	8	9	Diff.
190	27 8754	8982	9211	9439	9667	9895	0123	0351	0578	0806	228
1	28 1033	1261	1488	1715	1942	2169	2396	2622	2849	3075	227
2	3301	3527	3753	3979	4205	4431	4656	4882	5107	5332	226
3	5557	5782	6007	6232	6456	6681	6905	7130	7354	7578	225
4	7802	8026	8249	8473	8696	8920	9143	9366	9589	9812	223
195	29 0035	0257	0480	0702	0925	1147	1369	1591	1813	2034	222
6	2256	2478	2699	2920	3141	3363	3584	3804	4025	4246	221
7	4466	4687	4907	5127	5347	5567	5787	6007	6226	6446	220
8	6665	6884	7104	7323	7542	7761	7979	8198	8416	8635	219
9	8853	9071	9289	9507	9725	9943	0161	0378	0595	0813	218
200	30 1030	1247	1464	1681	1898	2114	2331	2547	2764	2980	217
1	3196	3412	3628	3844	4059	4275	4491	4706	4921	5136	216
2	5351	5566	5781	5996	6211	6425	6639	6854	7068	7282	215
3	7496	7710	7924	8137	8351	8564	8778	8991	9204	9417	213
4	9630	9843	0056	0268	0481	0693	0906	1118	1330	1542	212
205	31 1754	1966	2177	2389	2600	2812	3023	3234	3445	3656	211
6	3867	4078	4289	4499	4710	4920	5130	5340	5551	5760	210
7	5970	6180	6390	6599	6809	7018	7227	7436	7646	7854	209
8	8063	8272	8481	8689	8898	9106	9314	9522	9730	9938	208
9	32 0146	0354	0562	0769	0977	1184	1391	1598	1805	2012	207
210	2219	2426	2633	2839	3046	3252	3458	3665	3871	4077	206
1	4282	4488	4694	4899	5105	5310	5516	5721	5926	6131	205
2	6336	6541	6745	6950	7155	7359	7563	7767	7972	8176	204
3	8380	8583	8787	8991	9194	9398	9601	9805	0008	0211	203
4	33 0414	0617	0819	1022	1225	1427	1630	1832	2034	2236	202

PROPORTIONAL PARTS.

Diff.	1	2	3	4	5	6	7	8	9
225	22.5	45.0	67.5	90.0	112.5	135.0	157.5	180.0	202.5
224	22.4	44.8	67.2	89.6	112.0	134.4	156.8	179.2	201.6
223	22.3	44.6	66.9	89.2	111.5	133.8	156.1	178.4	200.7
222	22.2	44.4	66.6	88.8	111.0	133.2	155.4	177.6	199.8
221	22.1	44.2	66.3	88.4	110.5	132.6	154.7	176.8	198.9
220	22.0	44.0	66.0	88.0	110.0	132.0	154.0	176.0	198.0
219	21.9	43.8	65.7	87.6	109.5	131.4	153.3	175.2	197.1
218	21.8	43.6	65.4	87.2	109.0	130.8	152.6	174.4	196.2
217	21.7	43.4	65.1	86.8	108.5	130.2	151.9	173.6	195.3
216	21.6	43.2	64.8	86.4	108.0	129.6	151.2	172.8	194.4
215	21.5	43.0	64.5	86.0	107.5	129.0	150.5	172.0	193.5
214	21.4	42.8	64.2	85.6	107.0	128.4	149.8	171.2	192.6
213	21.3	42.6	63.9	85.2	106.5	127.8	149.1	170.4	191.7
212	21.2	42.4	63.6	84.8	106.0	127.2	148.4	169.6	190.8
211	21.1	42.2	63.3	84.4	105.5	126.6	147.7	168.8	189.9
210	21.0	42.0	63.0	84.0	105.0	126.0	147.0	168.0	189.0
209	20.9	41.8	62.7	83.6	104.5	125.4	146.3	167.2	188.1
208	20.8	41.6	62.4	83.2	104.0	124.8	145.6	166.4	187.2
207	20.7	41.4	62.1	82.8	103.5	124.2	144.9	165.6	186.3
206	20.6	41.2	61.8	82.4	103.0	123.6	144.2	164.8	185.4
205	20.5	41.0	61.5	82.0	102.5	123.0	143.5	164.0	184.5
204	20.4	40.8	61.2	81.6	102.0	122.4	142.8	163.2	183.6
203	20.3	40.6	60.9	81.2	101.5	121.8	142.1	162.4	182.7
202	20.2	40.4	60.6	80.8	101.0	121.2	141.4	161.6	181.8

Table VII.—*Continued*

N.	0	1	2	3	4	5	6	7	8	9	Diff.
215	33 2438	2640	2842	3044	3246	3447	3649	3850	4051	4253	202
6	4454	4655	4856	5057	5257	5458	5658	5859	6059	6260	201
7	6460	6660	6860	7060	7260	7459	7659	7858	8058	8257	200
8	8456	8656	8855	9054	9253	9451	9650	9849	0047	0246	199
9	34 0444	0642	0841	1039	1237	1435	1632	1830	2028	2225	198
220	2423	2620	2817	3014	3212	3409	3606	3802	3999	4196	197
1	4392	4589	4785	4981	5178	5374	5570	5766	5962	6157	196
2	6353	6549	6744	6939	7135	7330	7525	7720	7915	8110	195
3	8305	8500	8694	8889	9083	9278	9472	9666	9860	0054	194
4	35 0248	0442	0636	0829	1023	1216	1410	1603	1796	1989	193
225	2183	2375	2568	2761	2954	3147	3339	3532	3724	3916	193
6	4108	4301	4493	4685	4876	5068	5260	5452	5643	5834	192
7	6026	6217	6408	6599	6790	6981	7172	7363	7554	7744	191
8	7935	8125	8316	8506	8696	8886	9076	9266	9456	9646	190
9	9835	0025	0215	0404	0593	0783	0972	1161	1350	1539	189
230	36 1728	1917	2105	2294	2482	2671	2859	3048	3236	3424	188
1	3612	3800	3988	4176	4363	4551	4739	4926	5113	5301	188
2	5488	5675	5862	6049	6236	6423	6610	6796	6983	7169	187
3	7356	7542	7729	7915	8101	8287	8473	8659	8845	9030	186
4	9216	9401	9587	9772	9958	0143	0328	0513	0698	0883	185
235	37 1068	1253	1437	1622	1806	1991	2175	2360	2544	2728	184
6	2912	3096	3280	3464	3647	3831	4015	4198	4382	4565	184
7	4748	4932	5115	5298	5481	5664	5846	6029	6212	6394	183
8	6577	6759	6942	7124	7306	7488	7670	7852	8034	8216	182
9	8398	8580	8761	8943	9124	9306	9487	9668	9849	0030	181
	38										

PROPORTIONAL PARTS.

Diff.	1	2	3	4	5	6	7	8	9
202	20.2	40.4	60.6	80.8	101.0	121.2	141.4	161.6	181.8
201	20.1	40.2	60.3	80.4	100.5	120.6	140.7	160.8	180.9
200	20.0	40.0	60.0	80.0	100.0	120.0	140.0	160.0	180.0
199	19.9	39.8	59.7	79.6	99.5	119.4	139.3	159.2	179.1
198	19.8	39.6	59.4	79.2	99.0	118.8	138.6	158.4	178.2
197	19.7	39.4	59.1	78.8	98.5	118.2	137.9	157.6	177.3
196	19.6	39.2	58.8	78.4	98.0	117.6	137.2	156.8	176.4
195	19.5	39.0	58.5	78.0	97.5	117.0	136.5	156.0	175.5
194	19.4	38.8	58.2	77.6	97.0	116.4	135.8	155.2	174.6
193	19.3	38.6	57.9	77.2	96.5	115.8	135.1	154.4	173.7
192	19.2	38.4	57.6	76.8	96.0	115.2	134.4	153.6	172.8
191	19.1	38.2	57.3	76.4	95.5	114.6	133.7	152.8	171.9
190	19.0	38.0	57.0	76.0	95.0	114.0	133.0	152.0	171.0
189	18.9	37.8	56.7	75.6	94.5	113.4	132.3	151.2	170.1
188	18.8	37.6	56.4	75.2	94.0	112.8	131.6	150.4	169.2
187	18.7	37.4	56.1	74.8	93.5	112.2	130.9	149.6	168.3
186	18.6	37.2	55.8	74.4	93.0	111.6	130.2	148.8	167.4
185	18.5	37.0	55.5	74.0	92.5	111.0	129.5	148.0	166.5
184	18.4	36.8	55.2	73.6	92.0	110.4	128.8	147.2	165.6
183	18.3	36.6	54.9	73.2	91.5	109.8	128.1	146.4	164.7
182	18.2	36.4	54.6	72.8	91.0	109.2	127.4	145.6	163.8
181	18.1	36.2	54.3	72.4	90.5	108.6	126.7	144.8	162.9
180	18.0	36.0	54.0	72.0	90.0	108.0	126.0	144.0	162.0
179	17.9	35.8	53.7	71.6	89.5	107.4	125.3	143.2	161.1

TABLE VII.—*Continued*

N.	0	1	2	3	4	5	6	7	8	9	Diff.
240	38 0211	0392	0573	0754	0934	1115	1296	1476	1656	1837	181
1	2017	2197	2377	2557	2737	2917	3097	3277	3456	3636	180
2	3815	3995	4174	4353	4533	4712	4891	5070	5249	5428	179
3	5606	5785	5964	6142	6321	6499	6677	6856	7034	7212	178
4	7390	7568	7746	7924	8101	8279	8456	8634	8811	8989	178
245	9166	9343	9520	9698	9875	0051	0228	0405	0582	0759	177
6	39 0935	1112	1288	1464	1641	1817	1993	2169	2345	2521	176
7	2697	2873	3048	3224	3400	3575	3751	3926	4101	4277	176
8	4452	4627	4802	4977	5152	5326	5501	5676	5850	6025	175
9	6199	6374	6548	6722	6896	7071	7245	7419	7592	7766	174
250	7940	8114	8287	8461	8634	8808	8981	9154	9328	9501	173
1	9674	9847	0020	0192	0365	0538	0711	0883	1056	1228	173
2	40 1401	1573	1745	1917	2089	2261	2433	2605	2777	2949	172
3	3121	3292	3464	3635	3807	3978	4149	4320	4492	4663	171
4	4834	5005	5176	5346	5517	5688	5858	6029	6199	6370	171
255	6540	6710	6881	7051	7221	7391	7561	7731	7901	8070	170
6	8240	8410	8579	8749	8918	9087	9257	9426	9595	9764	169
7	9933	0102	0271	0440	0609	0777	0946	1114	1283	1451	169
8	41 1620	1788	1956	2124	2293	2461	2629	2796	2964	3132	168
9	3300	3467	3635	3803	3970	4137	4305	4472	4639	4806	167
260	4973	5140	5307	5474	5641	5808	5974	6141	6308	6474	167
1	6641	6807	6973	7139	7306	7472	7638	7804	7970	8135	166
2	8301	8467	8633	8798	8964	9129	9295	9460	9625	9791	165
3	9956	0121	0286	0451	0616	0781	0945	1110	1275	1439	165
4	42 1604	1768	1933	2097	2261	2426	2590	2754	2918	3082	164
265	3246	3410	3574	3737	3901	4065	4228	4392	4555	4718	164
6	4882	5045	5208	5371	5534	5697	5860	6023	6186	6349	163
7	6511	6674	6836	6999	7161	7324	7486	7648	7811	7973	162
8	8135	8297	8459	8621	8783	8944	9106	9268	9429	9591	162
9	9752	9914	0075	0236	0398	0559	0720	0881	1042	1203	161
	43										

PROPORTIONAL PARTS.

Diff.	1	2	3	4	5	6	7	8	9
178	17.8	35.6	53.4	71.2	89.0	106.8	124.6	142.4	160.2
177	17.7	35.4	53.1	70.8	88.5	106.2	123.9	141.6	159.3
176	17.6	35.2	52.8	70.4	88.0	105.6	123.2	140.8	158.4
175	17.5	35.0	52.5	70.0	87.5	105.0	122.5	140.0	157.5
174	17.4	34.8	52.2	69.6	87.0	104.4	121.8	139.2	156.6
173	17.3	34.6	51.9	69.2	86.5	103.8	121.1	138.4	155.7
172	17.2	34.4	51.6	68.8	86.0	103.2	120.4	137.6	154.8
171	17.1	34.2	51.3	68.4	85.5	102.6	119.7	136.8	153.9
170	17.0	34.0	51.0	68.0	85.0	102.0	119.0	136.0	153.0
169	16.9	33.8	50.7	67.6	84.5	101.4	118.3	135.2	152.1
168	16.8	33.6	50.4	67.2	84.0	100.8	117.6	134.4	151.2
167	16.7	33.4	50.1	66.8	83.5	100.2	116.9	133.6	150.3
166	16.6	33.2	49.8	66.4	83.0	99.6	116.2	132.8	149.4
165	16.5	33.0	49.5	66.0	82.5	99.0	115.5	132.0	148.5
164	16.4	32.8	49.2	65.6	82.0	98.4	114.8	131.2	147.6
163	16.3	32.6	48.9	65.2	81.5	97.8	114.1	130.4	146.7
162	16.2	32.4	48.5	64.8	81.0	97.2	113.4	129.6	145.8
161	16.1	32.2	48.3	64.4	80.5	96.6	112.7	128.8	144.9

No. 270
Log. 431

Table VII.—*Continued*

No. 299
Log. 476

N.	0	1	2	3	4	5	6	7	8	9	Diff.
270	43 1364	1525	1685	1846	2007	2167	2328	2488	2649	2809	161
1	2969	3130	3290	3450	3610	3770	3930	4090	4249	4409	160
2	4569	4729	4888	5048	5207	5367	5526	5685	5844	6004	159
3	6163	6322	6481	6640	6799	6957	7116	7275	7433	7592	159
4	7751	7909	8067	8226	8384	8542	8701	8859	9017	9175	158
275	9333	9491	9648	9806	9964	0122	0279	0437	0594	0752	158
6	44 0909	1066	1224	1381	1538	1695	1852	2009	2166	2323	157
7	2480	2637	2793	2950	3106	3263	3419	3576	3732	3889	157
8	4045	4201	4357	4513	4669	4825	4981	5137	5293	5449	156
9	5604	5760	5915	6071	6226	6382	6537	6692	6848	7003	155
280	7158	7313	7468	7623	7778	7933	8088	8242	8397	8552	155
1	8706	8861	9015	9170	9324	9478	9633	9787	9941	0095	154
2	45 0249	0403	0557	0711	0865	1018	1172	1326	1479	1633	154
3	1786	1940	2093	2247	2400	2553	2706	2859	3012	3165	153
4	3318	3471	3624	3777	3930	4082	4235	4387	4540	4692	153
285	4845	4997	5150	5302	5454	5606	5758	5910	6062	6214	152
6	6366	6518	6670	6821	6973	7125	7276	7428	7579	7731	152
7	7882	8033	8184	8336	8487	8638	8789	8940	9091	9242	151
8	9392	9543	9694	9845	9995	0146	0296	0447	0597	0748	151
9	46 0898	1048	1198	1348	1499	1649	1799	1948	2098	2248	150
290	2398	2548	2697	2847	2997	3146	3296	3445	3594	3744	150
1	3893	4042	4191	4340	4490	4639	4788	4936	5085	5234	149
2	5383	5532	5680	5829	5977	6126	6274	6423	6571	6719	149
3	6868	7016	7164	7312	7460	7608	7756	7904	8052	8200	148
4	8347	8495	8643	8790	8938	9085	9233	9380	9527	9675	148
295	9822	9969	0116	0263	0410	0557	0704	0851	0998	1145	147
6	47 1292	1438	1585	1732	1878	2025	2171	2318	2464	2610	146
7	2756	2903	3049	3195	3341	3487	3633	3779	3925	4071	146
8	4216	4362	4508	4653	4799	4944	5090	5235	5381	5526	146
9	5671	5816	5962	6107	6252	6397	6542	6687	6832	6976	145

PROPORTIONAL PARTS.

Diff.	1	2	3	4	5	6	7	8	9
161	16.1	32.2	48.3	64.4	80.5	96.6	112.7	128.8	144.9
160	16.0	32.0	48.0	64.0	80.0	96.0	112.0	128.0	144.0
159	15.9	31.8	47.7	63.6	79.5	95.4	111.3	127.2	143.1
158	15.8	31.6	47.4	63.2	79.0	94.8	110.6	126.4	142.2
157	15.7	31.4	47.1	62.8	78.5	94.2	109.9	125.6	141.3
156	15.6	31.2	46.8	62.4	78.0	93.6	109.2	124.8	140.4
155	15.5	31.0	46.5	62.0	77.5	93.0	108.5	124.0	139.5
154	15.4	30.8	46.2	61.6	77.0	92.4	107.8	123.2	138.6
153	15.3	30.6	45.9	61.2	76.5	91.8	107.1	122.4	137.7
152	15.2	30.4	45.6	60.8	76.0	91.2	106.4	121.6	136.8
151	15.1	30.2	45.3	60.4	75.5	90.6	105.7	120.8	135.9
150	15.0	30.0	45.0	60.0	75.0	90.0	105.0	120.0	135.0
149	14.9	29.8	44.7	59.6	74.5	89.4	104.3	119.2	134.1
148	14.8	29.6	44.4	59.2	74.0	88.8	103.6	118.4	133.2
147	14.7	29.4	44.1	58.8	73.5	88.2	102.9	117.6	132.3
146	14.6	29.2	43.8	58.4	73.0	87.6	102.2	116.8	131.4
145	14.5	29.0	43.5	58.0	72.5	87.0	101.5	116.0	130.5
144	14.4	28.8	43.2	57.6	72.0	86.4	100.8	115.2	129.6
143	14.3	28.6	42.9	57.2	71.5	85.8	100.1	114.4	128.7
142	14.2	28.4	42.6	56.8	71.0	85.2	99.4	113.6	127.8
141	14.1	28.2	42.3	56.4	70.5	84.6	98.7	112.8	126.9
140	14.0	28.0	42.0	56.0	70.0	84.0	98.0	112.0	126.0

TABLE VII.—*Continued*

N.	0	1	2	3	4	5	6	7	8	9	Diff.
300	47 7121	7266	7411	7555	7700	7844	7989	8133	8278	8422	145
1	8566	8711	8855	8999	9143	9287	9431	9575	9719	9863	144
2	48 0007	0151	0294	0438	0582	0725	0869	1012	1156	1299	144
3	1443	1586	1729	1872	2016	2159	2302	2445	2588	2731	143
4	2874	3016	3159	3302	3445	3587	3730	3872	4015	4157	143
305	4300	4442	4585	4727	4869	5011	5153	5295	5437	5579	142
6	5721	5863	6005	6147	6289	6430	6572	6714	6855	6997	142
7	7138	7280	7421	7563	7704	7845	7986	8127	8269	8410	141
8	8551	8692	8833	8974	9114	9255	9396	9537	9677	9818	141
9	9958	0099	0239	0380	0520	0661	0801	0941	1081	1222	140
310	49 1362	1502	1642	1782	1922	2062	2201	2341	2481	2621	140
1	2760	2900	3040	3179	3319	3458	3597	3737	3876	4015	139
2	4155	4294	4433	4572	4711	4850	4989	5128	5267	5406	139
3	5544	5683	5822	5960	6099	6238	6376	6515	6653	6791	139
4	6930	7068	7206	7344	7483	7621	7759	7897	8035	8173	138
315	8311	8448	8586	8724	8862	8999	9137	9275	9412	9550	138
6	9687	9824	9962	0099	0236	0374	0511	0648	0785	0922	137
7	50 1059	1196	1333	1470	1607	1744	1880	2017	2154	2291	137
8	2427	2564	2700	2837	2973	3109	3246	3382	3518	3655	136
9	3791	3927	4063	4199	4335	4471	4607	4743	4878	5014	136
320	5150	5286	5421	5557	5693	5828	5964	6099	6234	6370	136
1	6505	6640	6776	6911	7046	7181	7316	7451	7586	7721	135
2	7856	7991	8126	8260	8395	8530	8664	8799	8934	9068	135
3	9203	9337	9471	9606	9740	9874	0009	0143	0277	0411	134
4	51 0545	0679	0813	0947	1081	1215	1349	1482	1616	1750	134
325	1883	2017	2151	2284	2418	2551	2684	2818	2951	3084	133
6	3218	3351	3484	3617	3750	3883	4016	4149	4282	4415	133
7	4548	4681	4813	4946	5079	5211	5344	5476	5609	5741	133
8	5874	6006	6139	6271	6403	6535	6668	6800	6932	7064	132
9	7196	7328	7460	7592	7724	7855	7987	8119	8251	8382	132
330	8514	8646	8777	8909	9040	9171	9303	9434	9566	9697	131
1	9828	9959	0090	0221	0353	0484	0615	0745	0876	1007	131
2	52 1138	1269	1400	1530	1661	1792	1922	2053	2183	2314	131
3	2444	2575	2705	2835	2966	3096	3226	3356	3486	3616	130
4	3746	3876	4006	4136	4266	4396	4526	4656	4785	4915	130
335	5045	5174	5304	5434	5563	5693	5822	5951	6081	6210	129
6	6339	6469	6598	6727	6856	6985	7114	7243	7372	7501	129
7	7630	7759	7888	8016	8145	8274	8402	8531	8660	8788	129
8	8917	9045	9174	9302	9430	9559	9687	9815	9943	0072	128
9	53 0200	0328	0456	0584	0712	0840	0968	1096	1223	1351	128

PROPORTIONAL PARTS.

Diff.	1	2	3	4	5	6	7	8	9
139	13.9	27.8	41.7	55.6	69.5	83.4	97.3	111.2	125.1
138	13.8	27.6	41.4	55.2	69.0	82.8	96.6	110.4	124.2
137	13.7	27.4	41.1	54.8	68.5	82.2	95.9	109.6	123.3
136	13.6	27.2	40.8	54.4	68.0	81.6	95.2	108.8	122.4
135	13.5	27.0	40.5	54.0	67.5	81.0	94.5	108.0	121.5
134	13.4	26.8	40.2	53.6	67.0	80.4	93.8	107.2	120.6
133	13.3	26.6	39.9	53.2	66.5	79.8	93.1	106.4	119.7
132	13.2	26.4	39.6	52.8	66.0	79.2	92.4	105.6	118.8
131	13.1	26.2	39.3	52.4	65.5	78.6	91.7	104.8	117.9
130	13.0	26.0	39.0	52.0	65.0	78.0	91.0	104.0	117.0
129	12.9	25.8	38.7	51.6	64.5	77.4	90.3	103.2	116.1
128	12.8	25.6	38.4	51.2	64.0	76.8	89.6	102.4	115.2
127	12.7	25.4	38.1	50.8	63.5	76.2	88.9	101.6	114.3

No. 340
Log. 531

TABLE VII.—*Continued*

No. 379
Log. 579

N.	0	1	2	3	4	5	6	7	8	9	Diff.
340	53 1479	1607	1734	1862	1990	2117	2245	2372	2500	2627	128
1	2754	2882	3009	3136	3264	3391	3518	3645	3772	3899	127
2	4026	4153	4280	4407	4534	4661	4787	4914	5041	5167	127
3	5294	5421	5547	5674	5800	5927	6053	6180	6306	6432	126
4	6558	6685	6811	6937	7063	7189	7315	7441	7567	7693	126
345	7819	7945	8071	8197	8322	8448	8574	8699	8825	8951	126
6	9076	9202	9327	9452	9578	9703	9829	9954	0079	0204	125
7	54 0329	0455	0580	0705	0830	0955	1080	1205	1330	1454	125
8	1579	1704	1829	1953	2078	2203	2327	2452	2576	2701	125
9	2825	2950	3074	3199	3323	3447	3571	3696	3820	3944	124
350	4068	4192	4316	4440	4564	4688	4812	4936	5060	5183	124
1	5307	5431	5555	5678	5802	5925	6049	6172	6296	6419	124
2	6543	6666	6789	6913	7036	7159	7282	7405	7529	7652	123
3	7775	7898	8021	8144	8267	8389	8512	8635	8758	8881	123
4	9003	9126	9249	9371	9494	9616	9739	9861	9984	0106	123
355	55 0228	0351	0473	0595	0717	0840	0962	1084	1206	1328	122
6	1450	1572	1694	1816	1938	2060	2181	2303	2425	2547	122
7	2668	2790	2911	3033	3155	3276	3398	3519	3640	3762	121
8	3883	4004	4126	4247	4368	4489	4610	4731	4852	4973	121
9	5094	5215	5336	5457	5578	5699	5820	5940	6061	6182	121
360	6303	6423	6544	6664	6785	6905	7026	7146	7267	7387	120
1	7507	7627	7748	7868	7988	8108	8228	8349	8469	8589	120
2	8709	8829	8948	9068	9188	9308	9428	9548	9667	9787	120
3	9907	0026	0146	0265	0385	0504	0624	0743	0863	0982	119
4	56 1101	1221	1340	1459	1578	1698	1817	1936	2055	2174	119
365	2293	2412	2531	2650	2769	2887	3006	3125	3244	3362	119
6	3481	3600	3718	3837	3955	4074	4192	4311	4429	4548	119
7	4666	4784	4903	5021	5139	5257	5376	5494	5612	5730	118
8	5848	5966	6084	6202	6320	6437	6555	6673	6791	6909	118
9	7026	7144	7262	7379	7497	7614	7732	7849	7967	8084	118
370	8202	8319	8436	8554	8671	8788	8905	9023	9140	9257	117
1	9374	9491	9608	9725	9842	9959	0076	0193	0309	0426	117
2	57 0543	0660	0776	0893	1010	1126	1243	1359	1476	1592	117
3	1709	1825	1942	2058	2174	2291	2407	2523	2639	2755	116
4	2872	2988	3104	3220	3336	3452	3568	3684	3800	3915	116
375	4031	4147	4263	4379	4494	4610	4726	4841	4957	5072	116
6	5188	5303	5419	5534	5650	5765	5880	5996	6111	6226	115
7	6341	6457	6572	6687	6802	6917	7032	7147	7262	7377	115
8	7492	7607	7722	7836	7951	8066	8181	8295	8410	8525	115
9	8639	8754	8868	8983	9097	9212	9326	9441	9555	9669	114

PROPORTIONAL PARTS.

Diff.	1	2	3	4	5	6	7	8	9
128	12.8	25.6	38.4	51.2	64.0	76.8	89.6	102.4	115.2
127	12.7	25.4	38.1	50.8	63.5	76.2	88.9	101.6	114.3
126	12.6	25.2	37.8	50.4	63.0	75.6	88.2	100.8	113.4
125	12.5	25.0	37.5	50.0	62.5	75.0	87.5	100.0	112.5
124	12.4	24.8	37.2	49.6	62.0	74.4	86.8	99.2	111.6
123	12.3	24.6	36.9	49.2	61.5	73.8	86.1	98.4	110.7
122	12.2	24.4	36.6	48.8	61.0	73.2	85.4	97.6	109.8
121	12.1	24.2	36.3	48.4	60.5	72.6	84.7	96.8	108.9
120	12.0	24.0	36.0	48.0	60.0	72.0	84.0	96.0	108.0
119	11.9	23.8	35.7	47.6	59.5	71.4	83.3	95.2	107.1

No. 380
Log. 579

Table VII.—*Continued*

No. 414
Log. 617

N.	0	1	2	3	4	5	6	7	8	9	Diff.
380	57 9784	9898	0012	0126	0241	0355	0469	0583	0697	0811	114
1	58 0925	1039	1153	1267	1381	1495	1608	1722	1836	1950	
2	2063	2177	2291	2404	2518	2631	2745	2858	2972	3085	
3	3199	3312	3426	3539	3652	3765	3879	3992	4105	4218	
4	4331	4444	4557	4670	4783	4896	5009	5122	5235	5348	
385	5461	5574	5686	5799	5912	6024	6137	6250	6362	6475	113
6	6587	6700	6812	6925	7037	7149	7262	7374	7486	7599	
7	7711	7823	7935	8047	8160	8272	8384	8496	8608	8720	112
8	8832	8944	9056	9167	9279	9391	9503	9615	9726	9838	
9	9950	0061	0173	0284	0396	0507	0619	0730	0842	0953	
390	59 1065	1176	1287	1399	1510	1621	1732	1843	1955	2066	
1	2177	2288	2399	2510	2621	2732	2843	2954	3064	3175	111
2	3286	3397	3508	3618	3729	3840	3950	4061	4171	4282	
3	4393	4503	4614	4724	4834	4945	5055	5165	5276	5386	
4	5496	5606	5717	5827	5937	6047	6157	6267	6377	6487	
395	6597	6707	6817	6927	7037	7146	7256	7366	7476	7586	110
6	7695	7805	7914	8024	8134	8243	8353	8462	8572	8681	
7	8791	8900	9009	9119	9228	9337	9446	9556	9665	9774	
8	9883	9992	0101	0210	0319	0428	0537	0646	0755	0864	109
9	60 0973	1082	1191	1299	1408	1517	1625	1734	1843	1951	
400	2060	2169	2277	2386	2494	2603	2711	2819	2928	3036	
1	3144	3253	3361	3469	3577	3686	3794	3902	4010	4118	108
2	4226	4334	4442	4550	4658	4766	4874	4982	5089	5197	
3	5305	5413	5521	5628	5736	5844	5951	6059	6166	6274	
4	6381	6489	6596	6704	6811	6919	7026	7133	7241	7348	
405	7455	7562	7669	7777	7884	7991	8098	8205	8312	8419	107
6	8526	8633	8740	8847	8954	9061	9167	9274	9381	9488	
7	9594	9701	9808	9914	0021	0128	0234	0341	0447	0554	
8	61 0660	0767	0873	0979	1086	1192	1298	1405	1511	1617	
9	1723	1829	1936	2042	2148	2254	2360	2466	2572	2678	106
410	2784	2890	2996	3102	3207	3313	3419	3525	3630	3736	
1	3842	3947	4053	4159	4264	4370	4475	4581	4686	4792	
2	4897	5003	5108	5213	5319	5424	5529	5634	5740	5845	
3	5950	6055	6160	6265	6370	6476	6581	6686	6790	6895	105
4	7000	7105	7210	7315	7420	7525	7629	7734	7839	7943	

PROPORTIONAL PARTS.

Diff.	1	2	3	4	5	6	7	8	9
118	11.8	23.6	35.4	47.2	59.0	70.8	82.6	94.4	106.2
117	11.7	23.4	35.1	46.8	58.5	70.2	81.9	93.6	105.3
116	11.6	23.2	34.8	46.4	58.0	69.6	81.2	92.8	104.4
115	11.5	23.0	34.5	46.0	57.5	69.0	80.5	92.0	103.5
114	11.4	22.8	34.2	45.6	57.0	68.4	79.8	91.2	102.6
113	11.3	22.6	33.9	45.2	56.5	67.8	79.1	90.4	101.7
112	11.2	22.4	33.6	44.8	56.0	67.2	78.4	89.6	100.8
111	11.1	22.2	33.3	44.4	55.5	66.6	77.7	88.8	99.9
110	11.0	22.0	33.0	44.0	55.0	66.0	77.0	88.0	99.0
109	10.9	21.8	32.7	43.6	54.5	65.4	76.3	87.2	98.1
108	10.8	21.6	32.4	43.2	54.0	64.8	75.6	86.4	97.2
107	10.7	21.4	32.1	42.8	53.5	64.2	74.9	85.6	96.3
106	10.6	21.2	31.8	42.4	53.0	63.6	74.2	84.8	95.4
105	10.5	21.0	31.5	42.0	52.5	63.0	73.5	84.0	94.5
104	10.4	20.8	31.2	41.6	52.0	62.4	72.8	83.2	93.6

No. 415
Log. 618

Table VII.—*Continued*

No. 459
Log. 662

N.	0	1	2	3	4	5	6	7	8	9	Diff.
415	61 8048	8153	8257	8362	8466	8571	8676	8780	8884	8989	105
6	9093	9198	9302	9406	9511	9615	9719	9824	9928	0032	
7	62 0136	0240	0344	0448	0552	0656	0760	0864	0968	1072	104
8	1176	1280	1384	1488	1592	1695	1799	1903	2007	2110	
9	2214	2318	2421	2525	2628	2732	2835	2939	3042	3146	
420	3249	3353	3456	3559	3663	3766	3869	3973	4076	4179	
1	4282	4385	4488	4591	4695	4798	4901	5004	5107	5210	103
2	5312	5415	5518	5621	5724	5827	5929	6032	6135	6238	
3	6340	6443	6546	6648	6751	6853	6956	7058	7161	7263	
4	7366	7468	7571	7673	7775	7878	7980	8082	8185	8287	
425	8389	8491	8593	8695	8797	8900	9002	9104	9206	9308	102
6	9410	9512	9613	9715	9817	9919	0021	0123	0224	0326	
7	63 0428	0530	0631	0733	0835	0936	1038	1139	1241	1342	
8	1444	1545	1647	1748	1849	1951	2052	2153	2255	2356	
9	2457	2559	2660	2761	2862	2963	3064	3165	3266	3367	
430	3468	3569	3670	3771	3872	3973	4074	4175	4276	4376	101
1	4477	4578	4679	4779	4880	4981	5081	5182	5283	5383	
2	5484	5584	5685	5785	5886	5986	6087	6187	6287	6388	
3	6488	6588	6688	6789	6889	6989	7089	7189	7290	7390	
4	7490	7590	7690	7790	7890	7990	8090	8190	8290	8389	100
435	8489	8589	8689	8789	8888	8988	9088	9188	9287	9387	
6	9486	9586	9686	9785	9885	9984	0084	0183	0283	0382	
7	64 0481	0581	0680	0779	0879	0978	1077	1177	1276	1375	
8	1474	1573	1672	1771	1871	1970	2069	2168	2267	2366	
9	2465	2563	2662	2761	2860	2959	3058	3156	3255	3354	99
440	3453	3551	3650	3749	3847	3946	4044	4143	4242	4340	
1	4439	4537	4636	4734	4832	4931	5029	5127	5226	5324	
2	5422	5521	5619	5717	5815	5913	6011	6110	6208	6306	
3	6404	6502	6600	6698	6796	6894	6992	7089	7187	7285	98
4	7383	7481	7579	7676	7774	7872	7969	8067	8165	8262	
445	8360	8458	8555	8653	8750	8848	8945	9043	9140	9237	
6	9335	9432	9530	9627	9724	9821	9919	0016	0113	0210	
7	65 0308	0405	0502	0599	0696	0793	0890	0987	1084	1181	
8	1278	1375	1472	1569	1666	1762	1859	1956	2053	2150	97
9	2246	2343	2440	2536	2633	2730	2826	2923	3019	3116	
450	3213	3309	3405	3502	3598	3695	3791	3888	3984	4080	
1	4177	4273	4369	4465	4562	4658	4754	4850	4946	5042	
2	5138	5235	5331	5427	5523	5619	5715	5810	5906	6002	96
3	6098	6194	6290	6386	6482	6577	6673	6769	6864	6960	
4	7056	7152	7247	7343	7438	7534	7629	7725	7820	7916	
455	8011	8107	8202	8298	8393	8488	8584	8679	8774	8870	
6	8965	9060	9155	9250	9346	9441	9536	9631	9726	9821	
7	9916	0011	0106	0201	0296	0391	0486	0581	0676	0771	95
8	66 0865	0960	1055	1150	1245	1339	1434	1529	1623	1718	
9	1813	1907	2002	2096	2191	2286	2380	2475	2569	2663	

PROPORTIONAL PARTS.

Diff.	1	2	3	4	5	6	7	8	9
105	10.5	21.0	31.5	42.0	52.5	63.0	73.5	84.0	94.5
104	10.4	20.8	31.2	41.6	52.0	62.4	72.8	83.2	93.6
103	10.3	20.6	30.9	41.2	51.5	61.8	72.1	82.4	92.7
102	10.2	20.4	30.6	40.8	51.0	61.2	71.4	81.6	91.8
101	10.1	20.2	30.3	40.4	50.5	60.6	70.7	80.8	90.9
100	10.0	20.0	30.0	40.0	50.0	60.0	70.0	80.0	90.0
99	9.9	19.8	29.7	39.6	49.5	59.4	69.3	79.2	89.1

No. 460
Log. 662

TABLE VII.—*Continued*

No. 499
Log. 698

N.	0	1	2	3	4	5	6	7	8	9	Diff.
460	66 2758	2852	2947	3041	3135	3230	3324	3418	3512	3607	
1	3701	3795	3889	3983	4078	4172	4266	4360	4454	4548	
2	4642	4736	4830	4924	5018	5112	5206	5299	5393	5487	94
3	5581	5675	5769	5862	5956	6050	6143	6237	6331	6424	
4	6518	6612	6705	6799	6892	6986	7079	7173	7266	7360	
465	7453	7546	7640	7733	7826	7920	8013	8106	8199	8293	
6	8386	8479	8572	8665	8759	8852	8945	9038	9131	9224	
7	9317	9410	9503	9596	9689	9782	9875	9967	0060	0153	93
8	67 0246	0339	0431	0524	0617	0710	0802	0895	0988	1080	
9	1173	1265	1358	1451	1543	1636	1728	1821	1913	2005	
470	2098	2190	2283	2375	2467	2560	2652	2744	2836	2929	
1	3021	3113	3205	3297	3390	3482	3574	3666	3758	3850	
2	3942	4034	4126	4218	4310	4402	4494	4586	4677	4769	92
3	4861	4953	5045	5137	5228	5320	5412	5503	5595	5687	
4	5778	5870	5962	6053	6145	6236	6328	6419	6511	6602	
475	6694	6785	6876	6968	7059	7151	7242	7333	7424	7516	
6	7607	7698	7789	7881	7972	8063	8154	8245	8336	8427	
7	8518	8609	8700	8791	8882	8973	9064	9155	9246	9337	91
8	9428	9519	9610	9700	9791	9882	9973	0063	0154	0245	
9	68 0336	0426	0517	0607	0698	0789	0879	0970	1060	1151	
480	1241	1332	1422	1513	1603	1693	1784	1874	1964	2055	
1	2145	2235	2326	2416	2506	2596	2686	2777	2867	2957	
2	3047	3137	3227	3317	3407	3497	3587	3677	3767	3857	90
3	3947	4037	4127	4217	4307	4396	4486	4576	4666	4756	
4	4845	4935	5025	5114	5204	5294	5383	5473	5563	5652	
485	5742	5831	5921	6010	6100	6189	6279	6368	6458	6547	
6	6636	6726	6815	6904	6994	7083	7172	7261	7351	7440	
7	7529	7618	7707	7796	7886	7975	8064	8153	8242	8331	89
8	8420	8509	8598	8687	8776	8865	8953	9042	9131	9220	
9	9309	9398	9486	9575	9664	9753	9841	9930	0019	0107	
490	69 0196	0285	0373	0462	0550	0639	0728	0816	0905	0993	
1	1081	1170	1258	1347	1435	1524	1612	1700	1789	1877	
2	1965	2053	2142	2230	2318	2406	2494	2583	2671	2759	
3	2847	2935	3023	3111	3199	3287	3375	3463	3551	3639	88
4	3727	3815	3903	3991	4078	4166	4254	4342	4430	4517	
495	4605	4693	4781	4868	4956	5044	5131	5219	5307	5394	
6	5482	5569	5657	5744	5832	5919	6007	6094	6182	6269	
7	6356	6444	6531	6618	6706	6793	6880	6968	7055	7142	
8	7229	7317	7404	7491	7578	7665	7752	7839	7926	8014	
9	8100	8188	8275	8362	8449	8535	8622	8709	8796	8883	87

PROPORTIONAL PARTS.

Diff.	1	2	3	4	5	6	7	8	9
98	9.8	19.6	29.4	39.2	49.0	58.8	68.6	78.4	88.2
97	9.7	19.4	29.1	38.8	48.5	58.2	67.9	77.6	87.3
96	9.6	19.2	28.8	38.4	48.0	57.6	67.2	76.8	86.4
95	9.5	19.0	28.5	38.0	47.5	57.0	66.5	76.0	85.5
94	9.4	18.8	28.2	37.6	47.0	56.4	65.8	75.2	84.6
93	9.3	18.6	27.9	37.2	46.5	55.8	65.1	74.4	83.7
92	9.2	18.4	27.6	36.8	46.0	55.2	64.4	73.6	82.8
91	9.1	18.2	27.3	36.4	45.5	54.6	63.7	72.8	81.9
90	9.0	18.0	27.0	36.0	45.0	54.0	63.0	72.0	81.0
89	8.9	17.8	26.7	35.6	44.5	53.4	62.3	71.2	80.1
88	8.8	17.6	26.4	35.2	44.0	52.8	61.6	70.4	79.2
87	8.7	17.4	26.1	34.8	43.5	52.2	60.9	69.6	78.3
86	8.6	17.2	25.8	34.4	43.0	51.6	60.2	68.8	77.4

No. 500 Log. 698 — TABLE VII.—*Continued* — No. 544 Log. 736

N.	0	1	2	3	4	5	6	7	8	9	Diff.
500	69 8970	9057	9144	9231	9317	9404	9491	9578	9664	9751	
1	9838	9924	0011	0098	0184	0271	0358	0444	0531	0617	
2	70 0704	0790	0877	0963	1050	1136	1222	1309	1395	1482	
3	1568	1654	1741	1827	1913	1999	2086	2172	2258	2344	
4	2431	2517	2603	2689	2775	2861	2947	3033	3119	3205	
505	3291	3377	8463	3549	3635	3721	3807	3893	3979	4065	86
6	4151	4236	4322	4408	4494	4579	4665	4751	4837	4922	
7	5008	5094	5179	5205	5350	5430	5522	5607	5693	5778	
8	5864	5949	6035	6120	6206	6291	6376	6462	6547	6632	
9	6718	6803	6888	6974	7059	7144	7229	7315	7400	7485	
510	7570	7655	7740	7826	7911	7996	8081	8166	8251	8336	85
1	8421	8506	8591	8676	8761	8846	8931	9015	9100	9185	
2	9270	9355	9440	9524	9609	9694	9779	9863	9948	0033	
3	71 0117	0202	0287	0371	0456	0540	0625	0710	0794	0879	
4	0963	1048	1132	1217	1301	1385	1470	1554	1639	1723	
515	1807	1892	1976	2060	2144	2229	2313	2397	2481	2566	
6	2650	2734	2818	2902	2986	3070	3154	3238	3323	3407	84
7	3491	3575	3659	8742	3826	3910	3994	4078	4162	4246	
8	4330	4414	4497	4581	4665	4749	4833	4916	5000	5084	
9	5167	5251	5335	5418	5502	5586	5669	5753	5836	5920	
520	6003	6087	6170	6254	6337	6421	6504	6588	6671	6754	
1	6838	6921	7004	7088	7171	7254	7338	7421	7504	7587	
2	7671	7754	7837	7920	8003	8086	8169	8253	8336	8419	
3	8502	8585	8668	8751	8834	8917	9000	9083	9165	9248	83
4	9331	9414	9497	9580	9663	9745	9828	9911	9994	0077	
525	72 0159	0242	0325	0407	0490	0573	0655	0738	0821	0903	
6	0986	1068	1151	1233	1316	1398	1481	1563	1646	1728	
7	1811	1893	1975	2058	2140	2222	2305	2387	2469	2552	
8	2634	2716	2798	2881	2963	3045	3127	3209	3291	3374	82
9	3456	3538	3620	3702	3784	3866	3948	4030	4112	4194	
530	4276	4358	4440	4522	4604	4685	4767	4849	4931	5013	
1	5095	5176	5258	5340	5422	5503	5585	5667	5748	5830	
2	5912	5993	6075	6156	6238	6320	6401	6483	6564	6646	
3	6727	6809	6890	6972	7053	7134	7216	7297	7379	7460	
4	7541	7623	7704	7785	7866	7948	8029	8110	8191	8273	
535	8354	8435	8516	8597	8678	8759	8841	8922	9003	9084	81
6	9165	9246	9327	9408	9489	9570	9651	9732	9813	9893	
7	9974	0055	0136	0217	0298	0378	0459	0540	0621	0702	
8	73 0782	0863	0944	1024	1105	1186	1266	1347	1428	1508	
9	1589	1669	1750	1830	1911	1991	2072	2152	2233	2313	
540	2394	2474	2555	2635	2715	2796	2876	2956	3037	3117	
1	3197	3278	3358	3438	3518	3598	3679	3759	3839	3919	
2	3999	4079	4160	4240	4320	4400	4480	4560	4640	4720	80
3	4800	4880	4960	5040	5120	5200	5279	5359	5439	5519	
4	5599	5679	5759	5838	5918	5998	6078	6157	6237	6317	

PROPORTIONAL PARTS.

Diff.	1	2	3	4	5	6	7	8	9
87	8.7	17.4	26.1	34.8	43.5	52.2	60.9	69.6	78.3
86	8.6	17.2	25.8	34.4	43.0	51.6	60.2	68.8	77.4
85	8.5	17.0	25.5	34.0	42.5	51.0	59.5	68.0	76.5
84	8.4	16.8	25.2	33.6	42.0	50.4	58.8	67.2	75.6

TABLE VII.—*Continued*

N.	0	1	2	3	4	5	6	7	8	9	Diff.
545	73 6397	6476	6556	6635	6715	6795	6874	6954	7034	7113	
6	7193	7272	7352	7431	7511	7590	7670	7749	7829	7908	
7	7987	8067	8146	8225	8305	8384	8463	8543	8622	8701	
8	8781	8860	8939	9018	9097	9177	9256	9335	9414	9493	
9	9572	9651	9731	9810	9889	9968	0047	0126	0205	0284	79
550	74 0363	0442	0521	0600	0678	0757	0836	0915	0994	1073	
1	1152	1230	1309	1388	1467	1546	1624	1703	1782	1860	
2	1939	2018	2096	2175	2254	2332	2411	2489	2568	2647	
3	2725	2804	2882	2961	3039	3118	3196	3275	3353	3431	
4	3510	3588	3667	3745	3823	3902	3980	4058	4136	4215	
555	4293	4371	4449	4528	4606	4684	4762	4840	4919	4997	
6	5075	5153	5231	5309	5387	5465	5543	5621	5699	5777	78
7	5855	5933	6011	6089	6167	6245	6323	6401	6479	6556	
8	6634	6712	6790	6868	6945	7023	7101	7179	7256	7334	
9	7412	7489	7567	7645	7722	7800	7878	7955	8033	8110	
560	8188	8266	8343	8421	8498	8576	8653	8731	8808	8885	
1	8963	9040	9118	9195	9272	9350	9427	9504	9582	9659	
2	9736	9814	9891	9968	0045	0123	0200	0277	0354	0431	
3	75 0508	0586	0663	0740	0817	0894	0971	1048	1125	1202	
4	1279	1356	1433	1510	1587	1664	1741	1818	1895	1972	
565	2048	2125	2202	2279	2356	2433	2509	2586	2663	2740	77
6	2816	2893	2970	3047	3123	3200	3277	3353	3430	3506	
7	3583	3660	3736	3813	3889	3966	4042	4119	4195	4272	
8	4348	4425	4501	4578	4654	4730	4807	4883	4960	5036	
9	5112	5189	5265	5341	5417	5494	5570	5646	5722	5799	
570	5875	5951	6027	6103	6180	6256	6332	6408	6484	6560	
1	6636	6712	6788	6864	6940	7016	7092	7168	7244	7320	76
2	7396	7472	7548	7624	7700	7775	7851	7927	8003	8079	
3	8155	8230	8306	8382	8458	8533	8609	8685	8761	8836	
4	8912	8988	9063	9139	9214	9290	9366	9441	9517	9592	
575	9668	9743	9819	9894	9970	0045	0121	0196	0272	0347	
6	76 0422	0498	0573	0649	0724	0799	0875	0950	1025	1101	
7	1176	1251	1326	1402	1477	1552	1627	1702	1778	1853	
8	1928	2003	2078	2153	2228	2303	2378	2453	2529	2604	
9	2679	2754	2829	2904	2978	3053	3128	3203	3278	3353	75
580	3428	3503	3578	3653	3727	3802	3877	3952	4027	4101	
1	4176	4251	4326	4400	4475	4550	4624	4699	4774	4848	
2	4923	4998	5072	5147	5221	5296	5370	5445	5520	5594	
3	5669	5743	5818	5892	5966	6041	6115	6190	6264	6338	
4	6413	6487	6562	6636	6710	6785	6859	6933	7007	7082	

PROPORTIONAL PARTS.

Diff.	1	2	3	4	5	6	7	8	9
83	8.3	16.6	24.9	33.2	41.5	49.8	58.1	66.4	74.7
82	8.2	16.4	24.6	32.8	41.0	49.2	57.4	65.6	73.8
81	8.1	16.2	24.3	32.4	40.5	48.6	56.7	64.8	72.9
80	8.0	16.0	24.0	32.0	40.0	48.0	56.0	64.0	72.0
79	7.9	15.8	23.7	31.6	39.5	47.4	55.3	63.2	71.1
78	7.8	15.6	23.4	31.2	39.0	46.8	54.6	62.4	70.2
77	7.7	15.4	23.1	30.8	38.5	46.2	53.9	61.6	69.3
76	7.6	15.2	22.8	30.4	38.0	45.6	53.2	60.8	68.4
75	7.5	15.0	22.5	30.0	37.5	45.0	52.5	60.0	67.5
74	7.4	14.8	22.2	29.6	37.0	44.4	51.8	59.2	66.6

TABLE VII.—*Continued*

N.	0	1	2	3	4	5	6	7	8	9	Diff.
585	76 7156	7230	7304	7379	7453	7527	7601	7675	7749	7823	74
6	7898	7972	8046	8120	8194	8268	8342	8416	8490	8564	
7	8638	8712	8786	8860	8934	9008	9082	9156	9230	9303	
8	9377	9451	9525	9599	9673	9746	9820	9894	9968	0042	
9	77 0115	0189	0263	0336	0410	0484	0557	0631	0705	0778	
590	0852	0926	0999	1073	1146	1220	1293	1367	1440	1514	
1	1587	1661	1734	1808	1881	1955	2028	2102	2175	2248	
2	2322	2395	2468	2542	2615	2688	2762	2835	2908	2981	
3	3055	3128	3201	3274	3348	3421	3494	3567	3640	3713	
4	3786	3860	3933	4006	4079	4152	4225	4298	4371	4444	73
595	4517	4590	4663	4736	4809	4882	4955	5028	5100	5173	
6	5246	5319	5392	5465	5538	5610	5683	5756	5829	5902	
7	5974	6047	6120	6193	6265	6338	6411	6483	6556	6629	
8	6701	6774	6846	6919	6992	7064	7137	7209	7282	7354	
9	7427	7499	7572	7644	7717	7789	7862	7934	8006	8079	
600	8151	8224	8296	8368	8441	8513	8585	8658	8730	8802	
1	8874	8947	9019	9091	9163	9236	9308	9380	9452	9524	
2	9596	9669	9741	9813	9885	9957	0029	0101	0173	0245	72
3	78 0317	0389	0461	0533	0605	0677	0749	0821	0893	0965	
4	1037	1109	1181	1253	1324	1396	1468	1540	1612	1684	
605	1755	1827	1899	1971	2042	2114	2186	2258	2329	2401	
6	2473	2544	2616	2688	2759	2831	2902	2974	3046	3117	
7	3189	3260	3332	3403	3475	3546	3618	3689	3761	3832	
8	3904	3975	4046	4118	4189	4261	4332	4403	4475	4546	
9	4617	4689	4760	4831	4902	4974	5045	5116	5187	5259	
610	5330	5401	5472	5543	5615	5686	5757	5828	5899	5970	71
1	6041	6112	6183	6254	6325	6396	6467	6538	6609	6680	
2	6751	6822	6893	6964	7035	7106	7177	7248	7319	7390	
3	7460	7531	7602	7673	7744	7815	7885	7956	8027	8098	
4	8168	8239	8310	8381	8451	8522	8593	8663	8734	8804	
615	8875	8946	9016	9087	9157	9228	9299	9369	9440	9510	
6	9581	9651	9722	9792	9863	9933	0004	0074	0144	0215	
7	79 0285	0356	0426	0496	0567	0637	0707	0778	0848	0918	
8	0988	1059	1129	1199	1269	1340	1410	1480	1550	1620	
9	1691	1761	1831	1901	1971	2041	2111	2181	2252	2322	
620	2392	2462	2532	2602	2672	2742	2812	2882	2952	3022	70
1	3092	3162	3231	3301	3371	3441	3511	3581	3651	3721	
2	3790	3860	3930	4000	4070	4139	4209	4279	4349	4418	
3	4488	4558	4627	4697	4767	4836	4906	4976	5045	5115	
4	5185	5254	5324	5393	5463	5532	5602	5672	5741	5811	
625	5880	5949	6019	6088	6158	6227	6297	6366	6436	6505	
6	6574	6644	6713	6782	6852	6921	6990	7060	7129	7198	
7	7268	7337	7406	7475	7545	7614	7683	7752	7821	7890	
8	7960	8029	8098	8167	8236	8305	8374	8443	8513	8582	
9	8651	8720	8789	8858	8927	8996	9065	9134	9203	9272	69

PROPORTIONAL PARTS.

Diff.	1	2	3	4	5	6	7	8	9
75	7.5	15.0	22.5	30.0	37.5	45.0	52.5	60.0	67.5
74	7.4	14.8	22.2	29.6	37.0	44.4	51.8	59.2	66.6
73	7.3	14.6	21.9	29.2	36.5	43.8	51.1	58.4	65.7
72	7.2	14.4	21.6	28.8	36.0	43.2	50.4	57.6	64.8
71	7.1	14.2	21.3	28.4	35.5	42.6	49.7	56.8	63.9
70	7.0	14.0	21.0	28.0	35.0	42.0	49.0	56.0	63.0
69	6.9	13.8	20.7	27.6	34.5	41.4	48.3	55.2	62.1

No. 630
Log. 799

TABLE VII.—*Continued*

No. 674
Log. 829

N.	0	1	2	3	4	5	6	7	8	9	Diff.
630	79 9341	9409	9478	9547	9616	9685	9754	9823	9892	9961	
1	80 0029	0098	0167	0236	0305	0373	0442	0511	0580	0648	
2	0717	0786	0854	0923	0992	1061	1129	1198	1266	1335	
3	1404	1472	1541	1609	1678	1747	1815	1884	1952	2021	
4	2089	2158	2226	2295	2363	2432	2500	2568	2637	2705	
635	2774	2842	2910	2979	3047	3116	3184	3252	3321	3389	
6	3457	3525	3594	3662	3730	3798	3867	3935	4003	4071	
7	4139	4208	4276	4344	4412	4480	4548	4616	4685	4753	
8	4821	4889	4957	5025	5093	5161	5229	5297	5365	5433	
9	5501	5569	5637	5705	5773	5841	5908	5976	6044	6112	**68**
640	80 6180	6248	6316	6384	6451	6519	6587	6655	6723	6790	
1	6858	6926	6994	7061	7129	7197	7264	7332	7400	7467	
2	7535	7603	7670	7738	7806	7873	7941	8008	8076	8143	
3	8211	8279	8346	8414	8481	8549	8616	8684	8751	8818	
4	8886	8953	9021	9088	9156	9223	9290	9358	9425	9492	
645	9560	9627	9694	9762	9829	9896	9964	0031	0098	0165	
6	81 0233	0300	0367	0434	0501	0569	0636	0703	0770	0837	
7	0904	0971	1039	1106	1173	1240	1307	1374	1441	1508	**67**
8	1575	1642	1709	1776	1843	1910	1977	2044	2111	2178	
9	2245	2312	2379	2445	2512	2579	2646	2713	2780	2847	
650	2913	2980	3047	3114	3181	3247	3314	3381	3448	3514	
1	3581	3648	3714	3781	3848	3914	3981	4048	4114	4181	
2	4248	4314	4381	4447	4514	4581	4647	4714	4780	4847	
3	4913	4980	5046	5113	5179	5246	5312	5378	5445	5511	
4	5578	5644	5711	5777	5843	5910	5976	6042	6109	6175	
655	6241	6308	6374	6440	6506	6573	6639	6705	6771	6838	
6	6904	6970	7036	7102	7169	7235	7301	7367	7433	7499	
7	7565	7631	7698	7764	7830	7896	7962	8028	8094	8160	
8	8226	8292	8358	8424	8490	8556	8622	8688	8754	8820	
9	8885	8951	9017	9083	9149	9215	9281	9346	9412	9478	**66**
660	9544	9610	9676	9741	9807	9873	9939	0004	0070	0136	
1	82 0201	0267	0333	0399	0464	0530	0595	0661	0727	0792	
2	0858	0924	0989	1055	1120	1186	1251	1317	1382	1448	
3	1514	1579	1645	1710	1775	1841	1906	1972	2037	2103	
4	2168	2233	2299	2364	2430	2495	2560	2626	2691	2756	
665	2822	2887	2952	3018	3083	3148	3213	3279	3344	3409	
6	3474	3539	3605	3670	3735	3800	3865	3930	3996	4061	
7	4126	4191	4256	4321	4386	4451	4516	4581	4646	4711	
8	4776	4841	4906	4971	5036	5101	5166	5231	5296	5361	**65**
9	5426	5491	5556	5621	5686	5751	5815	5880	5945	6010	
670	6075	6140	6204	6269	6334	6399	6464	6528	6593	6658	
1	6723	6787	6852	6917	6981	7046	7111	7175	7240	7305	
2	7369	7434	7499	7563	7628	7692	7757	7821	7886	7951	
3	8015	8080	8144	8209	8273	8338	8402	8467	8531	8595	
4	8660	8724	8789	8853	8918	8982	9046	9111	9175	9239	

PROPORTIONAL PARTS.

Diff.	1	2	3	4	5	6	7	8	9
68	6.8	13.6	20.4	27.2	34.0	40.8	47.6	54.4	61.2
67	6.7	13.4	20.1	26.8	33.5	40.2	46.9	53.6	60.3
66	6.6	13.2	19.8	26.4	33.0	39.6	46.2	52.8	59.4
65	6.5	13.0	19.5	26.0	32.5	39.0	45.5	52.0	58.5
64	6.4	12.8	19.2	25.6	32.0	38.4	44.8	51.2	57.6

No. 675 — Log. 829 | TABLE VII.—*Continued* | No. 719 — Log. 857

N.	0	1	2	3	4	5	6	7	8	9	Diff.
675	82 9304	9368	9432	9497	9561	9625	9690	9754	9818	9882	
6	9947	0011	0075	0139	0204	0268	0332	0396	0460	0525	
7	83 0589	0653	0717	0781	0845	0909	0973	1037	1102	1166	
8	1230	1294	1358	1422	1486	1550	1614	1678	1742	1806	64
9	1870	1934	1998	2062	2126	2189	2253	2317	2381	2445	
680	2509	2573	2637	2700	2764	2828	2892	2956	3020	3083	
1	3147	3211	3275	3338	3402	3466	3530	3593	3657	3721	
2	3784	3848	3912	3975	4039	4103	4166	4230	4294	4357	
3	4421	4484	4548	4611	4675	4739	4802	4866	4929	4993	
4	5056	5120	5183	5247	5310	5373	5437	5500	5564	5627	
685	5691	5754	5817	5881	5944	6007	6071	6134	6197	6261	
6	6324	6387	6451	6514	6577	6641	6704	6767	6830	6894	
7	6957	7020	7083	7146	7210	7273	7336	7399	7462	7525	
8	7588	7652	7715	7778	7841	7904	7967	8030	8093	8156	
9	8219	8282	8345	8408	8471	8534	8597	8660	8723	8786	63
690	8849	8912	8975	9038	9101	9164	9227	9289	9352	9415	
1	9478	9541	9604	9667	9729	9792	9855	9918	9981	0043	
2	84 0106	0169	0232	0294	0357	0420	0482	0545	0608	0671	
3	0733	0796	0859	0921	0984	1046	1109	1172	1234	1297	
4	1359	1422	1485	1547	1610	1672	1735	1797	1860	1922	
695	1985	2047	2110	2172	2235	2297	2360	2422	2484	2547	
6	2609	2672	2734	2796	2859	2921	2983	3046	3108	3170	
7	3233	3295	3357	3420	3482	3544	3606	3669	3731	3793	
8	3855	3918	3980	4042	4104	4166	4229	4291	4353	4415	
9	4477	4539	4601	4664	4726	4788	4850	4912	4974	5036	
700	5098	5160	5222	5284	5346	5408	5470	5532	5594	5656	62
1	5718	5780	5842	5904	5966	6028	6090	6151	6213	6275	
2	6337	6399	6461	6523	6585	6646	6708	6770	6832	6894	
3	6955	7017	7079	7141	7202	7264	7326	7388	7449	7511	
4	7573	7634	7696	7758	7819	7881	7943	8004	8066	8128	
705	8189	8251	8312	8374	8435	8497	8559	8620	8682	8743	
6	8805	8866	8928	8989	9051	9112	9174	9235	9297	9358	
7	9419	9481	9542	9604	9665	9726	9788	9849	9911	9972	
8	85 0033	0095	0156	0217	0279	0340	0401	0462	0524	0585	
9	0646	0707	0769	0830	0891	0952	1014	1075	1136	1197	
710	1258	1320	1381	1442	1503	1564	1625	1686	1747	1809	
1	1870	1931	1992	2053	2114	2175	2236	2297	2358	2419	
2	2480	2541	2602	2663	2724	2785	2846	2907	2968	3029	61
3	0090	0150	3211	3272	3333	3304	3455	3516	3577	3637	
4	3698	3759	3820	3881	3941	4002	4063	4124	4185	4245	
715	4306	4367	4428	4488	4549	4610	4670	4731	4792	4852	
6	4913	4974	5034	5095	5156	5216	5277	5337	5398	5459	
7	5519	5580	5640	5701	5761	5822	5882	5943	6003	6064	
8	6124	6185	6245	6306	6366	6427	6487	6548	6608	6668	
9	6729	6789	6850	6910	6970	7031	7091	7152	7212	7272	

PROPORTIONAL PARTS.

Diff.	1	2	3	4	5	6	7	8	9
65	6.5	13.0	19.5	26.0	32.5	39.0	45.5	52.0	58.5
64	6.4	12.8	19.2	25.6	32.0	38.4	44.8	51.2	57.6
63	6.3	12.6	18.9	25.2	31.5	37.8	44.1	50.4	56.7
62	6.2	12.4	18.6	24.8	31.0	37.2	43.4	49.6	55.8
61	6.1	12.2	18.3	24.4	30.5	36.6	42.7	48.8	54.9
60	6.0	12.0	18.0	24.0	30.0	36.0	42.0	48.0	54.0

No. 720
Log. 857

TABLE VII.—*Continued*

No. 764
Log. 883

N.	0	1	2	3	4	5	6	7	8	9	Diff.
720	85 7332	7393	7453	7513	7574	7634	7694	7755	7815	7875	
1	7935	7995	8056	8116	8176	8236	8297	8357	8417	8477	
2	8537	8597	8657	8718	8778	8838	8898	8958	9018	9078	
3	9138	9198	9258	9318	9379	9439	9499	9559	9619	9679	60
4	9739	9799	9859	9918	9978	0038	0098	0158	0218	0278	
725	86 0338	0398	0458	0518	0578	0637	0697	0757	0817	0877	
6	0937	0996	1056	1116	1176	1236	1295	1355	1415	1475	
7	1534	1594	1654	1714	1773	1833	1893	1952	2012	2072	
8	2131	2191	2251	2310	2370	2430	2489	2549	2608	2668	
9	2728	2787	2847	2906	2966	3025	3085	3144	3204	3263	
730	3323	3382	3442	3501	3561	3620	3680	3739	3799	3858	
1	3917	3977	4036	4096	4155	4214	4274	4333	4392	4452	
2	4511	4570	4630	4689	4748	4808	4867	4926	4985	5045	
3	5104	5163	5222	5282	5341	5400	5459	5519	5578	5637	
4	5696	5755	5814	5874	5933	5992	6051	6110	6169	6228	
735	6287	6346	6405	6465	6524	6583	6642	6701	6760	6819	
6	6878	6937	6996	7055	7114	7173	7232	7291	7350	7409	59
7	7467	7526	7585	7644	7703	7762	7821	7880	7939	7998	
8	8056	8115	8174	8233	8292	8350	8409	8468	8527	8586	
9	8644	8703	8762	8821	8879	8938	8997	9056	9114	9173	
740	9232	9290	9349	9408	9466	9525	9584	9642	9701	9760	
1	9818	9877	9935	9994	0053	0111	0170	0228	0287	0345	
2	87 0404	0462	0521	0579	0638	0696	0755	0813	0872	0930	
3	0989	1047	1106	1164	1223	1281	1339	1398	1456	1515	
4	1573	1631	1690	1748	1806	1865	1923	1981	2040	2098	
745	2156	2215	2273	2331	2389	2448	2506	2564	2622	2681	
6	2739	2797	2855	2913	2972	3030	3088	3146	3204	3262	
7	3321	3379	3437	3495	3553	3611	3669	3727	3785	3844	
8	3902	3960	4018	4076	4134	4192	4250	4308	4366	4424	
9	4482	4540	4598	4656	4714	4772	4830	4888	4945	5003	58
750	5061	5119	5177	5235	5293	5351	5409	5466	5524	5582	
1	5640	5698	5756	5813	5871	5929	5987	6045	6102	6160	
2	6218	6276	6333	6391	6449	6507	6564	6622	6680	6737	
3	6795	6853	6910	6968	7026	7083	7141	7199	7256	7314	
4	7371	7429	7487	7544	7602	7659	7717	7774	7832	7889	
755	7947	8004	8062	8119	8177	8234	8292	8349	8407	8464	
6	8522	8579	8637	8694	8752	8809	8866	8924	8981	9039	
7	9096	9153	9211	9268	9325	9383	9440	9497	9555	9612	
8	9669	9726	9784	9841	9898	9956	0013	0070	0127	0185	
9	88 0242	0299	0356	0413	0471	0528	0585	0642	0699	0756	
760	0814	0871	0928	0985	1042	1099	1156	1213	1271	1328	
1	1385	1442	1499	1556	1613	1670	1727	1784	1841	1898	
2	1955	2012	2069	2126	2183	2240	2297	2354	2411	2468	57
3	2525	2581	2638	2695	2752	2809	2866	2923	2980	3037	
4	3093	3150	3207	3264	3321	3377	3434	3491	3548	3605	

PROPORTIONAL PARTS.

Diff.	1	2	3	4	5	6	7	8	9
59	5.9	11.8	17.7	23.6	29.5	35.4	41.3	47.2	53.1
58	5.8	11.6	17.4	23.2	29.0	34.8	40.6	46.4	52.2
57	5.7	11.4	17.1	22.8	28.5	34.2	39.9	45.6	51.3
56	5.6	11.2	16.8	22.4	28.0	33.6	39.2	44.8	50.4

No. 765 Log. 883 Table VII.—*Continued* No. 809 Log. 908

N.	0	1	2	3	4	5	6	7	8	9	Diff.
765	88 3661	3718	3775	3832	3888	3945	4002	4059	4115	4172	
6	4229	4285	4342	4399	4455	4512	4569	4625	4682	4739	
7	4795	4852	4909	4965	5022	5078	5135	5192	5248	5305	
8	5361	5418	5474	5531	5587	5644	5700	5757	5813	5870	
9	5926	5983	6039	6096	6152	6209	6265	6321	6378	6434	
770	6491	6547	6604	6660	6716	6773	6829	6885	6942	6998	
1	7054	7111	7167	7223	7280	7336	7392	7449	7505	7561	
2	7617	7674	7730	7786	7842	7898	7955	8011	8067	8123	
3	8179	8236	8292	8348	8404	8460	8516	8573	8629	8685	
4	8741	8797	8853	8909	8965	9021	9077	9134	9190	9246	
775	9302	9358	9414	9470	9526	9582	9638	9694	9750	9806	56
6	9862	9918	9974	0030	0086	0141	0197	0253	0309	0365	
7	89 0421	0477	0533	0589	0645	0700	0756	0812	0868	0924	
8	0980	1035	1091	1147	1203	1259	1314	1370	1426	1482	
9	1537	1593	1649	1705	1760	1816	1872	1928	1983	2039	
780	2095	2150	2206	2262	2317	2373	2429	2484	2540	2595	
1	2651	2707	2762	2818	2873	2929	2985	3040	3096	3151	
2	3207	3262	3318	3373	3429	3484	3540	3595	3651	3706	
3	3762	3817	3873	3928	3984	4039	4094	4150	4205	4261	
4	4316	4371	4427	4482	4538	4593	4648	4704	4759	4814	
785	4870	4925	4980	5036	5091	5146	5201	5257	5312	5367	
6	5423	5478	5533	5588	5644	5699	5754	5809	5864	5920	
7	5975	6030	6085	6140	6195	6251	6306	6361	6416	6471	
8	6526	6581	6636	6692	6747	6802	6857	6912	6967	7022	
9	7077	7132	7187	7242	7297	7352	7407	7462	7517	7572	
790	7627	7682	7737	7792	7847	7902	7957	8012	8067	8122	55
1	8176	8231	8286	8341	8396	8451	8506	8561	8615	8670	
2	8725	8780	8835	8890	8944	8999	9054	9109	9164	9218	
3	9273	9328	9383	9437	9492	9547	9602	9656	9711	9766	
4	9821	9875	9930	9985	0039	0094	0149	0203	0258	0312	
795	90 0367	0422	0476	0531	0586	0640	0695	0749	0804	0859	
6	0913	0968	1022	1077	1131	1186	1240	1295	1349	1404	
7	1458	1513	1567	1622	1676	1731	1785	1840	1894	1948	
8	2003	2057	2112	2166	2221	2275	2329	2384	2438	2492	
9	2547	2601	2655	2710	2764	2818	2873	2927	2981	3036	
800	3090	3144	3199	3253	3307	3361	3416	3470	3524	3578	
1	3633	3687	3741	3795	3849	3904	3958	4012	4066	4120	
2	4174	4229	4283	4337	4391	4445	4499	4553	4607	4661	
3	4716	4770	4824	4878	4932	4986	5040	5094	5148	5202	
4	5256	5310	5364	5418	5472	5526	5580	5634	5688	5742	54
805	5796	5850	5904	5958	6012	6066	6119	6173	6227	6281	
6	6335	6389	6443	6497	6551	6604	6658	6712	6766	6820	
7	6874	6927	6981	7035	7089	7143	7196	7250	7304	7358	
8	7411	7465	7519	7573	7626	7680	7734	7787	7841	7895	
9	7949	8002	8056	8110	8163	8217	8270	8324	8378	8431	

PROPORTIONAL PARTS.

Diff.	1	2	3	4	5	6	7	8	9
57	5.7	11.4	17.1	22.8	28.5	34.2	39.9	45.6	51.3
56	5.6	11.2	16.8	22.4	28.0	33.6	39.2	44.8	50.4
55	5.5	11.0	16.5	22.0	27.5	33.0	38.5	44.0	49.5
54	5.4	10.8	16.2	21.6	27.0	32.4	37.8	43.2	48.6

No. 810
Log. 908

TABLE VII.—*Continued*

No. 854
Log. 931

N.	0	1	2	3	4	5	6	7	8	9	Diff.
810	90 8485	8539	8592	8646	8699	8753	8807	8860	8914	8967	
1	9021	9074	9128	9181	9235	9289	9342	9396	9449	9503	
2	9556	9610	9663	9716	9770	9823	9877	9930	9984	0037	
3	91 0091	0144	0197	0251	0304	0358	0411	0464	0518	0571	
4	0624	0678	0731	0784	0838	0891	0944	0998	1051	1104	
815	1158	1211	1264	1317	1371	1424	1477	1530	1584	1637	
6	1690	1743	1797	1850	1903	1956	2009	2063	2116	2169	
7	2222	2275	2328	2381	2435	2488	2541	2594	2647	2700	
8	2753	2806	2859	2913	2966	3019	3072	3125	3178	3231	
9	3284	3337	3390	3443	3496	3549	3602	3655	3708	3761	53
820	3814	3867	3920	3973	4026	4079	4132	4184	4237	4290	
1	4343	4396	4449	4502	4555	4608	4660	4713	4766	4819	
2	4872	4925	4977	5030	5083	5136	5189	5241	5294	5347	
3	5400	5453	5505	5558	5611	5664	5716	5769	5822	5875	
4	5927	5980	6033	6085	6138	6191	6243	6296	6349	6401	
825	6454	6507	6559	6612	6664	6717	6770	6822	6875	6927	
6	6980	7033	7085	7138	7190	7243	7295	7348	7400	7453	
7	7506	7558	7611	7663	7716	7768	7820	7873	7925	7978	
8	8030	8083	8135	8188	8240	8293	8345	8397	8450	8502	
9	8555	8607	8659	8712	8764	8816	8869	8921	8973	9026	
830	9078	9130	9183	9235	9287	9340	9392	9444	9496	9549	
1	9601	9653	9706	9758	9810	9862	9914	9967	0019	0071	
2	92 0123	0176	0228	0280	0332	0384	0436	0489	0541	0593	
3	0645	0697	0749	0801	0853	0906	0958	1010	1062	1114	
4	1166	1218	1270	1322	1374	1426	1478	1530	1582	1634	52
835	1686	1738	1790	1842	1894	1946	1998	2050	2102	2154	
6	2206	2258	2310	2362	2414	2466	2518	2570	2622	2674	
7	2725	2777	2829	2881	2933	2985	3037	3089	3140	3192	
8	3244	3296	3348	3399	3451	3503	3555	3607	3658	3710	
9	3762	3814	3865	3917	3969	4021	4072	4124	4176	4228	
840	4279	4331	4383	4434	4486	4538	4589	4641	4693	4744	
1	4796	4848	4899	4951	5003	5054	5106	5157	5209	5261	
2	5312	5364	5415	5467	5518	5570	5621	5673	5725	5776	
3	5828	5879	5931	5982	6034	6085	6137	6188	6240	6291	
4	6342	6394	6445	6497	6548	6600	6651	6702	6754	6805	
845	6857	6908	6959	7011	7062	7114	7165	7216	7268	7319	
6	7370	7422	7473	7524	7576	7627	7678	7730	7781	7832	
7	7883	7935	7986	8037	8088	8140	8191	8242	8293	8345	
8	8396	8447	8498	8549	8601	8652	8703	8754	8805	8857	
9	8908	8959	9010	9061	9112	9163	9215	9266	9317	9368	
850	9419	9470	9521	9572	9623	9674	9725	9776	9827	9879	51
1	9930	9981	0032	0083	0134	0185	0236	0287	0338	0389	
2	93 0440	0491	0542	0592	0643	0694	0745	0796	0847	0898	
3	0949	1000	1051	1102	1153	1204	1254	1305	1356	1407	
4	1458	1509	1560	1610	1661	1712	1763	1814	1865	1915	

PROPORTIONAL PARTS.

Diff.	1	2	3	4	5	6	7	8	9
53	5.3	10.6	15.9	21.2	26.5	31.8	37.1	42.4	47.7
52	5.2	10.4	15.6	20.8	26.0	31.2	36.4	41.6	46.8
51	5.1	10.2	15.3	20.4	25.5	30.6	35.7	40.8	45.9
50	5.0	10.0	15.0	20.0	25.0	30.0	35.0	40.0	45.0

No. 855
Log. 931

Table VII.—*Continued*

No. 899
Log. 954

N.	0	1	2	3	4	5	6	7	8	9	Diff.
855	93 1966	2017	2068	2118	2169	2220	2271	2322	2372	2423	
6	2474	2524	2575	2626	2677	2727	2778	2829	2879	2930	
7	2981	3031	3082	3133	3183	3234	3285	3335	3386	3437	
8	3487	3538	3589	3639	3690	3740	3791	3841	3892	3943	
9	3993	4044	4094	4145	4195	4246	4296	4347	4397	4448	
860	4498	4549	4599	4650	4700	4751	4801	4852	4902	4953	
1	5003	5054	5104	5154	5205	5255	5306	5356	5406	5457	
2	5507	5558	5608	5658	5709	5759	5809	5860	5910	5960	
3	6011	6061	6111	6162	6212	6262	6313	6363	6413	6463	
4	6514	6564	6614	6665	6715	6765	6815	6865	6916	6966	
865	7016	7066	7116	7167	7217	7267	7317	7367	7418	7468	
6	7518	7568	7618	7668	7718	7769	7819	7869	7919	7969	
7	8019	8069	8119	8169	8219	8269	8320	8370	8420	8470	50
8	8520	8570	8620	8670	8720	8770	8820	8870	8920	8970	
9	9020	9070	9120	9170	9220	9270	9320	9369	9419	9469	
870	9519	9569	9619	9669	9719	9769	9819	9869	9918	9968	
1	94 0018	0068	0118	0168	0218	0267	0317	0367	0417	0467	
2	0516	0566	0616	0666	0716	0765	0815	0865	0915	0964	
3	1014	1064	1114	1163	1213	1263	1313	1362	1412	1462	
4	1511	1561	1611	1660	1710	1760	1809	1859	1909	1958	
875	2008	2058	2107	2157	2207	2256	2306	2355	2405	2455	
6	2504	2554	2603	2653	2702	2752	2801	2851	2901	2950	
7	3000	3049	3099	3148	3198	3247	3297	3346	3396	3445	
8	3495	3544	3593	3643	3692	3742	3791	3841	3890	3939	
9	3989	4038	4088	4137	4186	4236	4285	4335	4384	4433	
880	4483	4532	4581	4631	4680	4729	4779	4828	4877	4927	
1	4976	5025	5074	5124	5173	5222	5272	5321	5370	5419	
2	5469	5518	5567	5616	5665	5715	5764	5813	5862	5912	
3	5961	6010	6059	6108	6157	6207	6256	6305	6354	6403	
4	6452	6501	6551	6600	6649	6698	6747	6796	6845	6894	
885	6943	6992	7041	7090	7139	7189	7238	7287	7336	7385	
6	7434	7483	7532	7581	7630	7679	7728	7777	7826	7875	49
7	7924	7973	8022	8070	8119	8168	8217	8266	8315	8364	
8	8413	8462	8511	8560	8608	8657	8706	8755	8804	8853	
9	8902	8951	8999	9048	9097	9146	9195	9244	9292	9341	
890	9390	9439	9488	9536	9585	9634	9683	9731	9780	9829	
1	9878	9926	9975	0024	0073	0121	0170	0219	0267	0316	
2	95 0365	0414	0462	0511	0560	0608	0657	0706	0754	0803	
3	0851	0900	0949	0997	1046	1095	1143	1192	1240	1289	
4	1338	1386	1435	1483	1532	1580	1629	1677	1726	1775	
895	1823	1872	1920	1969	2017	2066	2114	2163	2211	2260	
6	2308	2356	2405	2453	2502	2550	2599	2647	2696	2744	
7	2792	2841	2889	2938	2986	3034	3083	3131	3180	3228	
8	3276	3325	3373	3421	3470	3518	3566	3615	3663	3711	
9	3760	3808	3856	3905	3953	4001	4049	4098	4146	4194	

PROPORTIONAL PARTS.

Diff.	1	2	3	4	5	6	7	8	9
51	5.1	10.2	15.3	20.4	25.5	30.6	35.7	40.8	45.9
50	5.0	10.0	15.0	20.0	25.0	30.0	35.0	40.0	45.0
49	4.9	9.8	14.7	19.6	24.5	29.4	34.3	39.2	44.1
48	4.8	9.6	14.4	19.2	24.0	28.8	33.6	38.4	43.2

No. 900
Log. 954

TABLE VII.—*Continued*

No. 944
Log. 975

N.	0	1	2	3	4	5	6	7	8	9	Diff.
900	95 4243	4291	4339	4387	4435	4484	4532	4580	4628	4677	
1	4725	4773	4821	4869	4918	4966	5014	5062	5110	5158	
2	5207	5255	5303	5351	5399	5447	5495	5543	5592	5640	
3	5688	5736	5784	5832	5880	5928	5976	6024	6072	6120	
4	6168	6216	6265	6313	6361	6409	6457	6505	6553	6601	48
905	6649	6697	6745	6793	6840	6888	6936	6984	7032	7080	
6	7128	7176	7224	7272	7320	7368	7416	7464	7512	7559	
7	7607	7655	7703	7751	7799	7847	7894	7942	7990	8038	
8	8086	8134	8181	8229	8277	8325	8373	8421	8468	8516	
9	8564	8612	8659	8707	8755	8803	8850	8898	8946	8994	
910	9041	9089	9137	9185	9232	9280	9328	9375	9423	9471	
1	9518	9566	9614	9661	9709	9757	9804	9852	9900	9947	
2	9995	0042	0090	0138	0185	0233	0280	0328	0376	0423	
3	96 0471	0518	0566	0613	0661	0709	0756	0804	0851	0899	
4	0946	0994	1041	1089	1136	1184	1231	1279	1326	1374	
915	1421	1469	1516	1563	1611	1658	1706	1753	1801	1848	
6	1895	1943	1990	2038	2085	2132	2180	2227	2275	2322	
7	2369	2417	2464	2511	2559	2606	2653	2701	2748	2795	
8	2843	2890	2937	2985	3032	3079	3126	3174	3221	3268	
9	3316	3363	3410	3457	3504	3552	3599	3646	3693	3741	
920	3788	3835	3882	3929	3977	4024	4071	4118	4165	4212	
1	4260	4307	4354	4401	4448	4495	4542	4590	4637	4684	
2	4731	4778	4825	4872	4919	4966	5013	5061	5108	5155	
3	5202	5249	5296	5343	5390	5437	5484	5531	5578	5625	
4	5672	5719	5766	5813	5860	5907	5954	6001	6048	6095	47
925	6142	6189	6236	6283	6329	6376	6423	6470	6517	6564	
6	6611	6658	6705	6752	6799	6845	6892	6939	6986	7033	
7	7080	7127	7173	7220	7267	7314	7361	7408	7454	7501	
8	7548	7595	7642	7688	7735	7782	7829	7875	7922	7969	
9	8016	8062	8109	8156	8203	8249	8296	8343	8390	8436	
930	8483	8530	8576	8623	8670	8716	8763	8810	8856	8903	
1	8950	8996	9043	9090	9136	9183	9229	9276	9323	9369	
2	9416	9463	9509	9556	9602	9649	9695	9742	9789	9835	
3	9882	9928	9975	0021	0068	0114	0161	0207	0254	0300	
4	97 0347	0393	0440	0486	0533	0579	0626	0672	0719	0765	
935	0812	0858	0904	0951	0997	1044	1090	1137	1183	1229	
6	1276	1322	1369	1415	1461	1508	1554	1601	1647	1693	
7	1740	1786	1832	1879	1925	1971	2018	2064	2110	2157	
8	2203	2249	2295	2342	2388	2434	2481	2527	2573	2619	
9	2666	2712	2758	2804	2851	2897	2943	2989	3035	3082	
940	3128	3174	3220	3266	3313	3359	3405	3451	3497	3543	
1	3590	3636	3682	3728	3774	3820	3866	3913	3959	4005	
2	4051	4097	4143	4189	4235	4281	4327	4374	4420	4466	
3	4512	4558	4604	4650	4696	4742	4788	4834	4880	4926	
4	4972	5018	5064	5110	5156	5202	5248	5294	5340	5386	46

PROPORTIONAL PARTS.

Diff.	1	2	3	4	5	6	7	8	9
47	4.7	9.4	14.1	18.8	23.5	28.2	32.9	37.6	42.3
46	4.6	9.2	13.8	18.4	23.0	27.6	32.2	36.8	41.4

No. 945 | Log. 975

No. 989 | Log. 995

TABLE VII.—*Continued*

N.	0	1	2	3	4	5	6	7	8	9	Diff.
945	97 5432	5478	5524	5570	5616	5662	5707	5753	5799	5845	
6	5891	5937	5983	6029	6075	6121	6167	6212	6258	6304	
7	6350	6396	6442	6488	6533	6579	6625	6671	6717	6763	
8	6808	6854	6900	6946	6992	7037	7083	7129	7175	7220	
9	7266	7312	7358	7403	7449	7495	7541	7586	7632	7678	
950	7724	7769	7815	7861	7906	7952	7998	8043	8089	8135	
1	8181	8226	8272	8317	8363	8409	8454	8500	8546	8591	
2	8637	8683	8728	8774	8819	8865	8911	8956	9002	9047	
3	9093	9138	9184	9230	9275	9321	9366	9412	9457	9503	
4	9548	9594	9639	9685	9730	9776	9321	9867	9912	9958	
955	98 0003	0049	0094	0140	0185	0231	0276	0322	0367	0412	
6	0458	0503	0549	0594	0640	0685	0730	0776	0821	0867	
7	0912	0957	1003	1048	1093	1139	1184	1229	1275	1320	
8	1366	1411	1456	1501	1547	1592	1637	1683	1728	1773	
9	1819	1864	1909	1954	2000	2045	2090	2135	2181	2226	
960	2271	2316	2362	2407	2452	2497	2543	2588	2633	2678	
1	2723	2769	2814	2859	2904	2949	2994	3040	3085	3130	
2	3175	3220	3265	3310	3356	3401	3446	3491	3536	3581	
3	3626	3671	3716	3762	3807	3852	3897	3942	3987	4032	
4	4077	4122	4167	4212	4257	4302	4347	4392	4437	4482	
965	4527	4572	4617	4662	4707	4752	4797	4842	4887	4932	45
6	4977	5022	5067	5112	5157	5202	5247	5292	5337	5382	
7	5426	5471	5516	5561	5606	5651	5696	5741	5786	5830	
8	5875	5920	5965	6010	6055	6100	6144	6189	6234	6279	
9	6324	6369	6413	6458	6503	6548	6593	6637	6682	6727	
970	6772	6817	6861	6906	6951	6996	7040	7085	7130	7175	
1	7219	7264	7309	7353	7398	7443	7488	7532	7577	7622	
2	7666	7711	7756	7800	7845	7890	7934	7979	8024	8068	
3	8113	8157	8202	8247	8291	8336	8381	8425	8470	8514	
4	8559	8604	8648	8693	8737	8782	8826	8871	8916	8960	
975	9005	9049	9094	9138	9183	9227	9272	9316	9361	9405	
6	9450	9494	9539	9583	9628	9672	9717	9761	9806	9850	
7	9895	9939	9983	0028	0072	0117	0161	0206	0250	0294	
8	99 0339	0383	0428	0472	0516	0561	0605	0650	0694	0738	
9	0783	0827	0871	0916	0960	1004	1049	1093	1137	1182	
980	1226	1270	1315	1359	1403	1448	1492	1536	1580	1625	
1	1669	1713	1758	1802	1846	1890	1935	1979	2023	2067	
2	2111	2156	2200	2244	2288	2333	2377	2421	2405	2509	
3	2554	2598	2642	2686	2730	2774	2819	2863	2907	2951	
4	2995	3039	3083	3127	3172	3216	3260	3304	3348	3392	
985	3436	3480	3524	3568	3613	3657	3701	3745	3789	3833	
6	3877	3921	3965	4009	4053	4097	4141	4185	4229	4273	
7	4317	4361	4405	4449	4493	4537	4581	4625	4669	4713	44
8	4757	4801	4845	4889	4933	4977	5021	5065	5108	5152	
9	5196	5240	5284	5328	5372	5416	5460	5504	5547	5591	

PROPORTIONAL PARTS.

Diff.	1	2	3	4	5	6	7	8	9
46	4.6	9.2	13.8	13.4	23.0	27.6	32.2	36.8	41.4
45	4.5	9.0	13.5	18.0	22.5	27.0	31.5	36.0	40.5
44	4.4	8.8	13.2	17.6	22.0	26.4	30.8	35.2	39.6
43	4.3	8.6	12.9	17.2	21.5	25.8	30.1	34.4	38.7

TABLE VII.—*Concluded*

N.	0	1	2	3	4	5	6	7	8	9	Diff.
990	99 5635	5679	5723	5767	5811	5854	5898	5942	5986	6030	
1	6074	6117	6161	6205	6249	6293	6337	6380	6424	6468	44
2	6512	6555	6599	6643	6687	6731	6774	6818	6862	6906	
3	6949	6993	7037	7080	7124	7168	7212	7255	7299	7343	
4	7386	7430	7474	7517	7561	7605	7648	7692	7736	7779	
995	7823	7867	7910	7954	7998	8041	8085	8129	8172	8216	
6	8259	8303	8347	8390	8434	8477	8521	8564	8608	8652	
7	8695	8739	8782	8826	8869	8913	8956	9000	9043	9087	
8	9131	9174	9218	9261	9305	9348	9392	9435	9479	9522	
9	9565	9609	9652	9696	9739	9783	9826	9870	9913	9957	43

TABLE VIII(a).—VALUES OF S, T, AND C IN TABLE VIII, PAGES 251 AND 252

If we were to plot the values of the logarithmic functions given in Table VIII as ordinates and corresponding minutes as abscissas it would be found that the points for each function were on a curve with variable radius. It would further be noted that the curves for sines, tangents, and cotangents were of comparatively small radii when the angles were small; that the curves for cosines, cotangents, and tangents of angles near 90° respectively had the same shape as the curves for sines, tangents, and cotangents of the complements of the angles; and that other than the portions of the curves just mentioned were nearly straight lines for short distances.

When seconds are involved it will be sufficiently accurate to interpolate in the ordinary manner between adjacent values in the tables—or in other words, to assume that the curve joining two adjacent points is a straight line—for all functions between 2° and 88°, and also for sines of angles between 88° and 90° and for cosines of angles between 0° and 2°. The values in the columns headed S, T, and C provide a means (1) of accurately determining for any given angle between 0° and 2° the logarithmic sine, tangent, or cotangent, and for any given angle between 88° and 90°, the logarithmic cosine, cotangent, or tangent; or (2) for a given value of the logarithmic sine, tangent, or cotangent of accurately determining the angle when it lies between 0° and 2°, and for any given value of the cosine, cotangent, or tangent, the angle when it lies between 88° and 90°.

Given: angle. Required: logarithmic function.

$$\left.\begin{aligned} \log\sin\alpha &= \log\alpha\ (\text{in seconds}) + S \\ \log\tan\alpha &= \log\alpha\ (\text{in seconds}) + T \\ \log\cot\alpha &= C - \log\alpha\ (\text{in seconds}) \end{aligned}\right\} \text{In which } \alpha \text{ is less than } 2°.$$

$$\left.\begin{aligned} \log\cos\beta &= \log(90° - \beta)\ (\text{in seconds}) + S \\ \log\tan\beta &= C - \log(90° - \beta)\ (\text{in seconds}) \\ \log\cot\beta &= \log(90° - \beta)\ (\text{in seconds}) + T \end{aligned}\right\} \text{In which } \beta \text{ lies between } 88° \text{ and } 90°$$

Given: logarithmic function. Required: angle.

$$\left.\begin{aligned} \log\alpha\ (\text{in seconds}) &= \log\sin\alpha - S \\ &= \log\tan\alpha - T \\ &= C - \log\cot\alpha \end{aligned}\right\} \text{In which } \alpha \text{ is less than } 2°.$$

$$\left.\begin{aligned} \log(90° - \beta)\ (\text{in seconds}) &= \log\cos\beta - S \\ &= C - \log\tan\beta \\ &= \log\cot\beta - T \end{aligned}\right\} \text{In which } \beta \text{ lies between } 88° \text{ and } 90°.$$

For examples, see following page.

TABLE VIII(a).—*Continued*

EXAMPLES

Given: angle. *Required: logarithmic function.*	*Given: logarithmic function.* *Required: angle.*
When $\left\{\begin{matrix}\alpha \\ 90°-\beta\end{matrix}\right\} = 19'\,22'' = 1162''$ $\log 1162'' = 3.065206$ S (for 19′) $= 4.685573$ $\left.\begin{matrix}\log\sin 19'\,22'' \\ \log\cos 89°\,40'\,38''\end{matrix}\right\} = 7.750779$	When $\left\{\begin{matrix}\log\sin\alpha \\ \log\cos\beta\end{matrix}\right\} = 7.750779$ S (for 19′) $= 4.685573$ $\log\left\{\begin{matrix}\alpha \\ 90°-\beta\end{matrix}\right\}$ (in seconds) $= 3.065206$ or $\begin{matrix}\alpha = 1162'' = 19'\,22'' \\ \beta = 89°\,40'\,38''\end{matrix}$
When $\left\{\begin{matrix}\alpha \\ 90°-\beta\end{matrix}\right\} = 23'\,21'' = 1401''$ $\log 1401'' = 3.146438$ C (for 23′) $= 15.314419$ $\left.\begin{matrix}\log\cot 23'\,21'' \\ \log\tan 89°\,36'\,39''\end{matrix}\right\} = 12.167981$	When $\left\{\begin{matrix}\log\cot\alpha \\ \log\tan\beta\end{matrix}\right\} = 12.167981$ C (for 23′) $= 15.314419$ $\log\left\{\begin{matrix}\alpha \\ 90°-\beta\end{matrix}\right\}$ (in seconds) $= 3.146438$ or $\begin{matrix}\alpha = 1401'' = 23'\,21'' \\ \beta = 89°\,36'\,39''\end{matrix}$

TABLE VIII.—LOGARITHMIC SINES, COSINES, TANGENTS, AND COTANGENTS

0° 179°

″	′	Sine.	S.*	T.*	Tang.	Cotang.	C.*	D. 1″.	Cosine.	′
			4.685				15.314			
0	0	Inf. neg.	575	575	Inf. neg.	Inf. pos.	425		10.00 0000	60
60	1	6.46 3726	575	575	6.46 3726	13.53 6274	425		0000	59
120	2	.76 4756	575	575	.76 4756	.23 5244	425		0000	58
180	3	6.94 0847	575	575	6.94 0847	13.05 9153	425		0000	57
240	4	7.06 5786	575	575	7.06 5786	12.93 4214	425		0000	56
300	5	7.16 2696	575	575	7.16 2696	12.83 7304	425	.02	10.00 0000	55
360	6	.24 1877	575	575	.24 1878	.75 8122	425	.00	9.99 9999	54
420	7	.30 8824	575	575	.30 8825	.69 1175	425	.00	9999	53
480	8	.36 6816	574	576	.36 6817	.63 3183	424	.00	9999	52
540	9	.41 7968	574	576	.41 7970	.58 2030	424	.02	9999	51
600	10	7.46 3726	574	576	7.46 3727	12.53 6273	424	.00	9.99 9998	50
660	11	.50 5118	574	576	.50 5120	.49 4880	424	.02	9998	49
720	12	.54 2906	574	577	.54 2909	.45 7091	423	.00	9997	48
780	13	.57 7668	574	577	.57 7672	.42 2328	423	.02	9997	47
840	14	.60 9853	574	577	.60 9857	.39 0143	423	.00	9996	46
900	15	7.63 9816	573	578	7.63 9820	12.36 0180	422	.02	9.99 9996	45
960	16	.66 7845	573	578	.66 7849	.33 2151	422	.00	9995	44
1020	17	.69 4173	573	578	.69 4179	.30 5821	422	.02	9995	43
1080	18	.71 8997	573	579	.71 9003	.28 0997	421	.02	9994	42
1140	19	.74 2478	573	579	.74 2484	.25 7516	421	.00	9993	41
1200	20	7.76 4754	572	580	7.76 4761	12.23 5239	420	.02	9.99 9993	40
1260	21	.78 5943	572	580	.78 5951	.21 4049	420	.02	9992	39
1320	22	.80 6146	572	581	.80 6155	.19 3845	419	.02	9991	38
1380	23	.82 5451	572	581	.82 5460	.17 4540	419	.02	9990	37
1440	24	.84 3934	571	582	.84 3944	.15 6056	418	.00	9989	36
1500	25	7.86 1662	571	583	7.86 1674	12.13 8326	417	.02	9.99 9989	35
1560	26	.87 8695	571	583	.87 8708	.12 1292	417	.02	9988	34
1620	27	.89 5085	570	584	.89 5099	.10 4901	416	.02	9987	33
1680	28	.91 0879	570	584	.91 0894	.08 9106	416	.02	9986	32
1740	29	.92 6119	570	585	.92 6134	.07 3866	415	.03	9985	31
1800	30	7.94 0842	569	586	7.94 0858	12.05 9142	414	.02	9.99 9983	30
1860	31	.95 5082	569	587	.95 5100	.04 4900	413	.02	9982	29
1920	32	.96 8870	569	587	.96 8889	.03 1111	413	.02	9981	28
1980	33	.98 2233	568	588	.98 2253	.01 7747	412	.02	9980	27
2040	34	7.99 5198	568	589	7.99 5219	12.00 4781	411	.03	9979	26
2100	35	8.00 7787	567	590	8.00 7809	11.99 2191	410	.02	9.99 9977	25
2160	36	.02 0021	567	591	.02 0044	.97 9956	409	.02	9976	24
2220	37	.03 1919	566	592	.03 1945	.96 8055	408	.03	9975	23
2280	38	.04 3501	566	593	.04 3527	.95 6473	407	.02	9973	22
2340	39	.05 4781	566	593	.05 4809	.94 5191	407	.02	9972	21
2400	40	8.06 5576	565	594	8.06 5806	11.93 4194	406	.03	9.99 9971	20
2460	41	.07 6500	565	595	.07 6531	.92 3469	405	.02	9969	19
2520	42	.08 6965	564	596	.08 6997	.91 3003	404	.03	9968	18
2580	43	.09 7183	564	598	.09 7217	.90 2783	402	.03	9966	17
2640	44	.10 7167	563	599	.10 7203	.89 2797	401	.02	9964	16
2700	45	8.11 6926	562	600	8.11 6963	11.88 3037	400	.03	9.99 9963	15
2760	46	.12 6471	562	601	.12 6510	.87 3490	399	.03	9961	14
2820	47	.13 5810	561	602	.13 5851	.86 4149	398	.02	9959	13
2880	48	.14 4953	561	603	.14 4996	.85 5004	397	.03	9958	12
2940	49	.15 3907	560	604	.15 3952	.84 6048	396	.03	9956	11
3000	50	8.16 2681	560	605	8.16 2727	11.83 7273	395	.03	9.99 9954	10
3060	51	.17 1280	559	607	.17 1328	.82 8672	393	.03	9952	9
3120	52	.17 9713	558	608	.17 9763	.82 0237	392	.03	9950	8
3180	53	.18 7985	558	609	.18 8036	.81 1964	391	.03	9948	7
3240	54	.19 6102	557	611	.19 6156	.80 3844	389	.03	9946	6
3300	55	8.20 4070	556	612	8.20 4126	11.79 5874	388	.03	9.99 9944	5
3360	56	.21 1895	556	613	.21 1953	.78 8047	387	.03	9942	4
3420	57	.21 9581	555	615	.21 9641	.78 0359	385	.03	9940	3
3480	58	.22 7134	554	616	.22 7195	.77 2805	384	.03	9938	2
3540	59	.23 4557	554	618	.23 4621	.76 5379	382	.03	9936	1
3600	60	8.24 1855	553	619	8.24 1921	11.75 8079	381		9.99 9934	0
			4.685				15.314			
″	′	Cosine.	S.*	T.*	Cotang.	Tang.	C.*	D. 1″.	Sine.	′

90° 89°

* For use of S, T, and C see Table VIII(a), p. 249.

1° TABLE VIII.—*Continued* 178°

″	′	Sine.	S.*	T.*	Tang.	Cotang.	C.*	D. 1″.	Cosine.	′
			4.685				15.314			
3600	**0**	**8.24 1855**	553	619	**8.24 1921**	**11.75 8079**	381	.03	**9.99 9934**	**60**
3660	1	.24 9033	552	620	.24 9102	.75 0898	380	.05	9932	59
3720	2	.25 6094	551	622	.25 6165	.74 3835	378	.03	9929	58
3780	3	.26 3042	551	623	.26 3115	.73 6885	377	.03	9927	57
3840	4	.26 9881	550	625	.26 9956	.73 0044	375	.05	9925	56
3900	**5**	**8.27 6614**	549	627	**8.27 6691**	**11.72 3309**	373	.03	**9.99 9922**	**55**
3960	6	.28 3243	548	628	.28 3323	.71 6677	372	.03	9920	54
4020	7	.28 9773	547	630	.28 9856	.71 0144	370	.05	9918	53
4080	8	.29 6207	546	632	.29 6292	.70 3708	368	.03	9915	52
4140	9	.30 2546	546	633	.30 2634	.69 7366	367	.05	9913	51
4200	**10**	**8.30 8794**	545	635	**8.30 8884**	**11.69 1116**	365	.05	**9.99 9910**	**50**
4260	11	.31 4954	544	637	.31 5046	.68 4954	363	.03	9907	49
4320	12	.32 1027	543	638	.32 1122	.67 8878	362	.05	9905	48
4380	13	.32 7016	542	640	.32 7114	.67 2886	360	.05	9902	47
4440	14	.33 2924	541	642	.33 3025	.66 6975	358	.03	9899	46
4500	**15**	**8.33 8753**	540	644	**8.33 8856**	**11.66 1144**	356	.05	**9.99 9897**	**45**
4560	16	.34 4504	539	646	.34 4610	.65 5390	354	.05	9894	44
4620	17	.35 0181	539	648	.35 0289	.64 9711	352	.05	9891	43
4680	18	.35 5783	538	649	.35 5895	.64 4105	351	.05	9888	42
4740	19	.36 1315	537	651	.36 1430	.63 8570	349	.05	9885	41
4800	**20**	**8.36 6777**	536	653	**8.36 6895**	**11.63 3105**	347	.05	**9.99 9882**	**40**
4860	21	.37 2171	535	655	.37 2292	.62 7708	345	.05	9879	39
4920	22	.37 7499	534	657	.37 7622	.62 2378	343	.05	9876	38
4980	23	.38 2762	533	659	.38 2889	.61 7111	341	.05	9873	37
5040	24	.38 7962	532	661	.38 8092	.61 1908	339	.05	9870	36
5100	**25**	**8.39 3101**	531	663	**8.39 3234**	**11.60 6766**	337	.05	**9.99 9867**	**35**
5160	26	.39 8179	530	666	.39 8315	.60 1685	334	.05	9864	34
5220	27	.40 3199	529	668	.40 3338	.59 6662	332	.05	9861	33
5280	28	.40 8161	527	670	.40 8304	.59 1696	330	.07	9858	32
5340	29	.41 3068	526	672	.41 3213	.58 6787	328	.05	9854	31
5400	**30**	**8.41 7919**	525	674	**8.41 8068**	**11.58 1932**	326	.05	**9.99 9851**	**30**
5460	31	.42 2717	524	676	.42 2869	.57 7131	324	.07	9848	29
5520	32	.42 7462	523	679	.42 7618	.57 2382	321	.05	9844	28
5580	33	.43 2156	522	681	.43 2315	.56 7685	319	.05	9841	27
5640	34	.43 6800	521	683	.43 6962	.56 3038	317	.07	9838	26
5700	**35**	**8.44 1394**	520	685	**8.44 1560**	**11.55 8440**	315	.05	**9.99 9834**	**25**
5760	36	.44 5941	518	688	.44 6110	.55 3890	312	.07	9831	24
5820	37	.45 0440	517	690	.45 0613	.54 9387	310	.05	9827	23
5880	38	4893	516	693	5070	4930	307	.07	9824	22
5940	39	.45 9301	515	695	.45 9481	.54 0519	305	.07	9820	21
6000	**40**	**8.46 3665**	514	697	**8.46 3849**	**11.53 6151**	303	.05	**9.99 9816**	**20**
6060	41	.46 7985	512	700	.46 8172	.53 1828	300	.07	9813	19
6120	42	.47 2263	511	702	.47 2454	.52 7546	298	.07	9809	18
6180	43	.47 6498	510	705	.47 6693	.52 3307	295	.07	9805	17
6240	44	.48 0693	509	707	.48 0892	.51 9108	293	.07	9801	16
6300	**45**	**8.48 4848**	507	710	**8.48 5050**	**11.51 4950**	290	.05	**9.99 9797**	**15**
6360	46	.48 8963	506	713	.48 9170	.51 0830	287	.07	9794	14
6420	47	.49 3040	505	715	.49 3250	.50 0750	285	.07	9790	13
6480	48	.49 7078	503	718	.49 7293	.50 2707	282	.07	9786	12
6540	49	.50 1080	502	720	.50 1298	.49 8702	280	.07	9782	11
6600	**50**	**8.50 5045**	501	723	**8.50 5267**	**11.49 4733**	277	.07	**9.99 9778**	**10**
6660	51	.50 8974	499	726	.50 9200	.49 0800	274	.08	9774	9
6720	52	.51 2867	498	729	.51 3098	.48 6902	271	.07	9769	8
6780	53	.51 6726	497	731	.51 6961	.48 3039	269	.07	9765	7
6840	54	.52 0551	495	734	.52 0790	.47 9210	266	.07	9761	6
6900	**55**	**8.52 4343**	494	737	**8.52 4586**	**11.47 5414**	263	.07	**9.99 9757**	**5**
6960	56	.52 8102	492	740	.52 8349	.47 1651	260	.03	9753	4
7020	57	.53 1828	491	743	.53 2080	.46 7920	257	.07	9748	3
7080	58	5523	490	745	5779	4221	255	.07	9744	2
7140	59	.53 9186	488	748	.53 9447	.46 0553	252	.08	9740	1
7200	**60**	**8.54 2819**	487	751	**8.54 3084**	**11.45 6916**	249		**9.99 9735**	**0**
			4.685				15.314			
″	′	Cosine.	S.*	T.*	Cotang.	Tang.	C.*	D. 1″.	Sine.	′

91° 88°

* For use of *S*, *T*, and *C* see Table VIII (a), p. 249.

2° TABLE VIII.—*Continued* 177°

′	Sine.	D. 1″.	Cosine.	D.1″.	Tang.	D.1″.	Cotang.	′
0	**8.54 2819**	60.05	**9.99 9735**	.07	**8.54 3084**	60.12	**11.45 6916**	60
1	6422	59.55	9731	.08	.54 6691	59.62	.45 3309	59
2	.54 9995	59.07	9726	.07	.55 0268	59.15	.44 9732	58
3	.55 3539	58.58	9722	.08	3817	58.65	6183	57
4	.55 7054	58.10	9717	.07	.55 7336	58.20	.44 2664	56
5	**8.56 0540**	57.65	**9.99 9713**	.08	**8.56 0828**	57.72	**11.43 9172**	55
6	3999	57.20	9708	.07	4291	57.27	5709	54
7	.56 7431	56.75	9704	.08	.56 7727	56.83	.43 2273	53
8	.57 0836	56.30	9699	.08	.57 1137	56.38	.42 8863	52
9	4214	55.87	9694	.08	4520	55.95	5480	51
10	**8.57 7566**	55.43	**9.99 9689**	.07	**8.57 7877**	55.52	**11.42 2123**	50
11	.58 0892	55.02	9685	.08	.58 1208	55.10	.41 8792	49
12	4193	54.60	9680	.08	4514	54.68	5486	48
13	.58 7469	54.20	9675	.08	.58 7795	54.27	.41 2205	47
14	.59 0721	53.78	9670	.08	.59 1051	53.87	.40 8949	46
15	**8.59 3948**	53.40	**9.99 9665**	.08	**8.59 4283**	53.48	**11.40 5717**	45
16	.59 7152	53.00	9660	.08	.59 7492	53.08	.40 2508	44
17	.60 0332	52.62	9655	.08	.60 0677	52.70	.39 9323	43
18	3489	52.23	9650	.08	3839	52.32	6161	42
19	6623	51.85	9645	.08	.60 6978	51.93	.39 3022	41
20	**8.60 9734**	51.48	**9.99 9640**	.08	**8.61 0094**	51.58	**11.38 9906**	40
21	.61 2823	51.13	9635	.10	3189	51.22	6811	39
22	5891	50.77	9629	.08	6262	50.85	3738	38
23	.61 8937	50.42	9624	.08	.61 9313	50.50	.38 0687	37
24	.62 1962	50.05	9619	.08	.62 2343	50.15	.37 7657	36
25	**8.62 4965**	49.72	**9.99 9614**	.10	**8.62 5352**	49.80	**11.37 4648**	35
26	.62 7948	49.38	9608	.08	.62 8340	49.47	.37 1660	34
27	.63 0911	49.05	9603	.10	.63 1308	49.13	.36 8692	33
28	3854	48.70	9597	.08	4256	48.80	5744	32
29	6776	48.40	9592	.10	.63 7184	48.48	.36 2816	31
30	**8.63 9680**	48.05	**9.99 9586**	.08	**8.64 0093**	48.15	**11.35 9907**	30
31	.64 2563	47.75	9581	.10	2982	47.85	7018	29
32	5428	47.43	9575	.08	5853	47.52	4147	28
33	.64 8274	47.13	9570	.10	.64 8704	47.22	.35 1296	27
34	.65 1102	46.82	9564	.10	.65 1537	46.92	.34 8463	26
35	**8.65 3911**	46.52	**9.99 9558**	.08	**8.65 4352**	46.62	**11.34 5648**	25
36	6702	46.22	9553	.10	7149	46.32	2851	24
37	.65 9475	45.92	9547	.10	.65 9928	46.02	.34 0072	23
38	.66 2230	45.63	9541	.10	.66 2689	45.73	.33 7311	22
39	4968	45.35	9535	.10	5433	45.45	4567	21
40	**8.66 7689**	45.07	**9.99 9529**	.08	**8.66 8160**	45.17	**11.33 1840**	20
41	.67 0393	44.78	9524	.10	.67 0870	44.88	.32 9130	19
42	3080	44.52	9518	.10	3563	44.60	6437	18
43	5751	44.23	9512	.10	6239	44.35	3761	17
44	.67 8405	43.97	9506	.10	.67 8900	44.07	.32 1100	16
45	**8.68 1043**	43.70	**9.99 9500**	.12	**8.68 1544**	43.80	**11.31 8456**	15
46	3665	43.45	9493	.10	4172	43.53	5828	14
47	6272	43.18	9487	.10	6784	43.28	3216	13
48	.68 8863	42.92	9481	.10	.68 9381	43.03	.31 0619	12
49	.69 1438	42.67	9475	.10	.69 1963	42.77	.30 8037	11
50	**8.69 3998**	42.42	**9.99 9469**	.10	**8.69 4529**	42.53	**11.30 5471**	10
51	6543	42.17	9463	.12	7081	42.27	2919	9
52	.69 9073	41.93	9456	.10	.69 9617	42.03	.30 0383	8
53	.70 1589	41.68	9450	.12	.70 2139	41.78	.29 7861	7
54	4090	41.45	9443	.10	4646	41.57	5354	6
55	**8.70 6577**	41.20	**9.99 9437**	.10	**8.70 7140**	41.30	**11.29 2860**	5
56	.70 9049	40.97	9431	.12	.70 9618	41.08	.29 0382	4
57	.71 1507	40.75	9424	.10	.71 2083	40.85	.28 7917	3
58	3952	40.52	9418	.12	4534	40.63	5466	2
59	6383	40.28	9411	.12	6972	40.40	3028	1
60	**8.71 8800**		**9.99 9404**		**8.71 9396**		**11.28 0604**	0
′	Cosine.	D.1″.	Sine.	D.1″.	Cotang.	D.1″.	Tang.	′

92° 87°

3° TABLE VIII.—*Continued* 176°

′	Sine.	D.1″.	Cosine.	D.1″.	Tang.	D.1″.	Cotang.	′
0	**8.71 8800**	40.07	**9.99 9404**	.10	**8.71 9396**	40.17	**11.28 0604**	**60**
1	.72 1204	39.85	9398	.12	.72 1806	39.97	.27 8194	59
2	3595	39.62	9391	.12	4204	39.73	5796	58
3	5972	39.42	9384	.10	6588	39.52	3412	57
4	.72 8337	39.18	9378	.12	.72 8959	39.30	.27 1041	56
5	**8.73 0688**	38.98	**9.99 9371**	.12	**8.73 1317**	39.10	**11.26 8683**	**55**
6	3027	38.78	9364	.12	3663	38.88	6337	54
7	5354	38.55	9357	.12	5996	38.68	4004	53
8	7667	38.37	9350	.12	.73 8317	38.48	.26 1683	52
9	.73 9969	38.17	9343	.12	.74 0626	38.27	.25 9374	51
10	**8.74 2259**	37.95	**9.99 9336**	.12	**8.74 2922**	38.08	**11.25 7078**	**50**
11	4536	37.77	9329	.12	5207	37.87	4793	49
12	6802	37.55	9322	.12	7479	37.68	2521	48
13	.74 9055	37.37	9315	.12	.74 9740	37.48	.25 0260	47
14	.75 1297	37.18	9308	.12	.75 1989	37.30	.24 8011	46
15	**8.75 3528**	36.98	**9.99 9301**	.12	**8.75 4227**	37.10	**11.24 5773**	**45**
16	5747	36.80	9294	.12	6453	36.92	3547	44
17	.75 7955	36.60	9287	.13	.75 8668	36.73	.24 1332	43
18	.76 0151	36.43	9279	.12	.76 0872	36.55	.23 9128	42
19	2337	36.23	9272	.12	3065	36.35	6935	41
20	**8.76 4511**	36.07	**9.99 9265**	.13	**8.76 5246**	36.18	**11.23 4754**	**40**
21	6675	35.88	9257	.12	7417	36.02	2583	39
22	.76 8828	35.70	9250	.13	.76 9578	35.82	.23 0422	38
23	.77 0970	35.52	9242	.12	.77 1727	35.65	.22 8273	37
24	3101	35.37	9235	.13	3866	35.48	6134	36
25	**8.77 5223**	35.17	**9.99 9227**	.12	**8.77 5995**	35.32	**11.22 4005**	**35**
26	7333	35.02	9220	.13	.77 8114	35.13	.22 1886	34
27	.77 9434	34.83	9212	.12	.78 0222	34.97	.21 9778	33
28	.78 1524	34.68	9205	.13	2320	34.80	7680	32
29	3605	34.50	9197	.13	4408	34.63	5592	31
30	**8.78 5675**	34.35	**9.99 9189**	.13	**8.78 6486**	34.47	**11.21 3514**	**30**
31	7736	34.18	9181	.12	.78 8554	34.32	.21 1446	29
32	.78 9787	34.02	9174	.13	.79 0613	34.15	.20 9387	28
33	.79 1828	33.85	9166	.13	2662	33.98	7338	27
34	3859	33.70	9158	.13	4701	33.83	5299	26
35	**8.79 5881**	33.55	**9.99 9150**	.13	**8.79 6731**	33.68	**11.20 3269**	**25**
36	7894	33.38	9142	.13	.79 8752	33.52	.20 1248	24
37	.79 9897	33.25	9134	.13	.80 0763	33.37	.19 9237	23
38	.80 1892	33.07	9126	.13	2765	33.22	7235	22
39	3876	32.93	9118	.13	4758	33.07	5242	21
40	**8.80 5852**	32.78	**9.99 9110**	.13	**8.80 6742**	32.92	**11.19 3258**	**20**
41	7819	32.63	9102	.13	.80 8717	32.77	.19 1283	19
42	.80 9777	32.48	9094	.13	.81 0683	32.63	.18 9317	18
43	.81 1726	32.35	9086	.15	2641	32.47	7359	17
44	3667	32.20	9077	.13	4589	32.33	5411	16
45	**8.81 5599**	32.05	**9.99 9069**	.13	**8.81 6529**	32.20	**11.18 3471**	**15**
46	7522	31.90	9061	.13	.81 8461	32.05	.18 1539	14
47	.81 9436	31.78	9053	.15	.82 0384	31.90	.17 9616	13
48	.82 1343	31.62	9044	.13	2298	31.78	7702	12
49	3240	31.50	9036	.15	4205	31.63	5795	11
50	**8.82 5130**	31.35	**9.99 9027**	.13	**8.82 6103**	31.48	**11.17 3897**	**10**
51	7011	31.22	9019	.15	7992	31.37	2008	9
52	.82 8884	31.08	9010	.13	.82 9874	31.23	.17 0126	8
53	.83 0749	30.97	9002	.15	.83 1748	31.08	.16 8252	7
54	2607	30.82	8993	.15	3613	30.97	6387	6
55	**8.83 4456**	30.68	**9.99 8984**	.13	**8.83 5471**	30.83	**11.16 4529**	**5**
56	6297	30.55	8976	.15	7321	30.70	2679	4
57	8130	30.43	8967	.15	.83 9163	30.58	.16 0837	3
58	.83 9956	30.30	8958	.13	.84 0998	30.45	.15 9002	2
59	.84 1774	30.18	8950	.15	2825	30.32	7175	1
60	**8.84 3585**		**9.99 8941**		**8.84 4644**		**11.15 5356**	**0**
′	Cosine.	D.1″.	Sine.	D.1″.	Cotang.	D.1″.	Tang.	′

93° 86°

4° **Table VIII.**—*Continued* 175°

′	Sine.	D.1″.	Cosine.	D.1″.	Tang.	D.1″.	Cotang.	′
0	8.84 3585	30.03	9.99 8941	.15	8.84 4644	30.18	11.15 5356	60
1	5387	29.93	8932	.15	6455	30.08	3545	59
2	7183	29.80	8923	.15	.84 8260	29.95	.15 1740	58
3	.84 8971	29.67	8914	.15	.85 0057	29.82	.14 9943	57
4	.85 0751	29.57	8905	.15	1846	29.70	8154	56
5	8.85 2525	29.43	9.99 8896	.15	8.85 3628	29.58	11.14 6372	55
6	4291	29.30	8887	.15	5403	29.47	4597	54
7	6049	29.20	8878	.15	7171	29.35	2829	53
8	7801	29.08	8869	.15	.85 8932	29.23	.14 1068	52
9	.85 9546	28.95	8860	.15	.86 0686	29.12	.13 9314	51
10	8.86 1283	28.85	9.99 8851	.17	8.86 2433	29.00	11.13 7567	50
11	3014	28.73	8841	.15	4173	28.88	5827	49
12	4738	28.62	8832	.15	5906	28.77	4094	48
13	6455	28.50	8823	.17	7632	28.65	2368	47
14	8165	28.38	8813	.15	.86 9351	28.55	.13 0649	46
15	8.86 9868	28.28	9.99 8804	.15	8.87 1064	28.43	11.12 8936	45
16	.87 1565	28.17	8795	.17	2770	28.32	7230	44
17	3255	28.05	8785	.15	4469	28.22	5531	43
18	4938	27.95	8776	.17	6162	28.12	3838	42
19	6615	27.83	8766	.15	7849	28.00	2151	41
20	8.87 8285	27.73	9.99 8757	.17	8.87 9529	27.88	11.12 0471	40
21	.87 9949	27.63	8747	.15	.88 1202	27.78	.11 8798	39
22	.88 1607	27.52	8738	.17	2869	27.68	7131	38
23	3258	27.42	8728	.17	4530	27.58	5470	37
24	4903	27.32	8718	.17	6185	27.47	3815	36
25	8.88 6542	27.20	9.99 8708	.15	8.88 7833	27.38	11.11 2167	35
26	8174	27.12	8699	.17	.88 9476	27.27	.11 0524	34
27	.88 9801	27.00	8689	.17	.89 1112	27.17	.10 8888	33
28	.89 1421	26.90	8679	.17	2742	27.07	7258	32
29	3035	26.80	8669	.17	4366	26.97	5634	31
30	8.89 4643	26.72	9.99 8659	.17	8.89 5984	26.87	11.10 4016	30
31	6246	26.60	8649	.17	7596	26.78	2404	29
32	7842	26.50	8639	.17	.89 9203	26.67	.10 0797	28
33	.89 9432	26.42	8629	.17	.90 0803	26.58	.09 9197	27
34	.90 1017	20.32	8619	.17	2398	26.48	7602	26
35	8.90 2596	26.22	9.99 8609	.17	8.90 3987	26.38	11.09 6013	25
36	4169	26.12	8599	.17	5570	26.28	4430	24
37	5736	26.02	8589	.18	7147	26.20	2853	23
38	7297	25.93	8578	.17	.90 8719	26.10	.09 1281	22
39	.90 8853	25.85	8568	.17	.91 0285	20.02	.08 9715	21
40	8.91 0404	25.75	9.99 8558	.17	8.91 1846	25.92	11.08 8154	20
41	1949	25.65	8548	.18	3401	25.83	6599	19
42	3488	25.57	8537	.17	4951	25.73	5049	18
43	5022	25.47	8527	.18	6495	25.63	3505	17
44	6550	25.38	8516	.17	8034	25.57	1966	16
45	8.91 8073	25.30	9.99 8506	.18	8.91 9568	25.47	11.08 0432	15
46	.91 9591	25.20	8495	.17	.92 1096	25.38	.07 8904	14
47	.92 1103	25.12	8485	.18	2619	25.28	7381	13
48	2610	25.03	8474	.17	4136	25.22	5864	12
49	4112	24.95	8464	.18	5649	25.12	4351	11
50	8.92 5609	24.85	9.99 8453	.18	8.92 7156	25.03	11.07 2844	10
51	7100	24.78	8442	.18	.92 8658	24.95	.07 1342	9
52	.92 8587	24.68	8431	.17	.93 0155	24.87	.06 9845	8
53	.93 0068	24.60	8421	.18	1647	24.78	8353	7
54	1544	24.52	8410	.18	3134	24.70	6866	6
55	8.93 3015	24.43	9.99 8399	.18	8.93 4616	24.62	11.06 5384	5
56	4481	24.35	8388	.18	6093	24.53	3907	4
57	5942	24.27	8377	.18	7565	24.45	2435	3
58	7398	24.20	8366	.18	.93 9032	24.37	.06 0968	2
59	.93 8850	24.10	8355	.18	.94 0494	24.30	.05 9506	1
60	8.94 0296		9.99 8344		8.94 1952		11.05 8048	0
′	Cosine.	D.1″.	Sine.	D.1″.	Cotang.	D.1″.	Tang.	′

94° 85°

5° TABLE VIII.—*Continued* 174°

′	Sine.	D.1″.	Cosine.	D.1″.	Tang.	D.1″.	Cotang.	′
0	8.94 0296	24.03	9.99 8344	.18	8.94 1952	24.20	11.05 8048	60
1	1738	23.93	8333	.18	3404	24.13	6596	59
2	3174	23.87	8322	.18	4852	24.05	5148	58
3	4606	23.80	8311	.18	6295	23.98	3705	57
4	6034	23.70	8300	.18	7734	23.90	2266	56
5	8.94 7456	23.63	9.99 8289	.20	8.94 9168	23.82	11.05 0832	55
6	.94 8874	23.55	8277	.18	.95 0597	23.73	.04 9403	54
7	.95 0287	23.48	8266	.18	2021	23.67	7979	53
8	1696	23.40	8255	.20	3441	23.58	6559	52
9	3100	23.32	8243	.18	4856	23.52	5144	51
10	8.95 4499	23.25	9.99 8232	.20	8.95 6267	23.45	11.04 3733	50
11	5894	23.17	8220	.18	7674	23.35	2326	49
12	7284	23.10	8209	.20	.95 9075	23.30	.04 0925	48
13	.95 8670	23.03	8197	.18	.96 0473	23.22	.03 9527	47
14	.96 0052	22.95	8186	.20	1866	23.15	8134	46
15	8.96 1429	22.87	9.99 8174	.18	8.96 3255	23.07	11.03 6745	45
16	2801	22.82	8163	.20	4639	23.00	5361	44
17	4170	22.73	8151	.20	6019	22.92	3981	43
18	5534	22.65	8139	.18	7394	22.87	2606	42
19	6893	22.60	8128	.20	.96 8766	22.78	.03 1234	41
20	8.96 8249	22.52	9.99 8116	.20	8.97 0133	22.72	11.02 9867	40
21	.96 9600	22.45	8104	.20	1496	22.65	8504	39
22	.97 0947	22.37	8092	.20	2855	22.57	7145	38
23	2289	22.32	8080	.20	4209	22.52	5791	37
24	3628	22.23	8068	.20	5560	22.43	4440	36
25	8.97 4962	22.18	9.99 8056	.20	8.97 6906	22.37	11.02 3094	35
26	6293	22.10	8044	.20	8248	22.30	1752	34
27	7619	22.03	8032	.20	.97 9586	22.25	.02 0414	33
28	.97 8941	21.97	8020	.20	.98 0921	22.17	.01 9079	32
29	.98 0259	21.90	8008	.20	2251	22.10	7749	31
30	8.98 1573	21.83	9.99 7996	.20	8.98 3577	22.03	11.01 6423	30
31	2883	21.77	7984	.20	4899	21.97	5101	29
32	4189	21.72	7972	.22	6217	21.92	3783	28
33	5491	21.63	7959	.20	7532	21.83	2468	27
34	6789	21.57	7947	.20	.98 8842	21.78	.01 1158	26
35	8.98 8083	21.52	9.99 7935	.22	8.99 0149	21.70	11.00 9851	25
36	.98 9374	21.43	7922	.20	1451	21.65	8549	24
37	.99 0660	21.38	7910	.22	2750	21.58	7250	23
38	1943	21.32	7897	.20	4045	21.53	5955	22
39	3222	21.25	7885	.22	5337	21.45	4663	21
40	8.99 4497	21.18	9.99 7872	.20	8.99 6624	21.40	11.00 3376	20
41	5768	21.13	7860	.22	7908	21.33	2092	19
42	7036	21.05	7847	.20	8.99 9188	21.28	11.00 0812	18
43	8299	21.02	7835	.22	9.00 0465	21.22	10.99 9535	17
44	8.99 9560	20.93	7822	.22	1738	21.15	8262	16
45	9.00 0816	20.88	9.99 7809	.20	9.00 3007	21.08	10.99 6993	15
46	2069	20.82	7797	.22	4272	21.03	5728	14
47	3318	20.75	7784	.22	5534	20.97	4466	13
48	4563	20.70	7771	.22	6792	20.92	3208	12
49	5805	20.65	7758	.22	8047	20.85	1953	11
50	9.00 7044	20.57	9.99 7745	.22	9.00 9298	20.80	10.99 0702	10
51	8278	20.53	7732	.22	.01 0546	20.73	.98 9454	9
52	.00 9510	20.45	7719	.22	1790	20.68	8210	8
53	.01 0737	20.42	7706	.22	3031	20.62	6969	7
54	1962	20.33	7693	.22	4268	20.57	5732	6
55	9.01 3182	20.30	9.99 7680	.22	9.01 5502	20.50	10.98 4498	5
56	4400	20.22	7667	.22	6732	20.45	3268	4
57	5613	20.18	7654	.22	7959	20.40	2041	3
58	6824	20.12	7641	.22	.01 9183	20.33	.98 0817	2
59	8031	20.07	7628	.23	.02 0403	20.28	.97 9597	1
60	9.01 9235		9.99 7614		9.02 1620		10.97 8380	0
′	Cosine.	D.1″.	Sine.	D.1″.	Cotang.	D.1″.	Tang.	′

95° 84°

6° TABLE VIII.—*Continued* 173°

′	Sine.	D.1″.	Cosine.	D.1″.	Tang.	D.1″.	Cotang.	′
0	9.01 9235	20.00	9.99 7614	.22	9.02 1620	20.23	10.97 8380	60
1	.02 0435	19.95	7601	.22	2834	20.17	7166	59
2	1632	19.88	7588	.23	4044	20.12	5956	58
3	2825	19.85	7574	.22	5251	20.07	4749	57
4	4016	19.78	7561	.23	6455	20.00	3545	56
5	9.02 5203	19.72	9.99 7547	.22	9.02 7655	19.95	10.97 2345	55
6	6386	19.68	7534	.23	.02 8852	19.90	.97 1148	54
7	7567	19.62	7520	.22	.03 0046	19.85	.96 9954	53
8	8744	19.57	7507	.23	1237	19.80	8763	52
9	.02 9918	19.52	7493	.22	2425	19.73	7575	51
10	9.03 1089	19.47	9.99 7480	.23	9.03 3609	19.70	10.96 6391	50
11	2257	19.40	7466	.23	4791	19.63	5209	49
12	3421	19.35	7452	.22	5969	19.58	4031	48
13	4582	19.32	7439	.23	7144	19.53	2856	47
14	5741	19.25	7425	.23	8316	19.48	1684	46
15	9.03 6896	19.20	9.99 7411	.23	9.03 9485	19.43	10.96 0515	45
16	8048	19.15	7397	.23	.04 0651	19.37	.95 9349	44
17	.03 9197	19.08	7383	.23	1813	19.33	8187	43
18	.04 0342	19.05	7369	.23	2973	19.28	7027	42
19	1485	19.00	7355	.23	4130	19.23	5870	41
20	9.04 2625	18.95	9.99 7341	.23	9.04 5284	19.17	10.95 4716	40
21	3762	18.88	7327	.23	6434	19.13	3566	39
22	4895	18.85	7313	.23	7582	19.08	2418	38
23	6026	18.80	7299	.23	8727	19.03	1273	37
24	7154	18.75	7285	.23	.04 9869	18.98	.95 0131	36
25	9.04 8279	18.68	9.99 7271	.23	9.05 1008	18.93	10.94 8992	35
26	.04 9400	18.65	7257	.25	2144	18.88	7856	34
27	.05 0519	18.60	7242	.23	3277	18.83	6723	33
28	1635	18.57	7228	.23	4407	18.80	5593	32
29	2749	18.50	7214	.25	5535	18.73	4465	31
30	9.05 3859	18.45	9.99 7199	.23	9.05 6659	18.70	10.94 3341	30
31	4966	18.42	7185	.25	7781	18.65	2219	29
32	6071	18.35	7170	.23	.05 8900	18.60	.94 1100	28
33	7172	18.32	7156	.25	.06 0016	18.57	.93 9984	27
34	8271	18.27	7141	.23	1130	18.50	8870	26
35	9.05 9367	18.22	9.99 7127	.25	9.06 2240	18.47	10.93 7760	25
36	.06 0460	18.18	7112	.23	3348	18.42	6652	24
37	1551	18.13	7098	.25	4453	18.38	5547	23
38	2639	18.08	7083	.25	5556	18.32	4444	22
39	3724	18.03	7068	.25	6655	18.28	3345	21
40	9.06 4806	17.98	9.99 7053	.23	9.06 7752	18.25	10.93 2248	20
41	5885	17.95	7039	.25	8846	18.20	1154	19
42	6962	17.90	7024	.25	.06 9938	18.15	.93 0062	18
43	8036	17.85	7009	.25	.07 1027	18.10	.92 8973	17
44	.06 9107	17.82	6994	.25	2113	18.07	7887	16
45	9.07 0176	17.77	9.99 6979	.25	9.07 3197	18.02	10.92 6803	15
46	1242	17.73	6964	.25	4278	17.97	5722	14
47	2306	17.67	6949	.25	5356	17.93	4644	13
48	3366	17.63	6934	.25	6432	17.88	3568	12
49	4424	17.60	6919	.25	7505	17.85	2495	11
50	9.07 5480	17.55	9.99 6904	.25	9.07 8576	17.80	10.92 1424	10
51	6533	17.50	6889	.25	.07 9644	17.77	.92 0356	9
52	7583	17.47	6874	.27	.08 0710	17.72	.91 9290	8
53	8631	17.42	6858	.25	1773	17.67	8227	7
54	.07 9676	17.38	6843	.27	2833	17.63	7167	6
55	9.08 0719	17.33	9.99 6828	.27	9.08 3891	17.60	10.91 6109	5
56	1759	17.30	6812	.25	4947	17.55	5053	4
57	2797	17.25	6797	.25	6000	17.50	4000	3
58	3832	17.20	6782	.27	7050	17.47	2950	2
59	4864	17.17	6766	.25	8098	17.43	1902	1
60	9.08 5894		9.99 6751		9.08 9144		10.91 0856	0
′	Cosine.	D.1″.	Sine.	D.1″.	Cotang.	D.1″.	Tang.	′

96° 83°

7° TABLE VIII.—*Continued* 172°

′	Sine.	D.1″.	Cosine.	D.1″.	Tang.	D.1″.	Cotang.	′
0	**9.08 5894**	17.13	**9.99 6751**	.27	**9.08 9144**	17.38	**10.91 0856**	**60**
1	6922	17.08	6735	.25	.09 0187	17.35	.90 9813	59
2	7947	17.05	6720	.27	1228	17.30	8772	58
3	8970	17.00	6704	.27	2266	17.27	7734	57
4	.08 9990	16.97	6688	.25	3302	17.23	6698	56
5	**9.09 1003**	16.93	**9.99 6673**	.27	**9.09 4336**	17.18	**10.90 5664**	**55**
6	2024	16.88	6657	.27	5367	17.13	4633	54
7	3037	16.83	6641	.27	6395	17.12	3605	53
8	4047	16.82	6625	.25	7422	17.07	2578	52
9	5056	16.77	6610	.27	8446	17.03	1554	51
10	**9.09 6062**	16.72	**9.99 6594**	.27	**9.09 9468**	16.98	**10.90 0532**	**50**
11	7065	16.68	6578	.27	.10 0487	16.95	.89 9513	49
12	8066	16.65	6562	.27	1504	16.92	8496	48
13	.09 9065	16.62	6546	.27	2519	16.88	7481	47
14	.10 0062	16.57	6530	.27	3532	16.83	6468	46
15	**9.10 1056**	16.53	**9.99 6514**	.27	**9.10 4542**	16.80	**10.89 5458**	**45**
16	2048	16.48	6498	.27	5550	16.77	4450	44
17	3037	16.47	6482	.28	6556	16.72	3444	43
18	4025	16.42	6465	.27	7559	16.68	2441	42
19	5010	16.37	6449	.27	8560	16.65	1440	41
20	**9.10 5992**	16.35	**9.99 6433**	.27	**9.10 9559**	16.62	**10.89 0441**	**40**
21	6973	16.30	6417	.28	.11 0556	16.58	.88 9444	39
22	7951	16.27	6400	.27	1551	16.53	8449	38
23	8927	16.23	6384	.27	2543	16.50	7457	37
24	.10 9901	16.20	6368	.28	3533	16.47	6467	36
25	**9.11 0873**	16.15	**9.99 6351**	.27	**9.11 4521**	16.43	**10 88 5479**	**35**
26	1842	16.12	6335	.28	5507	16.40	4493	34
27	2809	16.08	6318	.27	6491	16.35	3509	33
28	3774	16.05	6302	.28	7472	16.33	2528	32
29	4737	16.02	6285	.27	8452	16.28	1548	31
30	**9.11 5698**	15.97	**9.99 6269**	.28	**9.11 9429**	16.25	**10.88 0571**	**30**
31	6656	15.95	6252	.28	.12 0404	16.22	.87 9596	29
32	7613	15.90	6235	.27	1377	16.18	8623	28
33	8567	15.87	6219	.28	2348	16.15	7652	27
34	.11 9519	15.83	6202	.28	3317	16.12	6683	26
35	**9.12 0469**	15.80	**9.99 6185**	.28	**9.12 4284**	16.08	**10.87 5716**	**25**
36	1417	15.75	6168	.28	5249	16.03	4751	24
37	2362	15.73	6151	.28	6211	16.02	3789	23
38	3306	15.70	6134	.28	7172	15.97	2828	22
39	4248	15.65	6117	.28	8130	15.95	1870	21
40	**9.12 5187**	15.63	**9.99 6100**	.28	**9.12 9087**	15.90	**10.87 0913**	**20**
41	6125	15.58	6083	.28	.13 0041	15.88	.86 9959	19
42	7060	15.55	6066	.28	0994	15.83	9006	18
43	7993	15.53	6049	.28	1944	15.82	8056	17
44	8925	15.48	6032	.28	2893	15.77	7107	16
45	**9.12 9854**	15.45	**9.99 6015**	.28	**9.13 3839**	15.75	**10.86 6161**	**15**
46	.13 0781	15.42	5998	.30	4784	15.70	5216	14
47	1706	15.40	5980	.28	5726	15.68	4274	13
48	2630	15.35	5963	.28	6667	15.63	3333	12
49	3551	15.32	5946	.30	7605	15.62	2395	11
50	**9.13 4470**	15.28	**9.99 5928**	.28	**9.13 8542**	15.57	**10.86 1458**	**10**
51	5387	15.27	5911	.28	.13 9476	15.55	.86 0524	9
52	6303	15.22	5894	.30	.14 0409	15.52	.85 9591	8
53	7216	15.20	5876	.28	1340	15.48	8660	7
54	8128	15.15	5859	.30	2269	15.45	7731	6
55	**9.13 9037**	15.12	**9.99 5841**	.30	**9.14 3196**	15.42	**10.85 6804**	**5**
56	.13 9944	15.10	5823	.28	4121	15.38	5879	4
57	.14 0850	15.07	5806	.30	5044	15.37	4956	3
58	1754	15.02	5788	.28	5966	15.32	4034	2
59	2655	15.00	5771	.30	6885	15.30	3115	1
60	**9.14 3555**		**9.99 5753**		**9.14 7803**		**10.85 2197**	**0**
′	Cosine.	D.1″.	Sine.	D.1″.	Cotang.	D.1″.	Tang.	′

97° 82°

8° Table VIII.—*Continued* 171°

′	Sine.	D.1″.	Cosine.	D.1″.	Tang.	D.1″.	Cotang.	′
0	9.14 3555	14.97	9.99 5753	.30	9.14 7803	15.25	10.85 2197	60
1	4453	14.93	5735	.30	8718	15.23	1282	59
2	5349	14.90	5717	.30	.14 9632	15.20	.85 0368	58
3	6243	14.88	5699	.30	.15 0544	15.17	.84 9456	57
4	7136	14.83	5681	.28	1454	15.15	8546	56
5	9.14 8026	14.82	9.99 5664	.30	9.15 2363	15.10	10.84 7637	55
6	8915	14.78	5646	.30	3269	15.08	6731	54
7	.14 9802	14.73	5628	.30	4174	15.05	5826	53
8	.15 0686	14.72	5610	.32	5077	15.02	4923	52
9	1569	14.70	5591	.30	5978	14.98	4022	51
10	9.15 2451	14.65	9.99 5573	.30	9.15 6877	14.97	10.84 3123	50
11	3330	14.63	5555	.30	7775	14.93	2225	49
12	4208	14.58	5537	.30	8671	14.90	1329	48
13	5083	14.57	5519	.30	.15 9565	14.87	.84 0435	47
14	5957	14.55	5501	.32	.16 0457	14.83	.83 9543	46
15	9.15 6830	14.50	9.99 5482	.30	9.16 1347	14.82	10.83 8653	45
16	7700	14.48	5464	.30	2236	14.78	7764	44
17	8569	14.43	5446	.32	3123	14.75	6877	43
18	.15 9435	14.43	5427	.30	4008	14.73	5992	42
19	.16 0301	14.38	5409	.32	4892	14.70	5108	41
20	9.16 1164	14.35	9.99 5390	.30	9.16 5774	14.67	10.83 4226	40
21	2025	14.33	5372	.32	6654	14.63	3346	39
22	2885	14.30	5353	.32	7532	14.62	2468	38
23	3743	14.28	5334	.30	8409	14.58	1591	37
24	4600	14.23	5316	.32	.16 9284	14.55	.83 0716	36
25	9.16 5454	14.22	9.99 5297	.32	9.17 0157	14.53	10.82 9843	35
26	6307	14.20	5278	.30	1029	14.50	8971	34
27	7159	14.15	5260	.32	1899	14.47	8101	33
28	8008	14.13	5241	.32	2767	14.45	7233	32
29	8856	14.10	5222	.32	3634	14.42	6366	31
30	9.16 9702	14.08	9.99 5203	.32	9.17 4499	14.38	10.82 5501	30
31	.17 0547	14.03	5184	.32	5362	14.37	4638	29
32	1389	14.02	5165	.32	6224	14.33	3776	28
33	2230	14.00	5146	.32	7084	14.30	2916	27
34	3070	13.97	5127	.32	7942	14.28	2058	26
35	9.17 3908	13.93	9.99 5108	.32	9.17 8799	14.27	10.82 1201	25
36	4744	13.90	5089	.32	.17 9655	14.22	.82 0345	24
37	5578	13.88	5070	.32	.18 0508	14.20	.81 9492	23
38	6411	13.85	5051	.32	1360	14.18	8640	22
39	7242	13.83	5032	.32	2211	14.13	7789	21
40	9.17 8072	13.80	9.99 5013	.33	9.18 3059	14.13	10.81 6941	20
41	8900	13.77	4993	.32	3907	14.08	6093	19
42	.17 9726	13.75	4974	.32	4752	14.08	5248	18
43	.18 0551	13.72	4955	.33	5597	14.03	4403	17
44	1374	13.70	4935	.32	6439	14.02	3561	16
45	9.18 2196	13.67	9.99 4916	.33	9.18 7280	14.00	10.81 2720	15
46	3016	13.63	4896	.32	8120	13.97	1880	14
47	3834	13.62	4877	.33	8958	13.93	1042	13
48	4651	13.58	4857	.32	.18 9794	13.92	.81 0206	12
49	5466	13.57	4838	.33	.19 0629	13.88	.80 9371	11
50	9.18 6280	13.53	9.99 4818	.33	9.19 1462	13.87	10.80 8538	10
51	7092	13.52	4798	.32	2294	13.83	7706	9
52	7903	13.48	4779	.33	3124	13.82	6876	8
53	8712	13.45	4759	.33	3953	13.78	6047	7
54	.18 9519	13.43	4739	.32	4780	13.77	5220	6
55	9.19 0325	13.42	9.99 4720	.33	9.19 5606	13.73	10.80 4394	5
56	1130	13.38	4700	.33	6430	13.72	3570	4
57	1933	13.35	4680	.33	7253	13.68	2747	3
58	2734	13.33	4660	.33	8074	13.67	1926	2
59	3534	13.30	4640	.33	8894	13.65	1106	1
60	9.19 4332		9.99 4620		9.19 9713		10.80 0287	0
′	Cosine.	D.1″.	Sine.	D.1″.	Cotang.	D.1″.	Tang.	′

98° 81°

9° TABLE VIII.—*Continued* 170°

′	Sine.	D.1″.	Cosine.	D.1″.	Tang.	D.1″.	Cotang.	′
0	**9.19 4332**	13.28	**9.99 4620**	.33	**9.19 9713**	13.60	**10.80 0287**	**60**
1	5129	13.27	4600	.33	.20 0529	13.60	.79 9471	59
2	5925	13.23	4580	.33	1345	13.57	8655	58
3	6719	13.20	4560	.33	2159	13.53	7841	57
4	7511	13.18	4540	.35	2971	13.52	7029	56
5	**9.19 8302**	13.15	**9.99 4519**	.33	**9.20 3782**	13.50	**10.79 6218**	**55**
6	9091	13.13	4499	.33	4592	13.47	5408	54
7	.19 9879	13.12	4479	.33	5400	13.45	4600	53
8	.20 0666	13.08	4459	.35	6207	13.43	3793	52
9	1451	13.05	4438	.33	7013	13.40	2987	51
10	**9.20 2234**	13.05	**9.99 4418**	.33	**9.20 7817**	13.37	**10.79 2183**	**50**
11	3017	13.00	4398	.35	8619	13.35	1381	49
12	3797	13.00	4377	.33	.20 9420	13.33	.79 0580	48
13	4577	12.95	4357	.35	.21 0220	13.30	.78 9780	47
14	5354	12.95	4336	.33	1018	13.28	8982	46
15	**9.20 6131**	12.92	**9.99 4316**	.35	**9.21 1815**	13.27	**10.78 8185**	**45**
16	6906	12.88	4295	.35	2611	13.23	7389	44
17	7679	12.88	4274	.33	3405	13.22	6595	43
18	8452	12.83	4254	.35	4198	13.18	5802	42
19	9222	12.83	4233	.35	4989	13.18	5011	41
20	**9.20 9992**	12.80	**9.99 4212**	.35	**9.21 5780**	13.13	**10.78 4220**	**40**
21	.21 0760	12.77	4191	.33	6568	13.13	3432	39
22	1526	12.75	4171	.35	7356	13.10	2644	38
23	2291	12.73	4150	.35	8142	13.07	1858	37
24	3055	12.72	4129	.35	8926	13.07	1074	36
25	**9.21 3818**	12.68	**9.99 4108**	.35	**9.21 9710**	13.03	**10.78 0290**	**35**
26	4579	12.65	4087	.35	.22 0492	13.00	.77 9508	34
27	5338	12.65	4066	.35	1272	13.00	8728	33
28	6097	12.62	4045	.35	2052	12.97	7948	32
29	6854	12.58	4024	.35	2830	12.95	7170	31
30	**9.21 7609**	12.57	**9.99 4003**	.35	**9.22 3607**	12.92	**10.77 6393**	**30**
31	8363	12.55	3982	.37	4382	12.90	5618	29
32	9116	12.53	3960	.35	5156	12.88	4844	28
33	.21 9868	12.50	3939	.35	5929	12.85	4071	27
34	.22 0618	12.48	3918	.35	6700	12.85	3300	26
35	**9.22 1367**	12.47	**9.99 3897**	.37	**9.22 7471**	12.80	**10.77 2529**	**25**
36	2115	12.43	3875	.35	8239	12.80	1761	24
37	2861	12.42	3854	.37	9007	12.77	0993	23
38	3606	12.38	3832	.35	.22 9773	12.77	.77 0227	22
39	4349	12.38	3811	.37	.23 0539	12.72	.76 9461	21
40	**9.22 5092**	12.35	**9.99 3789**	.35	**9.23 1302**	12.72	**10.76 8698**	**20**
41	5833	12.33	3768	.37	2065	12.68	7935	19
42	6573	12.30	3746	.35	2826	12.67	7174	18
43	7311	12.28	3725	.37	3586	12.65	6414	17
44	8048	12.27	3703	.37	4345	12.63	5655	16
45	**9.22 8784**	12.23	**9.99 3681**	.35	**9.23 5103**	12.60	**10.76 4897**	**15**
46	.22 9518	12.23	3660	.37	5859	12.58	4141	14
47	.23 0252	12.20	3638	.37	6614	12.57	3386	13
48	0984	12.18	3616	.37	7368	12.53	2632	12
49	1715	12.15	3594	.37	8120	12.53	1880	11
50	**9.23 2444**	12.13	**9.99 3572**	.37	**9.23 8872**	12.50	**10.76 1128**	**10**
51	3172	12.12	3550	.37	.23 9622	12.48	.76 0378	9
52	3899	12.10	3528	.37	.24 0371	12.45	.75 9629	8
53	4625	12.07	3506	.37	1118	12.45	8882	7
54	5349	12.07	3484	.37	1865	12.42	8135	6
55	**9.23 6073**	12.03	**9.99 3462**	.37	**9.24 2610**	12.40	**10.75 7390**	**5**
56	6795	12.00	3440	.37	3354	12.38	6646	4
57	7515	12.00	3418	.37	4097	12.37	5903	3
58	8235	11.97	3396	.37	4839	12.33	5161	2
59	8953	11.95	3374	.38	5579	12.33	4421	1
60	**9.23 9670**		**9.99 3351**		**9.24 6319**		**10.75 3681**	**0**
′	Cosine.	D.1″.	Sine.	D.1″.	Cotang.	D.1″.	Tang.	′

99° 80°

10° **Table VIII.**—*Continued* 169°

′	Sine.	D.1″.	Cosine.	D.1″.	Tang.	D.1″.	Cotang.	′
0	**9.23 9670**	11.93	**9.99 3351**	.37	**9.24 6319**	12.30	**10.75 3681**	**60**
1	.24 0386	11.92	3329	.37	7057	12.28	2943	59
2	1101	11.88	3307	.38	7794	12.27	2206	58
3	1814	11.87	3284	.37	8530	12.23	1470	57
4	2526	11.85	3262	.37	9264	12.23	0736	56
5	**9.24 3237**	11.83	**9.99 3240**	.38	**9.24 9998**	12.20	**10.75 0002**	**55**
6	3947	11.82	3217	.37	.25 0730	12.18	.74 9270	54
7	4656	11.78	3195	.38	1461	12.17	8539	53
8	5363	11.77	3172	.38	2191	12.15	7809	52
9	6069	11.77	3149	.37	2920	12.13	7080	51
10	**9.24 6775**	11.72	**9.99 3127**	.38	**9.25 3648**	12.10	**10.74 6352**	**50**
11	7478	11.72	3104	.38	4374	12.10	5626	49
12	8181	11.70	3081	.37	5100	12.07	4900	48
13	8883	11.67	3059	.38	5824	12.05	4176	47
14	.24 9583	11.65	3036	.38	6547	12.03	3453	46
15	**9.25 0282**	11.63	**9.99 3013**	.38	**9.25 7269**	12.02	**10.74 2731**	**45**
16	0980	11.62	2990	.38	7990	12.00	2010	44
17	1677	11.60	2967	.38	8710	11.98	1290	43
18	2373	11.57	2944	.38	.25 9429	11.95	.74 0571	42
19	3067	11.57	2921	.38	.26 0146	11.95	.73 9854	41
20	**9.25 3761**	11.53	**9.99 2898**	.38	**9.26 0863**	11.92	**10.73 9137**	**40**
21	4453	11.52	2875	.38	1578	11.90	8422	39
22	5144	11.50	2852	.38	2292	11.88	7708	38
23	5834	11.48	2829	.38	3005	11.87	6995	37
24	6523	11.47	2806	.38	3717	11.85	6283	36
25	**9.25 7211**	11.45	**9.99 2783**	.40	**9.26 4428**	11.83	**10.73 5572**	**35**
26	7898	11.42	2759	.38	5138	11.82	4862	34
27	8583	11.42	2736	.38	5847	11.80	4153	33
28	9268	11.38	2713	.38	6555	11.77	3445	32
29	.25 9951	11.37	2690	.40	7261	11.77	2739	31
30	**9.26 0633**	11.35	**9.99 2666**	.38	**9.26 7967**	11.73	**10.73 2033**	**30**
31	1314	11.33	2643	.40	8671	11.73	1329	29
32	1994	11.32	2619	.38	.26 9375	11.70	.73 0625	28
33	2673	11.30	2596	.40	.27 0077	11.70	.72 9923	27
34	3351	11.27	2572	.38	0779	11.67	9221	26
35	**9.26 4027**	11.27	**9.99 2549**	.40	**9.27 1479**	11.65	**10.72 8521**	**25**
36	4703	11.23	2525	.40	2178	11.63	7822	24
37	5377	11.23	2501	.38	2876	11.62	7124	23
38	6051	11.20	2478	.40	3573	11.60	6427	22
39	6723	11.20	2454	.40	4260	11.58	5731	21
40	**9.26 7395**	11.17	**9.99 2430**	.40	**9.27 4964**	11.57	**10.72 5036**	**20**
41	8065	11.15	2406	.40	5658	11.55	4342	19
42	8734	11.13	2382	.38	6351	11.53	3649	18
43	.26 9402	11.12	2359	.40	7043	11.52	2957	17
44	.27 0069	11.10	2335	.40	7734	11.50	2266	16
45	**9.27 0735**	11.08	**9.99 2311**	.40	**9.27 8424**	11.48	**10.72 1576**	**15**
46	1400	11.07	2287	.40	9113	11.47	0887	14
47	2064	11.03	2263	.40	.27 9801	11.45	.72 0199	13
48	2726	11.03	2239	.42	.28 0488	11.43	.71 9512	12
49	3388	11.02	2214	.40	1174	11.40	8826	11
50	**9.27 4049**	10.98	**9.99 2190**	.40	**9.28 1858**	11.40	**10.71 8142**	**10**
51	4708	10.98	2166	.40	2542	11.38	7458	9
52	5367	10.97	2142	.40	3225	11.37	6775	8
53	6025	10.93	2118	.42	3907	11.35	6093	7
54	6681	10.93	2093	.40	4588	11.33	5412	6
55	**9.27 7337**	10.90	**9.99 2069**	.42	**9.28 5268**	11.32	**10.71 4732**	**5**
56	7991	10.90	2044	.40	5947	11.28	4053	4
57	8645	10.87	2020	.40	6624	11.28	3376	3
58	9297	10.85	1996	.42	7301	11.27	2699	2
59	.27 9948	10.85	1971	.40	7977	11.25	2023	1
60	**9.28 0599**		**9.99 1947**		**9.28 8652**		**10.71 1348**	**0**
′	Cosine.	D.1″.	Sine.	D.1″.	Cotang.	D.1″.	Tang.	′

100° 79°

11° TABLE VIII.—*Continued* 168°

′	Sine.	D.1″.	Cosine.	D.1″.	Tang.	D.1″.	Cotang.	′
0	**9.28 0599**	10.82	**9.99 1947**	.42	**9.28 8652**	11.23	**10.71 1348**	**60**
1	1248	10.82	1922	.42	9326	11.22	0674	59
2	1897	10.78	1897	.40	.28 9999	11.20	.71 0001	58
3	2544	10.77	1873	.42	.29 0671	11.18	.70 9329	57
4	3190	10.77	1848	.42	1342	11.18	8658	56
5	**9.28 3836**	10.73	**9.99 1823**	.40	**9.29 2013**	11.15	**10.70 7987**	**55**
6	4480	10.73	1799	.42	2682	11.13	7318	54
7	5124	10.70	1774	.42	3350	11.12	6650	53
8	5766	10.70	1749	.42	4017	11.12	5983	52
9	6408	10.67	1724	.42	4684	11.08	5316	51
10	**9.28 7048**	10.67	**9.99 1699**	.42	**9.29 5349**	11.07	**10.70 4651**	**50**
11	7688	10.63	1674	.42	6013	11.07	3987	49
12	8326	10.63	1649	.42	6677	11.03	3323	48
13	8964	10.60	1624	.42	7339	11.03	2661	47
14	.28 9600	10.60	1599	.42	8001	11.02	1999	46
15	**9.29 0236**	10.57	**9.99 1574**	.42	**9.29 8662**	11.00	**10.70 1338**	**45**
16	0870	10.57	1549	.42	9322	10.97	0678	44
17	1504	10.55	1524	.43	.29 9980	10.97	.70 0020	43
18	2137	10.52	1498	.42	.30 0638	10.95	.69 9362	42
19	2768	10.52	1473	.42	1295	10.93	8705	41
20	**9.29 3399**	10.50	**9.99 1448**	.43	**9.30 1951**	10.93	**10.69 8049**	**40**
21	4029	10.48	1422	.42	2607	10.90	7393	39
22	4658	10.47	1397	.42	3261	10.88	6739	38
23	5286	10.45	1372	.43	3914	10.88	6086	37
24	5913	10.43	1346	.42	4567	10.85	5433	36
25	**9.29 6539**	10.42	**9.99 1321**	.43	**9.30 5218**	10.85	**10.69 4782**	**35**
26	7164	10.40	1295	.42	5869	10.83	4131	34
27	7788	10.40	1270	.43	6519	10.82	3481	33
28	8412	10.37	1244	.43	7168	10.80	2832	32
29	9034	10.35	1218	.42	7816	10.78	2184	31
30	**9.29 9655**	10.35	**9.99 1193**	.43	**9.30 8463**	10.77	**10.69 1537**	**30**
31	.30 0276	10.32	1167	.43	9109	10.75	0891	29
32	0895	10.32	1141	.43	.30 9754	10.75	.69 0246	28
33	1514	10.30	1115	.42	.31 0399	10.72	.68 9601	27
34	2132	10.27	1090	.43	1042	10.72	8958	26
35	**9.30 2748**	10.27	**9.99 1064**	.43	**9.31 1685**	10.70	**10.68 8315**	**25**
36	3364	10.25	1038	.43	2327	10.68	7673	24
37	3979	10.23	1012	.43	2968	10.67	7032	23
38	4593	10.23	0986	.43	3608	10.65	6392	22
39	5207	10.20	0960	.43	4247	10.63	5753	21
40	**9.30 5819**	10.18	**9.99 0934**	.43	**9.31 4885**	10.63	**10.68 5115**	**20**
41	6430	10.18	0908	.43	5523	10.60	4477	19
42	7041	10.15	0882	.45	6159	10.60	3841	18
43	7650	10.15	0855	.43	6795	10.58	3205	17
44	8259	10.13	0829	.43	7430	10.57	2570	16
45	**9.30 8867**	10.12	**9.99 0803**	.43	**9.31 8064**	10.55	**10.68 1936**	**15**
46	.30 9474	10.10	0777	.45	8697	10.55	1303	14
47	.31 0080	10.08	0750	.43	9330	10.52	0670	13
48	0685	10.07	0724	.45	.31 9961	10.52	.68 0039	12
49	1289	10.07	0697	.43	.32 0592	10.50	.67 9408	11
50	**9.31 1893**	10.03	**9.99 0671**	.43	**9.32 1222**	10.48	**10.67 8778**	**10**
51	2495	10.03	0645	.45	1851	10.47	8149	9
52	3097	10.02	0618	.45	2479	10.45	7521	8
53	3698	9.98	0591	.43	3106	10.45	6894	7
54	4297	10.00	0565	.45	3733	10.42	6267	6
55	**9.31 4897**	9.97	**9.99 0538**	.45	**9.32 4358**	10.42	**10.67 5642**	**5**
56	5495	9.95	0511	.43	4983	10.40	5017	4
57	6092	9.95	0485	.45	5607	10.40	4393	3
58	6689	9.92	0458	.45	6231	10.37	3769	2
59	7284	9.92	0431	.45	6853	10.37	3147	1
60	**9.31 7879**		**9.99 0404**		**9.32 7475**		**10.67 2525**	**0**
′	Cosine.	D.1″.	Sine.	D.1″.	Cotang.	D.1″.	Tang.	′

101° 78°

12° TABLE VIII.—*Continued* 167°

′	Sine.	D.1″.	Cosine.	D.1″.	Tang.	D.1″.	Cotang.	′
0	**9.31 7879**	9.90	**9.99 0404**	.43	**9.32 7475**	10.33	**10.67 2525**	**60**
1	8473	9.88	0378	.45	8095	10.33	1905	59
2	9066	9.87	0351	.45	8715	10.32	1285	58
3	.31 9658	9.85	0324	.45	9334	10.32	0666	57
4	.32 0249	9.85	0297	.45	.32 9953	10.28	.67 0047	56
5	**9.32 0840**	9.83	**9.99 0270**	.45	**9.33 0570**	10.28	**10.66 9430**	**55**
6	1430	9.82	0243	.47	1187	10.27	8813	54
7	2019	9.80	0215	.45	1803	10.25	8197	53
8	2607	9.78	0188	.45	2418	10.25	7582	52
9	3194	9.77	0161	.45	3033	10.22	6967	51
10	**9.32 3780**	9.77	**9.99 0134**	.45	**9.33 3646**	10.22	**10.66 6354**	**50**
11	4366	9.73	0107	.47	4259	10.20	5741	49
12	4950	9.73	0079	.45	4871	10.18	5129	48
13	5534	9.72	0052	.45	5482	10.18	4518	47
14	6117	9.72	.99 0025	.47	6093	10.15	3907	46
15	**9.32 6700**	9.68	**9.98 9997**	.45	**9.33 6702**	10.15	**10.66 3298**	**45**
16	7281	9.68	9970	.47	7311	10.13	2689	44
17	7862	9.67	9942	.45	7919	10.13	2081	43
18	8442	9.65	9915	.47	8527	10.10	1473	42
19	9021	9.63	9887	.45	9133	10.10	0867	41
20	**9.32 9599**	9.62	**9.98 9860**	.47	**9.33 9739**	10.08	**10.66 0261**	**40**
21	.33 0176	9.62	9832	.47	.34 0344	10.07	.65 9656	39
22	0753	9.60	9804	.45	0948	10.07	9052	38
23	1329	9.57	9777	.47	1552	10.05	8448	37
24	1903	9.58	9749	.47	2155	10.03	7845	36
25	**9.33 2478**	9.55	**9.98 9721**	.47	**9.34 2757**	10.02	**10.65 7243**	**35**
26	3051	9.55	9693	.47	3358	10.00	6642	34
27	3624	9.52	9665	.47	3958	10.00	6042	33
28	4195	9.53	9637	.45	4558	9.98	5442	32
29	4767	9.50	9610	.47	5157	9.97	4843	31
30	**9.33 5337**	9.48	**9.98 9582**	.48	**9.34 5755**	9.97	**10.65 4245**	**30**
31	5906	9.48	9553	.47	6353	9.93	3647	29
32	6475	9.47	9525	.47	6949	9.93	3051	28
33	7043	9.45	9497	.47	7545	9.93	2455	27
34	7610	9.43	9469	.47	8141	9.90	1859	26
35	**9.33 8176**	9.43	**9.98 9441**	.47	**9.34 8735**	9.90	**10.65 1265**	**25**
36	8742	9.42	9413	.47	9329	9.88	0671	24
37	9307	9.40	9385	.48	.34 9922	9.87	.65 0078	23
38	.33 9871	9.38	9356	.47	.35 0514	9.87	.64 9486	22
39	.34 0434	9.37	9328	.47	1106	9.85	8894	21
40	**9.34 0996**	9.37	**9.98 9300**	.48	**9.35 1697**	9.83	**10.64 8303**	**20**
41	1558	9.35	9271	.47	2287	9.82	7713	19
42	2119	9.33	9243	.48	2876	9.82	7124	18
43	2679	9.33	9214	.47	3465	9.80	6535	17
44	3239	9.30	9186	.48	4053	9.78	5947	16
45	**9.34 3797**	9.30	**9.98 9157**	.48	**9.35 4640**	9.78	**10.64 5360**	**15**
46	4355	9.28	9128	.47	5227	9.77	4773	14
47	4912	9.28	9100	.48	5813	9.75	4187	13
48	5469	9.25	9071	.48	6398	9.73	3602	12
49	6024	9.25	9042	.47	6982	9.73	3018	11
50	**9.34 6579**	9.25	**9.98 9014**	.48	**9.35 7566**	9.72	**10.64 2434**	**10**
51	7134	9.22	8985	.48	8149	9.70	1851	9
52	7687	9.22	8956	.48	8731	9.70	1269	8
53	8240	9.20	8927	.48	9313	9.67	0687	7
54	8792	9.18	8898	.48	.35 9893	9.68	.64 0107	6
55	**9.34 9343**	9.17	**9.98 8869**	.48	**9.36 0474**	9.65	**10.63 9526**	**5**
56	.34 9893	9.17	8840	.48	1053	9.65	8947	4
57	.35 0443	9.15	8811	.48	1632	9.63	8368	3
58	0992	9.13	8782	.48	2210	9.62	7790	2
59	1540	9.13	8753	.48	2787	9.62	7213	1
60	**9.35 2088**		**9.98 8724**		**9.36 3364**		**10.63 6636**	**0**
′	Cosine.	D.1″.	Sine.	D.1″.	Cotang.	D.1″.	Tang.	′
102°								77°

13° TABLE VIII.—*Continued* 166°

′	Sine.	D.1″.	Cosine.	D.1″.	Tang.	D.1″.	Cotang.	′
0	9.35 2088	9.12	9.98 8724	.48	9.36 3364	9.60	10.63 6636	60
1	2635	9.10	8695	.48	3940	9.58	6060	59
2	3181	9.08	8666	.50	4515	9.58	5485	58
3	3726	9.08	8636	.48	5090	9.57	4910	57
4	4271	9.07	8607	.48	5664	9.55	4336	56
5	9.35 4815	9.05	9.98 8578	.50	9.36 6237	9.55	10.63 3763	55
6	5358	9.05	8548	.48	6810	9.53	3190	54
7	5901	9.03	8519	.50	7382	9.52	2618	53
8	6443	9.02	8489	.48	7953	9.52	2047	52
9	6984	9.00	8460	.50	8524	9.50	1476	51
10	9.35 7524	9.00	9.98 8430	.48	9.36 9094	9.48	10.63 0906	50
11	8064	8.98	8401	.50	.36 9663	9.48	.63 0337	49
12	8603	8.97	8371	.48	.37 0232	9.45	.62 9768	48
13	9141	8.95	8342	.50	0799	9.47	9201	47
14	.35 9678	8.95	8312	.50	1367	9.43	8633	46
15	9.36 0215	8.95	9.98 8282	.50	9.37 1933	9.43	10.62 8067	45
16	0752	8.92	8252	.48	2499	9.42	7501	44
17	1287	8.92	8223	.50	3064	9.42	6936	43
18	1822	8.90	8193	.50	3629	9.40	6371	42
19	2356	8.88	8163	.50	4193	9.38	5807	41
20	9.36 2889	8.88	9.98 8133	.50	9.37 4756	9.38	10.62 5244	40
21	3422	8.87	8103	.50	5319	9.37	4681	39
22	3954	8.85	8073	.50	5881	9.35	4119	38
23	4485	8.85	8043	.50	6442	9.35	3558	37
24	5016	8.83	8013	.50	7003	9.33	2997	36
25	9.36 5546	8.82	9.98 7983	.50	9.37 7563	9.32	10.62 2437	35
26	6075	8.82	7953	.52	8122	9.32	1878	34
27	6604	8.78	7922	.50	8681	9.30	1319	33
28	7131	8.80	7892	.50	9239	9.30	0761	32
29	7659	8.77	7862	.50	.37 9797	9.28	.62 0203	31
30	9.36 8185	8.77	9.98 7832	.52	9.38 0354	9.27	10.61 9646	30
31	8711	8.75	7801	.50	0910	9.27	9090	29
32	9236	8.75	7771	.52	1466	9.23	8534	28
33	.36 9761	8.72	7740	.50	2020	9.25	7980	27
34	.37 0285	8.72	7710	.52	2575	9.23	7425	26
35	9.37 0808	8.70	9.98 7679	.50	9.38 3129	9.22	10.61 6871	25
36	1330	8.70	7649	.52	3682	9.20	6318	24
37	1852	8.68	7618	.50	4234	9.20	5766	23
38	2373	8.68	7588	.52	4786	9.18	5214	22
39	2894	8.67	7557	.52	5337	9.18	4663	21
40	9.37 3414	8.65	9.98 7526	.50	9.38 5888	9.17	10.61 4112	20
41	3933	8.65	7496	.52	6438	9.15	3562	19
42	4452	8.63	7465	.52	6987	9.15	3013	18
43	4970	8.62	7434	.52	7536	9.13	2464	17
44	5487	8.60	7403	.52	8084	9.12	1916	16
45	9.37 6003	8.60	9.98 7372	.52	9.38 8631	9.12	10.61 1369	15
46	6519	8.60	7341	.52	9178	9.10	0822	14
47	7035	8.57	7310	.52	.38 9724	9.10	.61 0276	13
48	7549	8.57	7279	.52	.39 0270	9.08	.60 9730	12
49	8063	8.57	7248	.52	0815	9.08	9185	11
50	9.37 8577	8.53	9.98 7217	.52	9.39 1360	9.05	10.60 8640	10
51	9089	8.53	7186	.52	1903	9.07	8097	9
52	.37 9601	8.53	7155	.52	2447	9.03	7553	8
53	.38 0113	8.52	7124	.53	2989	9.03	7011	7
54	0624	8.50	7092	.52	3531	9.03	6469	6
55	9.38 1134	8.48	9.98 7061	.52	9.39 4073	9.02	10.60 5927	5
56	1643	8.48	7030	.53	4614	9.00	5386	4
57	2152	8.48	6998	.52	5154	9.00	4846	3
58	2661	8.45	6967	.52	5694	8.98	4306	2
59	3168	8.45	6936	.53	6233	8.97	3767	1
60	9.38 3675		9.98 6904		9.39 6771		10.60 3229	0
′	Cosine.	D.1″.	Sine.	D.1″.	Cotang.	D.1″.	Tang.	′

103° 76°

14° TABLE VIII.—*Continued* 165°

′	Sine.	D.1″.	Cosine.	D.1″.	Tang.	D.1″.	Cotang.	′
0	**9.38 3675**	8.45	**9.98 6904**	.52	**9.39 6771**	8.97	**10.60 3229**	**60**
1	4182	8.42	6873	.53	7309	8.95	2691	59
2	4687	8.42	6841	.53	7846	8.95	2154	58
3	5192	8.42	6809	.52	8383	8.93	1617	57
4	5697	8.40	6778	.53	8919	8.93	1081	56
5	**9.38 6201**	8.38	**9.98 6746**	.53	**9.39 9455**	8.92	**10.60 0545**	**55**
6	6704	8.38	6714	.52	.39 9990	8.90	.60 0010	54
7	7207	8.37	6683	.53	.40 0524	8.90	.59 9476	53
8	7709	8.35	6651	.53	1058	8.88	8942	52
9	8210	8.35	6619	.53	1591	8.88	8409	51
10	**9.38 8711**	8.33	**9.98 6587**	.53	**9.40 2124**	8.87	**10.59 7876**	**50**
11	9211	8.33	6555	.53	2656	8.85	7344	49
12	.38 9711	8.32	6523	.53	3187	8.85	6813	48
13	.39 0210	8.30	6491	.53	3718	8.85	6282	47
14	0708	8.30	6459	.53	4249	8.82	5751	46
15	**9.39 1206**	8.28	**9.98 6427**	.53	**9.40 4778**	8.83	**10.59 5222**	**45**
16	1703	8.27	6395	.53	5308	8.80	4692	44
17	2199	8.27	6363	.53	5836	8.80	4164	43
18	2695	8.27	6331	.53	6364	8.80	3636	42
19	3191	8.23	6299	.55	6892	8.78	3108	41
20	**9.39 3685**	8.23	**9.98 6266**	.53	**9.40 7419**	8.77	**10.59 2581**	**40**
21	4179	8.23	6234	.53	7945	8.77	2055	39
22	4673	8.22	6202	.55	8471	8.75	1529	38
23	5166	8.20	6169	.53	8996	8.75	1004	37
24	5658	8.20	6137	.55	.40 9521	8.73	.59 0479	36
25	**9.39 6150**	8.18	**9.98 6104**	.53	**9.41 0045**	8.73	**10.58 9955**	**35**
26	6641	8.18	6072	.55	0569	8.72	9431	34
27	7132	8.15	6039	.53	1092	8.72	8908	33
28	7621	8.17	6007	.55	1615	8.70	8385	32
29	8111	8.15	5974	.53	2137	8.68	7863	31
30	**9.39 8600**	8.13	**9.98 5942**	.55	**9.41 2658**	8.68	**10.58 7342**	**30**
31	9088	8.12	5909	.55	3179	8.67	6821	29
32	.39 9575	8.12	5876	.55	3699	8.67	6301	28
33	.40 0062	8.12	5843	.53	4219	8.65	5781	27
34	0549	8.10	5811	.55	4738	8.65	5262	26
35	**9.40 1035**	8.08	**9.98 5778**	.55	**9.41 5257**	8.63	**10.58 4743**	**25**
36	1520	8.08	5745	.55	5775	8.63	4225	24
37	2005	8.07	5712	.55	6293	8.62	3707	23
38	2489	8.05	5679	.55	6810	8.60	3190	22
39	2972	8.05	5646	.55	7326	8.60	2674	21
40	**9.40 3455**	8.05	**9.98 5613**	.55	**9.41 7842**	8.60	**10.58 2158**	**20**
41	3938	8.03	5580	.55	8358	8.58	1642	19
42	4420	8.02	5547	.55	8873	8.57	1127	18
43	4901	8.02	5514	.57	9387	8.57	0613	17
44	5382	8.00	5480	.55	.41 9901	8.57	.58 0099	16
45	**9.40 5862**	7.98	**9.98 5447**	.55	**9.42 0415**	8.55	**10.57 9585**	**15**
46	6341	7.98	5414	.55	0927	8.55	9073	14
47	6820	7.98	5381	.57	1440	8.53	8560	13
48	7299	7.97	5347	.55	1952	8.52	8048	12
49	7777	7.95	5314	.57	2463	8.52	7537	11
50	**9.40 8254**	7.95	**9.98 5280**	.55	**9.42 2974**	8.50	**10.57 7026**	**10**
51	8731	7.93	5247	.57	3484	8.48	6516	9
52	9207	7.93	5213	.55	3993	8.50	6007	8
53	.40 9682	7.92	5180	.57	4503	8.47	5497	7
54	.41 0157	7.92	5146	.55	5011	8.47	4989	6
55	**9.41 0632**	7.90	**9.98 5113**	.57	**9.42 5519**	8.47	**10.57 4481**	**5**
56	1106	7.88	5079	.57	6027	8.45	3973	4
57	1579	7.88	5045	.57	6534	8.45	3466	3
58	2052	7.87	5011	.55	7041	8.43	2959	2
59	2524	7.87	4978	.57	7547	8.42	2453	1
60	**9.41 2996**		**9.98 4944**		**9.42 8052**		**10.57 1948**	**0**
′	Cosine.	D.1″.	Sine.	D.1″.	Cotang.	D.1″.	Tang.	′

104° 75°

15° TABLE VIII.—*Continued* 164°

′	Sine.	D.1″.	Cosine.	D.1″.	Tang.	D.1″.	Cotang.	′
0	9.41 2996	7.85	9.98 4944	.57	9.42 8052	8.43	10.57 1948	60
1	3467	7.85	4910	.57	8558	8.40	1442	59
2	3938	7.83	4876	.57	9062	8.40	0938	58
3	4408	7.83	4842	.57	.42 9566	8.40	.57 0434	57
4	4878	7.82	4808	.57	.43 0070	8.38	.56 9930	56
5	9.41 5347	7.80	9.98 4774	.57	9.43 0573	8.37	10.56 9427	55
6	5815	7.80	4740	.57	1075	8.37	8925	54
7	6283	7.80	4706	.57	1577	8.37	8423	53
8	6751	7.77	4672	.57	2079	8.35	7921	52
9	7217	7.78	4638	.58	2580	8.33	7420	51
10	9.41 7684	7.77	9.98 4603	.57	9.43 3080	8.33	10.56 6920	50
11	8150	7.75	4569	.57	3580	8.33	6420	49
12	8615	7.73	4535	.58	4080	8.32	5920	48
13	9079	7.75	4500	.57	4579	8.32	5421	47
14	.41 9544	7.72	4466	.57	5078	8.30	4922	46
15	9.42 0007	7.72	9.98 4432	.58	9.43 5576	8.28	10.56 4424	45
16	0470	7.72	4397	.57	6073	8.28	3927	44
17	0933	7.70	4363	.58	6570	8.28	3430	43
18	1395	7.70	4328	.57	7067	8.27	2933	42
19	1857	7.68	4294	.58	7563	8.27	2437	41
20	9.42 2318	7.67	9.98 4259	.58	9.43 8059	8.25	10.56 1941	40
21	2778	7.67	4224	.57	8554	8.23	1446	39
22	3238	7.65	4190	.58	9048	8.25	0952	38
23	3697	7.65	4155	.58	.43 9543	8.22	.56 0457	37
24	4156	7.65	4120	.58	.44 0036	8.22	.55 9964	36
25	9.42 4615	7.63	9.98 4085	.58	9.44 0529	8.22	10.55 9471	35
26	5073	7.62	4050	.58	1022	8.20	8978	34
27	5530	7.62	4015	.57	1514	8.20	8486	33
28	5987	7.60	3981	.58	2006	8.18	7994	32
29	6443	7.60	3946	.58	2497	8.18	7503	31
30	9.42 6899	7.58	9.98 3911	.60	9.44 2988	8.18	10.55 7012	30
31	7354	7.58	3875	.58	3479	8.15	6521	29
32	7809	7.57	3840	.58	3968	8.17	6032	28
33	8263	7.57	3805	.58	4458	8.15	5542	27
34	8717	7.55	3770	.58	4947	8.13	5053	26
35	9.42 9170	7.55	9.98 3735	.58	9.44 5435	8.13	10.55 4565	25
36	.42 9623	7.53	3700	.60	5923	8.13	4077	24
37	.43 0075	7.53	3664	.58	6411	8.12	3589	23
38	0527	7.52	3629	.58	6898	8.10	3102	22
39	0978	7.52	3594	.60	7384	8.10	2616	21
40	9.43 1429	7.50	9.98 3558	.58	9.44 7870	8.10	10.55 2130	20
41	1879	7.50	3523	.60	8356	8.08	1644	19
42	2329	7.48	3487	.58	8841	8.08	1159	18
43	2778	7.47	3452	.60	9326	8.07	0674	17
44	3226	7.48	3416	.58	.44 9810	8.07	.55 0190	16
45	9.43 3675	7.45	9.98 3381	.60	9.45 0294	8.05	10.54 9706	15
46	4122	7.45	3345	.60	0777	8.05	9223	14
47	4569	7.45	3309	.60	1260	8.05	8740	13
48	5016	7.43	3273	.58	1743	8.03	8257	12
49	5462	7.43	3238	.60	2225	8.02	7775	11
50	9.43 5908	7.42	9.98 3202	.60	9.45 2706	8.02	10.54 7294	10
51	6353	7.42	3166	.60	3187	8.02	6813	9
52	6798	7.40	3130	.60	3668	8.00	6332	8
53	7242	7.40	3094	.60	4148	8.00	5852	7
54	7686	7.38	3058	.60	4628	7.98	5372	6
55	9.43 8129	7.38	9.98 3022	.60	9.45 5107	7.98	10.54 4893	5
56	8572	7.37	2986	.60	5586	7.97	4414	4
57	9014	7.37	2950	.60	6064	7.97	3936	3
58	9456	7.35	2914	.60	6542	7.95	3458	2
59	.43 9897	7.35	2878	.60	7019	7.95	2981	1
60	9.44 0338		9.98 2842		9.45 7496		10.54 2404	0
′	Cosine.	D.1″.	Sine.	D.1″.	Cotang.	D.1″.	Tang.	′

105° 74°

16° **Table VIII.**—*Continued* 163°

′	Sine.	D.1″.	Cosine.	D.1″.	Tang.	D.1″.	Cotang.	′
0	**9.44 0338**	7.33	**9.98 2842**	.62	**9.45 7496**	7.95	**10.54 2504**	**60**
1	0778	7.33	2805	.60	7973	7.93	2027	59
2	1218	7.33	2769	.60	8449	7.93	1551	58
3	1658	7.30	2733	.62	8925	7.92	1075	57
4	2096	7.32	2696	.60	9400	7.92	0600	56
5	**9.44 2535**	7.30	**9.98 2660**	.60	**9.45 9875**	7.90	**10.54 0125**	**55**
6	2973	7.28	2624	.62	.46 0349	7.90	.53 9651	54
7	3410	7.28	2587	.60	0823	7.90	9177	53
8	3847	7.28	2551	.62	1297	7.88	8703	52
9	4284	7.27	2514	.62	1770	7.87	8230	51
10	**9.44 4720**	7.25	**9.98 2477**	.60	**9.46 2242**	7.88	**10.53 7758**	**50**
11	5155	7.25	2441	.62	2715	7.85	7285	49
12	5590	7.25	2404	.62	3186	7.87	6814	48
13	6025	7.23	2367	.60	3658	7.83	6342	47
14	6459	7.23	2331	.62	4128	7.85	5872	46
15	**9.44 6893**	7.22	**9.98 2294**	.62	**9.46 4599**	7.83	**10.53 5401**	**45**
16	7326	7.22	2257	.62	5069	7.83	4931	44
17	7759	7.20	2220	.62	5539	7.82	4461	43
18	8191	7.20	2183	.62	6008	7.82	3992	42
19	8623	7.18	2146	.62	6477	7.80	3523	41
20	**9.44 9054**	7.18	**9.98 2109**	.62	**9.46 6945**	7.80	**10.53 3055**	**40**
21	9485	7.17	2072	.62	7413	7.78	2587	39
22	.44 9915	7.17	2035	.62	7880	7.78	2120	38
23	.45 0345	7.17	1998	.62	8347	7.78	1653	37
24	0775	7.15	1961	.62	8814	7.77	1186	36
25	**9.45 1204**	7.13	**9.98 1924**	.63	**9.46 9280**	7.77	**10.53 0720**	**35**
26	1632	7.13	1886	.62	.46 9746	7.75	.53 0254	34
27	2060	7.13	1849	.62	.47 0211	7.75	.52 9789	33
28	2488	7.12	1812	.63	0676	7.75	9324	32
29	2915	7.12	1774	.62	1141	7.73	8859	31
30	**9.45 3342**	7.10	**9.98 1737**	.62	**9.47 1605**	7.73	**10.52 8395**	**30**
31	3768	7.10	1700	.63	2069	7.72	7931	29
32	4194	7.08	1662	.62	2532	7.72	7468	28
33	4619	7.08	1625	.63	2995	7.70	7005	27
34	5044	7.08	1587	.63	3457	7.70	6543	26
35	**9.45 5469**	7.07	**9.98 1549**	.62	**9.47 3919**	7.70	**10.52 6081**	**25**
36	5893	7.05	1512	.63	4381	7.68	5619	24
37	6316	7.05	1474	.63	4842	7.68	5158	23
38	6739	7.05	1436	.62	5303	7.67	4697	22
39	7162	7.03	1399	.63	5763	7.67	4237	21
40	**9.45 7584**	7.03	**9.98 1361**	.63	**9.47 6223**	7.67	**10.52 3777**	**20**
41	8006	7.02	1323	.63	6683	7.65	3317	19
42	8427	7.02	1285	.63	7142	7.65	2858	18
43	8848	7.00	1247	.63	7601	7.63	2399	17
44	9268	7.00	1209	.63	8059	7.63	1941	16
45	**9.45 9688**	7.00	**9.98 1171**	.63	**9.47 8517**	7.63	**10.52 1483**	**15**
46	.46 0108	6.98	1133	.63	8975	7.62	1025	14
47	0527	6.98	1095	.63	9432	7.62	0568	13
48	0946	6.97	1057	.63	.47 9889	7.60	.52 0111	12
49	1364	6.97	1019	.63	.48 0345	7.60	.51 9655	11
50	**9.46 1782**	6.95	**9.98 0981**	.65	**9.48 0801**	7.60	**10.51 9199**	**10**
51	2199	6.95	0942	.63	1257	7.58	8743	9
52	2616	6.93	0904	.63	1712	7.58	8288	8
53	3032	6.93	0866	.65	2167	7.57	7833	7
54	3448	6.93	0827	.63	2621	7.57	7379	6
55	**9.46 3864**	6.92	**9.98 0789**	.65	**9.48 3075**	7.57	**10.51 6925**	**5**
56	4279	6.92	0750	.63	3529	7.55	6471	4
57	4694	6.90	0712	.65	3982	7.55	6018	3
58	5108	6.90	0673	.63	4435	7.53	5565	2
59	5522	6.88	0635	.65	4887	7.53	5113	1
60	**9.46 5935**		**9.98 0596**		**9.48 5339**		**10.51 4661**	**0**
′	Cosine.	D.1″.	Sine.	D.1″.	Cotang.	D.1″.	Tang.	′

106° **73°**

17° TABLE VIII.—*Continued* 162°

′	Sine.	D.1″.	Cosine.	D.1″.	Tang.	D.1″.	Cotang.	′
0	**9.46 5935**	6.88	**9.98 0596**	.63	**9.48 5339**	7.53	**10.51 4661**	**60**
1	6348	6.88	0558	.65	5791	7.52	4209	59
2	6761	6.87	0519	.65	6242	7.52	3758	58
3	7173	6.87	0480	.63	6693	7.50	3307	57
4	7585	6.85	0442	.65	7143	7.50	2857	56
5	**9.46 7996**	6.85	**9.98 0403**	.65	**9.48 7593**	7.50	**10.51 2407**	**55**
6	8407	6.83	0364	.65	8043	7.48	1957	54
7	8817	6.83	0325	.65	8492	7.48	1508	53
8	9227	6.83	0286	.65	8941	7.48	1059	52
9	.46 9637	6.82	0247	.65	9390	7.47	0610	51
10	**9.47 0046**	6.82	**9.98 0208**	.65	**9.48 9838**	7.47	**10.51 0162**	**50**
11	0455	6.80	0169	.65	.49 0286	7.45	.50 9714	49
12	0863	6.80	0130	.65	0733	7.45	9267	48
13	1271	6.80	0091	.65	1180	7.45	8820	47
14	1679	6.78	0052	.67	1627	7.43	8373	46
15	**9.47 2086**	6.77	**9.98 0012**	.65	**9.49 2073**	7.43	**10.50 7927**	**45**
16	2492	6.77	.97 9973	.65	2519	7.43	7481	44
17	2898	6.77	9934	.65	2965	7.42	7035	43
18	3304	6.77	9895	.67	3410	7.40	6590	42
19	3710	6.75	9855	.65	3854	7.42	6146	41
20	**9.47 4115**	6.73	**9.97 9816**	.67	**9.49 4299**	7.40	**10.50 5701**	**40**
21	4519	6.73	9776	.65	4743	7.38	5257	39
22	4923	6.73	9737	.67	5186	7.40	4814	38
23	5327	6.72	9697	.65	5630	7.38	4370	37
24	5730	6.72	9658	.67	6073	7.37	3927	36
25	**9.47 6133**	6.72	**9.97 9618**	.65	**9.49 6515**	7.37	**10.50 3485**	**35**
26	6536	6.70	9579	.67	6957	7.37	3043	34
27	6938	6.70	9539	.67	7399	7.37	2601	33
28	7340	6.68	9499	.67	7841	7.35	2159	32
29	7741	6.68	9459	.65	8282	7.33	1718	31
30	**9.47 8142**	6.67	**9.97 9420**	.67	**9.49 8722**	7.35	**10.50 1278**	**30**
31	8542	6.67	9380	.67	9163	7.33	0837	29
32	8942	6.67	9340	.67	.49 9603	7.32	.50 0397	28
33	9342	6.65	9300	.67	.50 0042	7.32	.49 9958	27
34	.47 9741	6.65	9260	.67	0481	7.32	9519	26
35	**9.48 0140**	6.65	**9.97 9220**	.67	**9.50 0920**	7.32	**10.49 9080**	**25**
36	0539	6.63	9180	.67	1359	7.30	8641	24
37	0937	6.62	9140	.67	1797	7.30	8203	23
38	1334	6.62	9100	.68	2235	7.28	7765	22
39	1731	6.62	9059	.67	2672	7.28	7328	21
40	**9.48 2128**	6.62	**9.97 9019**	.67	**9.50 3109**	7.28	**10.49 6891**	**20**
41	2525	6.60	8979	.67	3546	7.27	6454	19
42	2921	6.58	8939	.68	3982	7.27	6018	18
43	3316	6.60	8898	.67	4418	7.27	5582	17
44	3712	6.58	8858	.68	4854	7.25	5146	16
45	**9.48 4107**	6.57	**9.97 8817**	.67	**9.50 5289**	7.25	**10.49 4711**	**15**
46	4501	6.57	8777	.67	5724	7.25	4276	14
47	4895	6.57	8737	.68	6159	7.23	3841	13
48	5289	6.55	8696	.68	6593	7.23	3407	12
49	5682	6.55	8655	.67	7027	7.23	2973	11
50	**9.48 6075**	6.53	**9.97 8615**	.68	**9.50 7460**	7.22	**10.49 2540**	**10**
51	6467	6.55	8574	.68	7893	7.22	2107	9
52	6860	6.52	8533	.67	8326	7.22	1674	8
53	7251	6.53	8493	.68	8759	7.20	1241	7
54	7643	6.52	8452	.68	9191	7.18	0809	6
55	**9.48 8034**	6.50	**9.97 8411**	.68	**9.50 9622**	7.20	**10.49 0378**	**5**
56	8424	6.50	8370	.68	.51 0054	7.18	.48 9946	4
57	8814	6.50	8329	.68	0485	7.18	9515	3
58	9204	6.48	8288	.68	0916	7.17	9084	2
59	9593	6.48	8247	.68	1346	7.17	8654	1
60	**9.48 9982**		**9.97 8206**		**9.51 1776**		**10.48 8224**	**0**
′	Cosine.	D.1″.	Sine.	D.1″.	Cotang.	D.1″.	Tang.	′

107° 72°

18° **Table VIII.**—*Continued* 161°

′	Sine.	D.1″.	Cosine.	D.1″.	Tang.	D.1″.	Cotang.	′
0	**9.48 9982**	6.48	**9.97 8206**	.68	**9.51 1776**	7.17	**10.48 8224**	**60**
1	.49 0371	6.47	8165	.68	2206	7.15	7794	59
2	0759	6.47	8124	.68	2635	7.15	7365	58
3	1147	6.47	8083	.68	3064	7.15	6936	57
4	1535	6.45	8042	.68	3493	7.13	6507	56
5	**9.49 1922**	6.43	**9.97 8001**	.70	**9.51 3921**	7.13	**10.48 6079**	**55**
6	2308	6.45	7959	.68	4349	7.13	5651	54
7	2695	6.43	7918	.68	4777	7.12	5223	53
8	3081	6.42	7877	.70	5204	7.12	4796	52
9	3466	6.42	7835	.08	5631	7.10	4369	51
10	**9.49 3851**	6.42	**9.97 7794**	.70	**9.51 6057**	7.12	**10.48 3934**	**50**
11	4236	6.42	7752	.68	6484	7.10	3516	49
12	4621	6.40	7711	.70	6910	7.08	3090	48
13	5005	6.38	7669	.68	7335	7.10	2665	47
14	5388	6.40	7628	.70	7761	7.08	2239	46
15	**9.49 5772**	6.37	**9.97 7586**	.70	**9.51 8186**	7.07	**10.48 1814**	**45**
16	6154	6.38	7544	.68	8610	7.07	1390	44
17	6537	6.37	7503	.70	9034	7.07	0966	43
18	6919	6.37	7461	.70	9458	7.07	0542	42
19	7301	6.35	7419	.70	.51 9882	7.05	.48 0118	41
20	**9.49 7682**	6.35	**9.97 7377**	.70	**9.52 0305**	7.05	**10.47 9695**	**40**
21	8064	6.33	7335	.70	0728	7.05	9272	39
22	8444	6.35	7293	.70	1151	7.03	8849	38
23	8825	6.32	7251	.70	1573	7.03	8427	37
24	9204	6.33	7209	.70	1995	7.03	8005	36
25	**9.49 9584**	6.32	**9.97 7167**	.70	**9.52 2417**	7.02	**10.47 7583**	**35**
26	.49 9963	6.32	7125	.70	2838	7.02	7162	34
27	.50 0342	6.32	7083	.70	3259	7.02	6741	33
28	0721	6.30	7041	.70	3680	7.00	6320	32
29	1099	6.28	6999	.70	4100	7.00	5900	31
30	**9.50 1476**	6.30	**9.97 6957**	.72	**9.52 4520**	7.00	**10.47 5480**	**30**
31	1854	6.28	6914	.70	4940	6.98	5060	29
32	2231	6.27	6872	.70	5359	6.98	4641	28
33	2607	6.28	6830	.72	5778	6.98	4222	27
34	2984	6.27	6787	.70	6197	6.97	3803	26
35	**9.50 3360**	6.25	**9.97 6745**	.72	**9.52 6615**	6.97	**10.47 3385**	**25**
36	3735	6.25	6702	.70	7033	6.97	2967	24
37	4110	6.25	6660	.72	7451	6.95	2549	23
38	4485	6.25	6617	.72	7868	6.95	2132	22
39	4860	6.23	6574	.70	8285	6.95	1715	21
40	**9.50 5234**	6.23	**9.97 6532**	.72	**9.52 8702**	6.95	**10.47 1298**	**20**
41	5608	6.22	6489	.72	9119	6.93	0881	19
42	5981	6.22	6446	.70	9535	6.93	0465	18
43	6354	6.22	6404	.72	.52 9951	6.92	.47 0049	17
44	6727	6.20	6361	.72	.53 0366	6.92	.46 9634	16
45	**9.50 7099**	6.20	**9.97 6318**	.72	**9.53 0781**	6.92	**10.46 9219**	**15**
46	7471	6.20	6275	.72	1196	6.92	8804	14
47	7843	6.18	6232	.72	1611	6.90	8389	13
48	8214	6.18	6189	.72	2025	6.90	7975	12
49	8585	6.18	6146	.72	2439	6.90	7561	11
50	**9.50 8956**	6.17	**9.97 6103**	.72	**9.53 2853**	6.88	**10.46 7147**	**10**
51	9326	6.17	6060	.72	3266	6.88	6734	9
52	.50 9696	6.15	6017	.72	3679	6.88	6321	8
53	.51 0065	6.15	5974	.73	4092	6.87	5908	7
54	0434	6.15	5930	.72	4504	6.87	5496	6
55	**9.51 0803**	6.15	**9.97 5887**	.72	**9.53 4916**	6.87	**10.46 5084**	**5**
56	1172	6.13	5844	.73	5328	6.85	4672	4
57	1540	6.12	5800	.72	5739	6.85	4261	3
58	1907	6.13	5757	.72	6150	6.85	3850	2
59	2275	6.12	5714	.73	6561	6.85	3429	1
60	**9.51 2642**		**9.97 5670**		**9.53 6972**		**10.46 3028**	**0**
′	Cosine.	D.1″.	Sine.	D.1″.	Cotang.	D.1″.	Tang.	′

108° 71°

19° TABLE VIII.—*Continued* 160°

′	Sine.	D.1″.	Cosine.	D.1″.	Tang.	D.1″.	Cotang.	′
0	**9.51 2642**	6.12	**9.97 5670**	.72	**9.53 6972**	6.83	**10.46 3028**	**60**
1	3009	6.10	5627	.73	7382	6.83	2618	59
2	3375	6.10	5583	.73	7792	6.83	2208	58
3	3741	6.10	5539	.72	8202	6.82	1798	57
4	4107	6.08	5496	.73	8611	6.82	1389	56
5	**9.51 4472**	6.08	**9.97 5452**	.73	**9.53 9020**	6.82	**10.46 0980**	**55**
6	4837	6.08	5408	.72	9429	6.80	0571	54
7	5202	6.07	5365	.73	.53 9837	6.80	.46 0163	53
8	5566	6.07	5321	.73	.54 0245	6.80	.45 9755	52
9	5930	6.07	5277	.73	0653	6.80	9347	51
10	**9.51 6294**	6.05	**9.97 5233**	.73	**9.54 1061**	6.78	**10 45 8939**	**50**
11	6657	6.05	5189	.73	1468	6.78	8532	49
12	7020	6.03	5145	.73	1875	6.77	8125	48
13	7382	6.05	5101	.73	2281	6.78	7719	47
14	7745	6.03	5057	.73	2688	6.77	7312	46
15	**9.51 8107**	6.02	**9.97 5013**	.73	**9.54 3094**	6.75	**10.45 6906**	**45**
16	8468	6.02	4969	.73	3499	6.77	6501	44
17	8829	6.02	4925	.75	3905	6.75	6095	43
18	9190	6.02	4880	.73	4310	6.75	5690	42
19	9551	6.00	4836	.73	4715	6.73	5285	41
20	**9.51 9911**	6.00	**9.97 4792**	.73	**9.54 5119**	6.75	**10.45 4881**	**40**
21	.52 0271	6.00	4748	.75	5524	6.73	4476	39
22	0631	5.98	4703	.73	5928	6.72	4072	38
23	0990	5.98	4659	.75	6331	6.73	3669	37
24	1349	5.97	4614	.73	6735	6.72	3265	36
25	**9.52 1707**	5.98	**9.97 4570**	.75	**9.54 7138**	6.70	**10.45 2862**	**35**
26	2066	5.97	4525	.73	7540	6.72	2460	34
27	2424	5.95	4481	.75	7943	6.70	2057	33
28	2781	5.95	4436	.75	8345	6.70	1655	32
29	3138	5.95	4391	.73	8747	6.70	1253	31
30	**9.52 3495**	5.95	**9.97 4347**	.75	**9.54 9149**	6.68	**10.45 0851**	**30**
31	3852	5.93	4302	.75	9550	6.68	0450	29
32	4208	5.93	4257	.75	.54 9951	6.68	.45 0049	28
33	4564	5.93	4212	.75	.55 0352	6.67	.44 9648	27
34	4920	5.92	4167	.75	0752	6.68	9248	26
35	**9.52 5275**	5.92	**9.97 4122**	.75	**9.55 1153**	6.65	**10.44 8847**	**25**
36	5630	5.90	4077	.75	1552	6.67	8448	24
37	5984	5.92	4032	.75	1952	6.65	8048	23
38	6339	5.90	3987	.75	2351	6.65	7649	22
39	6693	5.88	3942	.75	2750	6.65	7250	21
40	**9.52 7046**	5.90	**9.97 3897**	.75	**9.55 3149**	6.65	**10.44 6851**	**20**
41	7400	5.88	3352	.75	3548	6.63	6452	19
42	7753	5.87	3807	.77	3946	6.63	6054	18
43	8105	5.88	3761	.75	4344	6.62	5656	17
44	8458	5.87	3716	.75	4741	6.63	5259	16
45	**9.52 8810**	5.85	**9.97 3671**	.77	**9.55 5139**	6.62	**10.44 4861**	**15**
46	9161	5.87	3625	.75	5536	6.62	4464	14
47	9513	5.85	3580	.75	5933	6.60	4067	13
48	.52 9864	5.85	3535	.77	6329	6.60	3671	12
49	.53 0215	5.83	3489	.75	6725	6.60	3275	11
50	**9.53 0565**	5.83	**9.97 3444**	.77	**9.55 7121**	6.60	**10.44 2879**	**10**
51	0915	5.83	3398	.77	7517	6.60	2483	9
52	1265	5.82	3352	.75	7913	6.58	2087	8
53	1614	5.82	3307	.77	8308	6.58	1692	7
54	1963	5.82	3261	.77	8703	6.57	1297	6
55	**9.53 2312**	5.82	**9.97 3215**	.77	**9.55 9097**	6.57	**10.44 0903**	**5**
56	2661	5.80	3169	.75	9491	6.57	0509	4
57	3009	5.80	3124	.77	.55 9885	6.57	.44 0115	3
58	3357	5.78	3078	.77	.56 0279	6.57	.43 9721	2
59	3704	5.80	3032	.77	0673	6.55	9327	1
60	**9.53 4052**		**9.97 2986**		**9.56 1066**		**10.43 8934**	**0**
′	Cosine.	D.1″.	Sine.	D.1″.	Cotang.	D.1″.	Tang.	′

109° 70°

20° TABLE VIII.—*Continued* 159°

′	Sine.	D.1″.	Cosine.	D.1″.	Tang.	D.1″.	Cotang.	′
0	**9.53 4052**	5.78	**9.97 2986**	.77	**9.56 1066**	6.55	**10.43 8934**	**60**
1	4399	5.77	2940	.77	1459	6.53	8541	59
2	4745	5.78	2894	.77	1851	6.55	8149	58
3	5092	5.77	2848	.77	2244	6.53	7756	57
4	5438	5.75	2802	.78	2636	6.53	7364	56
5	**9.53 5783**	5.77	**9.97 2755**	.77	**9.56 3028**	6.52	**10.43 6972**	**55**
6	6129	5.75	2709	.77	3419	6.53	6581	54
7	6474	5.73	2663	.77	3811	6.52	6189	53
8	6818	5.75	2617	.78	4202	6.52	5798	52
9	7163	5.73	2570	.77	4593	6.50	5407	51
10	**9.53 7507**	5.73	**9.97 2524**	.77	**9.56 4983**	6.50	**10.43 5017**	**50**
11	7851	5.72	2478	.78	5373	6.50	4627	49
12	8194	5.73	2431	.77	5763	6.50	4237	48
13	8538	5.70	2385	.78	6153	6.48	3847	47
14	8880	5.72	2338	.78	6542	6.50	3458	46
15	**9.53 9223**	5.70	**9.97 2291**	.77	**9.56 6932**	6.47	**10.43 3068**	**45**
16	9565	5.70	2245	.78	7320	6.48	2680	44
17	.53 9907	5.68	2198	.78	7709	6.48	2291	43
18	.54 0249	5.68	2151	.77	8098	6.47	1902	42
19	0590	5.68	2105	.78	8486	6.45	1514	41
20	**9.54 0931**	5.68	**9.97 2058**	.78	**9.56 8873**	6.47	**10.43 1127**	**40**
21	1272	5.68	2011	.78	9261	6.45	0739	39
22	1613	5.67	1964	.78	.56 9648	6.45	.43 0352	38
23	1953	5.67	1917	.78	.57 0035	6.45	.42 9965	37
24	2293	5.65	1870	.78	0422	6.45	9578	36
25	**9.54 2632**	5.65	**9.97 1823**	.78	**9.57 0809**	6.43	**10.42 9191**	**35**
26	2971	5.65	1776	.78	1195	6.43	8805	34
27	3310	5.65	1729	.78	1581	6.43	8419	33
28	3649	5.63	1682	.78	1967	6.42	8033	32
29	3987	5.63	1635	.78	2352	6.43	7648	31
30	**9.54 4325**	5.63	**9.97 1588**	.80	**9.57 2738**	6.42	**10.42 7262**	**30**
31	4663	5.62	1540	.78	3123	6.40	6877	29
32	5000	5.63	1493	.78	3507	6.42	6493	28
33	5338	5.60	1446	.80	3892	6.40	6108	27
34	5674	5.62	1398	.78	4276	6.40	5724	26
35	**9.54 6011**	5.60	**9.97 1351**	.80	**9.57 4660**	6.40	**10.42 5340**	**25**
36	6347	5.60	1303	.78	5044	6.38	4956	24
37	6683	5.60	1256	.80	5427	6.38	4573	23
38	7019	5.58	1208	.78	5810	6.38	4190	22
39	7354	5.58	1161	.80	6193	6.38	3807	21
40	**9.54 7689**	5.58	**9.97 1113**	.78	**9.57 6576**	6.38	**10.42 3424**	**20**
41	8024	5.58	1066	.80	6959	6.37	3041	19
42	8359	5.57	1018	.80	7341	6.37	2659	18
43	8693	5.57	0970	.80	7723	6.35	2277	17
44	9027	5.55	0922	.80	8104	6.37	1896	16
45	**9.54 9360**	5.55	**9.97 0874**	.78	**9.57 8486**	6.35	**10.42 1514**	**15**
46	.54 9693	5.55	0827	.80	8867	6.35	1133	14
47	.55 0026	5.55	0779	.80	9248	6.35	0752	13
48	0359	5.55	0731	.80	57 9629	6.33	.42 0371	12
49	0692	5.53	0683	.80	58 0009	6.33	.41 9991	11
50	**9.55 1024**	5.53	**9.97 0635**	.82	**9.58 0389**	6.33	**10.41 9611**	**10**
51	1356	5.52	0586	.80	0769	6.33	9231	9
52	1687	5.52	0538	.80	1149	6.32	8851	8
53	2018	5.52	0490	.80	1528	6.32	8472	7
54	2349	5.52	0442	.80	1907	6.32	8093	6
55	**9.55 2680**	5.50	**9.97 0394**	.82	**9.58 2286**	6.32	**10.41 7714**	**5**
56	3010	5.52	0345	.80	2665	6.32	7335	4
57	3341	5.48	0297	.80	3044	6.30	6956	3
58	3670	5.50	0249	.82	3422	6.30	6578	2
59	4000	5.48	0200	.80	3800	6.28	6200	1
60	**9.55 4329**		**9.97 0152**		**9.58 4177**		**10.41 5823**	**0**
′	Cosine.	D.1″.	Sine.	D.1″.	Cotang.	D.1″.	Tang.	′

110° 69°

21° TABLE VIII.—*Continued* 158°

′	Sine.	D.1″.	Cosine.	D.1″.	Tang.	D.1″.	Cotang.	′
0	**9.55 4329**	5.48	**9.97 0152**	.82	**9.58 4177**	6.30	**10.41 5823**	**60**
1	4658	5.48	0103	.80	4555	6.28	5445	59
2	4987	5.47	0055	.82	4932	6.28	5068	58
3	5315	5.47	.97 0006	.82	5309	6.28	4691	57
4	5643	5.47	.96 9957	.80	5686	6.27	4314	56
5	**9.55 5971**	5.47	**9.96 9909**	.82	**9.58 6062**	6.28	**10.41 3938**	**55**
6	6299	5.45	9860	.82	6439	6.27	3561	54
7	6626	5.45	9811	.82	6815	6.25	3185	53
8	6953	5.45	9762	.80	7190	6.27	2810	52
9	7280	5.43	9714	.82	7566	6.25	2434	51
10	**9.55 7606**	5.43	**9.96 9665**	.82	**9.58 7941**	6.25	**10.41 2059**	**50**
11	7932	5.43	9616	.82	8316	6.25	1684	49
12	8258	5.42	9567	.82	8691	6.25	1309	48
13	8583	5.43	9518	.82	9066	6.23	0934	47
14	8909	5.42	9469	.82	9440	6.23	0560	46
15	**9.55 9234**	5.40	**9.96 9420**	.83	**9.58 9814**	6.23	**10.41 0186**	**45**
16	9558	5.42	9370	.82	.59 0188	6.23	.40 9812	44
17	.55 9883	5.40	9321	.82	0562	6.22	9438	43
18	.56 0207	5.40	9272	.82	0935	6.22	9065	42
19	0531	5.40	9223	.83	1308	6.22	8692	41
20	**9.56 0855**	5.38	**9.96 9173**	.82	**9.59 1681**	6.22	**10.40 8319**	**40**
21	1178	5.38	9124	.82	2054	6.20	7946	39
22	1501	5.38	9075	.83	2426	6.22	7574	38
23	1824	5.37	9025	.82	2799	6.20	7201	37
24	2146	5.37	8976	.83	3171	6.18	6829	36
25	**9.56 2468**	5.37	**9.96 8926**	.82	**9.59 3542**	6.20	**10.40 6458**	**35**
26	2790	5.37	8877	.83	3914	6.18	6086	34
27	3112	5.35	8827	.83	4285	6.18	5715	33
28	3433	5.37	8777	.82	4656	6.18	5344	32
29	3755	5.33	8728	.83	5027	6.18	4973	31
30	**9.56 4075**	5.35	**9.96 8678**	.83	**9.59 5398**	6.17	**10.40 4602**	**30**
31	4396	5.33	8628	.83	5768	6.17	4232	29
32	4716	5.33	8578	.83	6138	6.17	3862	28
33	5036	5.33	8528	.82	6508	6.17	3492	27
34	5356	5.33	8479	.83	6878	6.15	3122	26
35	**9.56 5676**	5.32	**9.96 8429**	.83	**9.59 7247**	6.15	**10.40 2753**	**25**
36	5995	5.32	8379	.83	7616	6.15	2384	24
37	6314	5.30	8329	.85	7985	6.15	2015	23
38	6632	5.32	8278	.83	8354	6.13	1646	22
39	6951	5.30	8228	.83	8722	6.15	1278	21
40	**9.56 7269**	5.30	**9.96 8178**	.83	**9.59 9091**	6.13	**10.40 0909**	**20**
41	7587	5.28	8128	.83	9459	6.13	0541	19
42	7904	5.30	8078	.85	.59 9827	6.12	.40 0173	18
43	8222	5.28	8027	.83	.60 0194	6.13	.39 9806	17
44	8539	5.28	7977	.83	0562	6.12	9438	16
45	**9.56 8856**	5.27	**9.96 7927**	.85	**9.60 0929**	6.12	**10.39 9071**	**15**
46	9172	5.27	7876	.83	1296	6.12	8704	14
47	9488	5.27	7826	.85	1663	6.10	8337	13
48	.56 9804	5.27	7775	.83	2029	6.10	7971	12
49	.57 0120	5.25	7725	.85	2395	6.10	7605	11
50	**9.57 0435**	5.27	**9.96 7674**	.83	**9.60 2761**	6.10	**10.39 7239**	**10**
51	0751	5.25	7624	.85	3127	6.10	6873	9
52	1066	5.23	7573	.85	3493	6.08	6507	8
53	1380	5.25	7522	.85	3858	6.08	6142	7
54	1695	5.23	7471	.83	4223	6.08	5777	6
55	**9.57 2009**	5.23	**9.96 7421**	.85	**9.60 4588**	6.08	**10.39 5412**	**5**
56	2323	5.22	7370	.85	4953	6.07	5047	4
57	2636	5.23	7319	.85	5317	6.08	4683	3
58	2950	5.22	7268	.85	5682	6.07	4318	2
59	3263	5.20	7217	.85	6046	6.07	3954	1
60	**9.57 3575**		**9.96 7166**		**9.60 6410**		**10.39 3590**	**0**
′	Cosine.	D.1″.	Sine.	D.1″.	Cotang.	D.1″.	Tang.	′

111° 68°

22° TABLE VIII.—*Continued* 157°

′	Sine.	D.1″.	Cosine.	D.1″.	Tang.	D.1″.	Cotang.	′
0	**9.57 3575**	5.22	**9.96 7166**	.85	**9.60 6410**	6.05	**10.39 3590**	**60**
1	3888	5.20	7115	.85	6773	6.07	3227	59
2	4200	5.20	7064	.85	7137	6.05	2863	58
3	4512	5.20	7013	.87	7500	6.05	2500	57
4	4824	5.20	6961	.85	7863	6.03	2137	56
5	**9.57 5136**	5.18	**9.96 6910**	.85	**9.60 8225**	6.05	**10.39 1775**	**55**
6	5447	5.18	6859	.85	8588	6.03	1412	54
7	5758	5.18	6808	.87	8950	6.03	1050	53
8	6069	5.17	6756	.85	9312	6.03	0688	52
9	6379	5.17	6705	.87	.00 9674	6.03	.30 0326	51
10	**9.57 6689**	5.17	**9.96 6653**	.85	**9.61 0036**	6.02	**10.38 9964**	**50**
11	6999	5.17	6602	.87	0397	6.03	9603	49
12	7309	5.15	6550	.85	0759	6.02	9241	48
13	7618	5.15	6499	.87	1120	6.00	8880	47
14	7927	5.15	6447	.87	1480	6.02	8520	46
15	**9.57 8236**	5.15	**9.96 6395**	.85	**9.61 1841**	6.00	**10.38 8159**	**45**
16	8545	5.13	6344	.87	2201	6.00	7799	44
17	8853	5.15	6292	.87	2561	6.00	7439	43
18	9162	5.13	6240	.87	2921	6.00	7079	42
19	9470	5.12	6188	.87	3281	6.00	6719	41
20	**9.57 9777**	5.13	**9.96 6136**	.85	**9.61 3641**	5.98	**10.38 6359**	**40**
21	.58 0085	5.12	6085	.87	4000	5.98	6000	39
22	0392	5.12	6033	.87	4359	5.98	5641	38
23	0699	5.10	5981	.87	4718	5.98	5282	37
24	1005	5.12	5929	.88	5077	5.97	4923	36
25	**9.58 1312**	5.10	**9.96 5876**	.87	**9.61 5435**	5.97	**10.38 4565**	**35**
26	1618	5.10	5824	.87	5793	5.97	4207	34
27	1924	5.08	5772	.87	6151	5.97	3849	33
28	2229	5.10	5720	.87	6509	5.97	3491	32
29	2535	5.08	5668	.88	6867	5.95	3133	31
30	**9.58 2840**	5.08	**9.96 5615**	.87	**9.61 7224**	5.97	**10.38 2776**	**30**
31	3145	5.07	5563	.87	7582	5.95	2418	29
32	3449	5.08	5511	.88	7939	5.93	2061	28
33	3754	5.07	5458	.87	8295	5.95	1705	27
34	4058	5.05	5406	.88	8652	5.93	1348	26
35	**9.58 4361**	5.07	**9.96 5353**	.87	**9.61 9008**	5.93	**10.38 0992**	**25**
36	4665	5.05	5301	.88	9364	5.93	0636	24
37	4968	5.07	5248	.88	.61 9720	5.93	.38 0280	23
38	5272	5.03	5195	.87	.62 0076	5.93	.37 9924	22
39	5574	5.05	5143	.88	0432	5.92	9568	21
40	**9.58 5877**	5.03	**9.96 5090**	.88	**9.62 0787**	5.92	**10.37 9213**	**20**
41	6179	5.05	5037	.88	1142	5.92	8858	19
42	6482	5.02	4984	.88	1497	5.92	8503	18
43	6783	5.03	4931	.87	1852	5.92	8148	17
44	7085	5.02	4879	.88	2207	5.90	7793	16
45	**9.58 7386**	5.03	**9.96 4826**	.88	**9.62 2561**	5.90	**10.37 7439**	**15**
46	7688	5.02	4773	.88	2915	5.90	7085	14
47	7989	5.00	4720	.90	3269	5.90	6731	13
48	8289	5.02	4666	.88	3623	5.88	6377	12
49	8590	5.00	4613	.88	3976	5.90	6024	11
50	**9.58 8890**	5.00	**9.96 4560**	.88	**9.62 4330**	5.88	**10.37 5670**	**10**
51	9190	4.98	4507	.88	4683	5.88	5317	9
52	9489	5.00	4454	.90	5036	5.87	4964	8
53	.58 9789	4.98	4400	.88	5388	5.88	4612	7
54	.59 0088	4.98	4347	.88	5741	5.87	4259	6
55	**9.59 0387**	4.98	**9.96 4294**	.90	**9.62 6093**	5.87	**10.37 3907**	**5**
56	0686	4.97	4240	.88	6445	5.87	3555	4
57	0984	4.97	4187	.90	6797	5.87	3203	3
58	1282	4.97	4133	.88	7149	5.87	2851	2
59	1580	4.97	4080	.90	7501	5.85	2499	1
60	**9.59 1878**		**9.96 4026**		**9.62 7852**		**10.37 2148**	**0**
′	Cosine.	D.1″.	Sine.	D.1″.	Cotang.	D.1″.	Tang.	′

112° 67°

23° TABLE VIII.—*Continued* 156°

′	Sine.	D.1″.	Cosine.	D.1″.	Tang.	D.1″.	Cotang.	′
0	9.59 1878	4.97	9.96 4026	.90	9.62 7825	5.85	10.37 2148	60
1	2176	4.95	3972	.88	8203	5.85	1797	59
2	2473	4.95	3919	.90	8554	5.85	1446	58
3	2770	4.95	3865	.90	8905	5.83	1095	57
4	3067	4.93	3811	.90	9255	5.85	0745	56
5	9.59 3363	4.93	9.96 3757	.88	9.62 9606	5.83	10.37 0394	55
6	3659	4.93	3704	.90	.62 9956	5.83	.37 0044	54
7	3955	4.93	3650	.90	.63 0306	5.83	.36 9694	53
8	4251	4.93	3596	.90	0656	5.82	9344	52
9	4547	4.92	3542	.90	1005	5.83	8995	51
10	9.59 4842	4.92	9.96 3488	.90	9.63 1355	5.82	10.36 8645	50
11	5137	4.92	3434	.92	1704	5.82	8296	49
12	5432	4.92	3379	.90	2053	5.82	7947	48
13	5727	4.90	3325	.90	2402	5.80	7598	47
14	6021	4.90	3271	.90	2750	5.82	7250	46
15	9.59 6315	4.90	9.96 3217	.90	9.63 3099	5.80	10.36 6901	45
16	6609	4.90	3163	.92	3447	5.80	6553	44
17	6903	4.88	3108	.90	3795	5.80	6205	43
18	7196	4.90	3054	.92	4143	5.78	5857	42
19	7490	4.88	2999	.90	4490	5.80	5510	41
20	9.59 7783	4.87	9.96 2945	.92	9.63 4838	5.78	10.36 5162	40
21	8075	4.88	2890	.90	5185	5.78	4815	39
22	8368	4.87	2836	.92	5532	5.78	4468	38
23	8660	4.87	2781	.90	5879	5.78	4121	37
24	8952	4.87	2727	.92	6226	5.77	3774	36
25	9.59 9244	4.87	9.96 2672	.92	9.63 6572	5.78	10.36 3428	35
26	9536	4.85	2617	.92	6919	5.77	3081	34
27	.59 9827	4.85	2562	.90	7265	5.77	2735	33
28	.60 0118	4.85	2508	.92	7611	5.75	2389	32
29	0409	4.85	2453	.92	7956	5.77	2044	31
30	9.60 0700	4.83	9.96 2398	.92	9 63 8302	5.75	10.36 1698	30
31	0990	4.83	2343	.92	8647	5.75	1353	29
32	1280	4.83	2288	.92	8992	5.75	1008	28
33	1570	4.83	2233	.92	9337	5.75	0663	27
34	1860	4.83	2178	.92	.63 9682	5.75	.36 0318	26
35	9.60 2150	4.82	9.96 2123	.93	9.64 0027	5.73	10.35 9973	25
36	2439	4.82	2067	.92	0371	5.75	9629	24
37	2728	4.82	2012	.92	0716	5.73	9284	23
38	3017	4.80	1957	.92	1060	5.73	8940	22
39	3305	4.82	1902	.93	1404	5.72	8596	21
40	9.60 3594	4.80	9.96 1846	.92	9.64 1747	5.73	10.35 8253	20
41	3882	4.80	1791	.93	2091	5.72	7909	19
42	4170	4.78	1735	.92	2434	5.72	7566	18
43	4457	4.80	1680	.93	2777	5.72	7223	17
44	4745	4.78	1624	.92	3120	5.72	6880	16
45	9.60 5032	4.78	9.96 1569	.93	9.64 3463	5.72	10.35 6537	15
46	5319	4.78	1513	.92	3806	5.70	6194	14
47	5606	4.77	1458	.93	4148	5.70	5852	13
48	5892	4.78	1402	.93	4490	5.70	5510	12
49	6179	4.77	1346	.93	4832	5.70	5168	11
50	9.60 6465	4.77	9.96 1290	.92	9.64 5174	5.70	10.35 4826	10
51	6751	4.75	1235	.93	5516	5.68	4484	9
52	7036	4.77	1179	.93	5857	5.70	4143	8
53	7322	4.75	1123	.93	6199	5.68	3801	7
54	7607	4.75	1067	.93	6540	5.68	3460	6
55	9.60 7892	4.75	9.96 1011	.93	9.64 6881	5.68	10.35 3119	5
56	8177	4.73	0955	.93	7222	5.67	2778	4
57	8461	4.73	0899	.93	7562	5.68	2438	3
58	8745	4.73	0843	.93	7903	5.67	2097	2
59	9029	4.73	0786	.95	8243	5.67	1757	1
60	9.60 9313		9.96 0730	.93	9.64 8583		10.35 1417	0
′	Cosine.	D.1″.	Sine.	D.1″.	Cotang.	D.1″.	Tang.	′

113° 66°

24° TABLE VIII.—*Continued* 155°

′	Sine.	D.1″.	Cosine.	D.1″.	Tang.	D.1″.	Cotang.	′
0	**9.60 9313**	4.73	**9.96 0730**	.93	**9.64 8583**	5.67	**10.35 1417**	**60**
1	9597	4.72	0674	.93	8923	5.67	1077	59
2	.60 9880	4.73	0618	.95	9263	5.65	0737	58
3	.61 0164	4.72	0561	.93	9602	5.67	0398	57
4	0447	4.70	0505	.95	.64 9942	5.65	.35 0058	56
5	**9.61 0729**	4.72	**9.96 0448**	.93	**9.65 0281**	5.65	**10.34 9719**	**55**
6	1012	4.70	0392	.95	0620	5.65	9380	54
7	1294	4.70	0335	.93	0959	5.63	9041	53
8	1576	4.70	0279	.95	1297	5.65	8703	52
9	1858	4.70	0222	.95	1636	5.63	8364	51
10	**9.61 2140**	4.68	**9.96 0165**	.93	**9.65 1974**	5.63	**10.34 8026**	**50**
11	2421	4.68	0109	.95	2312	5.63	7688	49
12	2702	4.68	.96 0052	.95	2650	5.63	7350	48
13	2983	4.68	.95 9995	.95	2988	5.63	7012	47
14	3264	4.68	9938	.93	3326	5.62	6674	46
15	**9.61 3545**	4.67	**9.95 9882**	.95	**9.65 3663**	5.62	**10.34 6337**	**45**
16	3825	4.67	9825	.95	4000	5.62	6000	44
17	4105	4.67	9768	.95	4337	5.62	5663	43
18	4385	4.67	9711	.95	4674	5.62	5326	42
19	4665	4.65	9654	.97	5011	5.62	4989	41
20	**9.61 4944**	4.65	**9.95 9596**	.95	**9.65 5348**	5.60	**10.34 4652**	**40**
21	5223	4.65	9539	.95	5684	5.60	4316	39
22	5502	4.65	9482	.95	6020	5.60	3980	38
23	5781	4.65	9425	.95	6356	5.60	3644	37
24	6060	4.63	9368	.97	6692	5.60	3308	36
25	**9.61 6338**	4.63	**9.95 9310**	.95	**9.65 7028**	5.60	**10.34 2972**	**35**
26	6616	4.63	9253	.97	7364	5.58	2636	34
27	6894	4.63	9195	.95	7699	5.58	2301	33
28	7172	4.63	9138	.97	8034	5.58	1966	32
29	7450	4.62	9080	.95	8369	5.58	1631	31
30	**9.61 7727**	4.62	**9.95 9023**	.97	**9.65 8704**	5.58	**10.34 1296**	**30**
31	8004	4.62	8965	.95	9039	5.57	0961	29
32	8281	4.62	8908	.97	9373	5.58	0627	28
33	8558	4.60	8850	.97	.65 9708	5.57	.34 0292	27
34	8834	4.60	8792	.97	.66 0042	5.57	.33 9958	26
35	**9.61 9110**	4.60	**9.95 8734**	.95	**9.66 0376**	5.57	**10.33 9624**	**25**
36	9386	4.60	8677	.97	0710	5.55	9290	24
37	9662	4.60	8619	.97	1043	5.57	8957	23
38	.61 9938	4.58	8561	.97	1377	5.55	8623	22
39	.62 0213	4.58	8503	.97	1710	5.55	8290	21
40	**9.62 0488**	4.58	**9.95 8445**	.97	**9.66 2043**	5.55	**10.33 7957**	**20**
41	0763	4.58	8387	.97	2376	5.55	7624	19
42	1038	4.58	8329	.97	2709	5.55	7291	18
43	1313	4.57	8271	.97	3042	5.55	6958	17
44	1587	4.57	8213	.98	3375	5.53	6625	16
45	**9.62 1861**	4.57	**9.95 8154**	.97	**9.66 3707**	5.53	**10.33 6293**	**15**
46	2135	4.57	8096	.97	4039	5.53	5961	14
47	2409	4.55	8038	.98	4371	5.53	5629	13
48	2682	4.57	7979	.97	4703	5.53	5297	12
49	2956	4.55	7921	.97	5035	5.52	4965	11
50	**9.62 3229**	4.55	**9.95 7863**	.98	**9.66 5366**	5.53	**10.33 4634**	**10**
51	3502	4.53	7804	.97	5698	5.52	4302	9
52	3774	4.55	7746	.98	6029	5.52	3971	8
53	4047	4.53	7687	.98	6360	5.52	3640	7
54	4319	4.53	7628	.97	6691	5.50	3309	6
55	**9.62 4591**	4.53	**9.95 7570**	.98	**9.66 7021**	5.52	**10.33 2979**	**5**
56	4863	4.53	7511	.98	7352	5.50	2648	4
57	5135	4.52	7452	.98	7682	5.52	2318	3
58	5406	4.52	7393	.97	8013	5.50	1987	2
59	5677	4.52	7335	.98	8343	5.50	1657	1
60	**9.62 5948**		**9.95 7276**		**9.66 8673**		**10.33 1327**	**0**
′	Cosine.	D.1″.	Sine.	D.1″.	Cotang.	D.1″.	Tang.	′

114° 65°

25° **TABLE VIII.—*Continued*** 154°

′	Sine.	D.1″.	Cosine.	D.1″.	Tang.	D.1″.	Cotang.	′
0	**9.62 5948**	4.52	**9.95 7276**	.98	**9.66 8673**	5.48	**10.33 1327**	**60**
1	6219	4.52	7217	.98	9002	5.50	0998	59
2	6490	4.50	7158	.98	9332	5.48	0668	58
3	6760	4.50	7099	.98	9661	5.50	0339	57
4	7030	4.50	7040	.98	.66 9991	5.48	.33 0009	56
5	**9.62 7300**	4.50	**9.95 6981**	1.00	**9.67 0320**	5.48	**10.32 9680**	**55**
6	7570	4.50	6921	.98	0649	5.47	9351	54
7	7840	4.48	6862	.98	0977	5.48	9023	53
8	8109	4.48	6803	.98	1306	5.48	8694	52
9	8378	4.48	6744	1.00	1635	5.47	8365	51
10	**9.62 8647**	4.48	**9.95 6684**	.98	**9.67 1963**	5.47	**10.32 8037**	**50**
11	8916	4.48	6625	.98	2291	5.47	7709	49
12	9185	4.47	6566	1.00	2619	5.47	7381	48
13	9453	4.47	6506	.98	2947	5.45	7053	47
14	9721	4.47	6447	1.00	3274	5.47	6726	46
15	**9.62 9989**	4.47	**9.95 6387**	1.00	**9.67 3602**	5.45	**10.32 6398**	**45**
16	.63 0257	4.45	6327	.98	3929	5.47	6071	44
17	0524	4.47	6268	1.00	4257	5.45	5743	43
18	0792	4.45	6208	1.00	4584	5.45	5416	42
19	1059	4.45	6148	.98	4911	5.43	5089	41
20	**9.63 1326**	4.45	**9.95 6089**	1.00	**9.67 5237**	5.45	**10.32 4763**	**40**
21	1593	4.43	6029	1.00	5564	5.43	4436	39
22	1859	4.43	5969	1.00	5890	5.45	4110	38
23	2125	4.45	5909	1.00	6217	5.43	3783	37
24	2392	4.43	5849	1.00	6543	5.43	3457	36
25	**9.63 2658**	4.42	**9.95 5789**	1.00	**9.67 6869**	5.42	**10.32 3131**	**35**
26	2923	4.43	5729	1.00	7194	5.43	2806	34
27	3189	4.42	5669	1.00	7520	5.43	2480	33
28	3454	4.42	5609	.98	7846	5.42	2154	32
29	3719	4.42	5548	1.00	8171	5.42	1829	31
30	**9.63 3984**	4.42	**9.95 5488**	1.00	**9.67 8496**	5.42	**10.32 1504**	**30**
31	4249	4.42	5428	1.00	8821	5.42	1179	29
32	4514	4.40	5368	1.02	9146	5.42	0854	28
33	4778	4.40	5307	1.00	9471	5.40	0529	27
34	5042	4.40	5247	1.02	.67 9795	5.42	.32 0205	26
35	**9.63 5306**	4.40	**9.95 5186**	1.00	**9.68 0120**	5.40	**10.31 9880**	**25**
36	5570	4.40	5126	1.02	0444	5.40	9556	24
37	5834	4.38	5065	1.00	0768	5.40	9232	23
38	6097	4.38	5005	1.02	1092	5.40	8908	22
39	6360	4.38	4944	1.02	1416	5.40	8584	21
40	**9.63 6623**	4.38	**9.95 4883**	1.00	**9.68 1740**	5.38	**10.31 8260**	**20**
41	6886	4.37	4823	1.02	2063	5.40	7937	19
42	7148	4.38	4762	1.02	2387	5.38	7613	18
43	7411	4.37	4701	1.02	2710	5.38	7290	17
44	7673	4.37	4640	1.02	3033	5.38	6967	16
45	**9.63 7935**	4.37	**9.95 4579**	1.02	**9.68 3356**	5.38	**10.31 6644**	**15**
46	8197	4.35	4518	1.02	3679	5.37	6321	14
47	8458	4.37	4457	1.02	4001	5.38	5999	13
48	8720	4.35	4396	1.02	4324	5.37	5676	12
49	8981	4.35	4335	1.02	4646	5.37	5354	11
50	**9.63 9242**	4.35	**9.95 4274**	1.02	**9.68 4968**	5.37	**10.31 5032**	**10**
51	9503	4.35	4213	1.02	5290	5.37	4710	9
52	.63 9764	4.33	4152	1.03	5612	5.37	4388	8
53	.64 0024	4.33	4090	1.02	5934	5.35	4066	7
54	0284	4.33	4029	1.02	6255	5.37	3745	6
55	**9.64 0544**	4.33	**9.95 3968**	1.03	**9.68 6577**	5.35	**10.31 3423**	**5**
56	0804	4.33	3906	1.02	6898	5.35	3102	4
57	1064	4.33	3845	1.03	7219	5.35	2781	3
58	1324	4.32	3783	1.02	7540	5.35	2460	2
59	1583	4.32	3722	1.03	7861	5.35	2139	1
60	**9.64 1842**		**9.95 3660**		**9.68 8182**		**10.31 1818**	**0**
′	Cosine.	D.1″.	Sine.	D.1″.	Cotang.	D.1″.	Tang.	′

115° **64°**

26° **Table VIII.**—*Continued* 153°

′	Sine.	D.1″.	Cosine.	D.1″.	Tang.	D.1″.	Cotang.	′
0	**9.64 1842**	4.32	**9.95 3660**	1.02	**9.68 8182**	5.33	**10.31 1818**	**60**
1	2101	4.32	3599	1.03	8502	5.32	1498	59
2	2360	4.30	3537	1.03	8823	5.33	1177	58
3	2618	4.32	3475	1.03	9143	5.33	0857	57
4	2877	4.30	3413	1.02	9463	5.33	0537	56
5	**9.64 3135**	4.30	**9.95 3352**	1.03	**9.68 9783**	5.33	**10.31 0217**	**55**
6	3393	4.28	3290	1.03	.69 0103	5.33	.30 9897	54
7	3650	4.30	3228	1.03	0423	5.32	9577	53
8	3908	4.28	3166	1.03	0742	5.33	9258	52
9	4165	4.30	3104	1.03	1062	5.32	8938	51
10	**9.64 4423**	4.28	**9.95 3042**	1.03	**9.69 1381**	5.32	**10.30 8619**	**50**
11	4680	4.27	2980	1.03	1700	5.32	8300	49
12	4936	4.28	2918	1.05	2019	5.32	7981	48
13	5193	4.28	2855	1.03	2338	5.30	7662	47
14	5450	4.27	2793	1.03	2656	5.32	7344	46
15	**9.64 5706**	4.27	**9.95 2731**	1.03	**9.69 2975**	5.30	**10.30 7025**	**45**
16	5962	4.27	2669	1.05	3293	5.32	6707	44
17	6218	4.27	2606	1.03	3612	5.30	6388	43
18	6474	4.25	2544	1.05	3930	5.30	6070	42
19	6729	4.25	2481	1.03	4228	5.30	5752	41
20	**9.64 6984**	4.27	**9.95 2419**	1.05	**9.69 4566**	5.28	**10.30 5434**	**40**
21	7240	4.23	2356	1.03	4883	5.30	5117	39
22	7494	4.25	2294	1.05	5201	5.28	4799	38
23	7749	4.25	2231	1.05	5518	5.30	4482	37
24	8004	4.23	2168	1.03	5836	5.28	4164	36
25	**9.64 8258**	4.23	**9.95 2103**	1.05	**9.69 6153**	5.28	**10.30 3847**	**35**
26	8512	4.23	2043	1.05	6470	5.28	3530	34
27	8766	4.23	1980	1.05	6787	5.27	3213	33
28	9020	4.23	1917	1.05	7103	5.28	2897	32
29	9274	4.22	1854	1.05	7420	5.27	2580	31
30	**9.64 9527**	4.23	**9.95 1791**	1.05	**9.69 7736**	5.28	**10.30 2264**	**30**
31	.64 9781	4.22	1728	1.05	8053	5.27	1947	29
32	.65 0034	4.22	1665	1.05	8369	5.27	1631	28
33	0287	4.20	1602	1.05	8685	5.27	1315	27
34	0539	4.22	1539	1.05	9001	5.25	0999	26
35	**9.65 0792**	4.20	**9.95 1476**	1.07	**9.69 9316**	5.27	**10.30 0684**	**25**
36	1044	4.22	1412	1.05	9632	5.25	0368	24
37	1297	4.20	1349	1.05	.69 9947	5.27	.30 0053	23
38	1549	4.18	1286	1.07	.70 0263	5.25	.29 9737	22
39	1800	4.20	1222	1.05	0578	5.25	9422	21
40	**9.65 2052**	4.20	**9.95 1159**	1.05	**9.70 0893**	5.25	**10.29 9107**	**20**
41	2304	4.18	1096	1.07	1208	5.25	8792	19
42	2555	4.18	1032	1.07	1523	5.23	8477	18
43	2806	4.18	0968	1.05	1837	5.25	8163	17
44	3057	4.18	0905	1.07	2152	5.23	7848	16
45	**9.65 3308**	4.17	**9.95 0841**	1.05	**9.70 2466**	5.25	**10.29 7534**	**15**
46	3558	4.17	0778	1.07	2781	5.23	7219	14
47	3808	4.18	0714	1.07	3095	5.23	6905	13
48	4059	4.17	0650	1.07	3409	5.22	6591	12
49	4309	4.15	0586	1.07	3722	5.23	6278	11
50	**9.65 4558**	4.17	**9.95 0522**	1.07	**9.70 4036**	5.23	**10.29 5964**	**10**
51	4808	4.17	0458	1.07	4350	5.22	5650	9
52	5058	4.15	0394	1.07	4663	5.22	5337	8
53	5307	4.15	0330	1.07	4976	5.23	5024	7
54	5556	4.15	0266	1.07	5290	5.22	4710	6
55	**9.65 5805**	4.15	**9.95 0202**	1.07	**9.70 5603**	5.22	**10.29 4397**	**5**
56	6054	4.13	0138	1.07	5916	5.20	4084	4
57	6302	4.15	0074	1.07	6228	5.22	3772	3
58	6551	4.13	.95 0010	1.08	6541	5.22	3459	2
59	6799	4.13	.94 9945	1.07	6854	5.20	3146	1
60	**9.65 7047**		**9.94 9881**		**9.70 7166**		**10.29 2834**	**0**
′	Cosine.	D.1″.	Sine.	D.1″.	Cotang.	D.1″.	Tang.	′

116° **63°**

27° TABLE VIII.—*Continued* 152°

′	Sine.	D.1″.	Cosine.	D.1″.	Tang.	D.1″.	Cotang.	′
0	9.65 7047	4.13	9.94 9881	1.08	9.70 7166	5.20	10.29 2834	60
1	7295	4.12	9816	1.07	7478	5.20	2522	59
2	7542	4.13	9752	1.07	7790	5.20	2210	58
3	7790	4.12	9688	1.08	8102	5.20	1898	57
4	8037	4.12	9623	1.08	8414	5.20	1586	56
5	9.65 8284	4.12	9.94 9558	1.07	9.70 8726	5.18	10.29 1274	55
6	8531	4.12	9494	1.08	9037	5.20	0963	54
7	8778	4.12	9429	1.08	9349	5.18	0651	53
8	9025	4.10	9364	1.07	9660	5.18	0340	52
9	9271	4.10	9300	1.08	9971	5.18	.29 0029	51
10	9.65 9517	4.10	9.94 9235	1.08	9.71 0282	5.18	10.28 9718	50
11	.65 9763	4.10	9170	1.08	0593	5.18	9407	49
12	.66 0009	4.10	9105	1.08	0904	5.18	9096	48
13	0255	4.10	9040	1.08	1215	5.17	8785	47
14	0501	4.08	8975	1.08	1525	5.18	8475	46
15	9.66 0746	4.08	9.94 8910	1.08	9.71 1836	5.17	10.28 8164	45
16	0991	4.08	8845	1.08	2146	5.17	7854	44
17	1236	4.08	8780	1.08	2456	5.17	7544	43
18	1481	4.08	8715	1.08	2766	5.17	7234	42
19	1726	4.07	8650	1.10	3076	5.17	6924	41
20	9.66 1970	4.07	9.94 8584	1.08	9.71 3386	5.17	10.28 6614	40
21	2214	4.08	8519	1.08	3696	5.15	6304	39
22	2459	4.07	8454	1.10	4005	5.15	5995	38
23	2703	4.05	8388	1.08	4314	5.17	5686	37
24	2946	4.07	8323	1.10	4624	5.15	5376	36
25	9.66 3190	4.05	9.94 8257	1.08	9.71 4933	5.15	10.28 5067	35
26	3433	4.07	8192	1.10	5242	5.15	4758	34
27	3677	4.05	8126	1.10	5551	5.15	4449	33
28	3920	4.05	8060	1.08	5860	5.13	4140	32
29	4163	4.05	7995	1.10	6168	5.15	3832	31
30	9.66 4406	4.03	9.94 7929	1.10	9.71 6477	5.13	10.28 3523	30
31	4648	4.05	7863	1.10	6785	5.13	3215	29
32	4891	4.03	7797	1.10	7093	5.13	2907	28
33	5133	4.03	7731	1.10	7401	5.13	2599	27
34	5375	4.03	7665	1.08	7709	5.13	2291	26
35	9.66 5617	4.03	9.94 7600	1.12	9.71 8017	5.13	10.28 1983	25
36	5859	4.02	7533	1.10	8325	5.13	1675	24
37	6100	4.03	7467	1.10	8633	5.12	1367	23
38	6342	4.02	7401	1.10	8940	5.13	1060	22
39	6583	4.02	7335	1.10	9248	5.12	0752	21
40	9.66 6824	4.02	9.94 7269	1.10	9.71 9555	5.12	10.28 0445	20
41	7065	4.00	7203	1.12	.71 9862	5.12	.28 0138	19
42	7305	4.02	7136	1.10	.72 0169	5.12	.27 9831	18
43	7546	4.00	7070	1.10	0476	5.12	9524	17
44	7786	4.02	7004	1.12	0783	5.10	9217	16
45	9.66 8027	4.00	9.94 6937	1.10	9.72 1089	5.12	10.27 8911	15
46	8267	3.98	6871	1.12	1396	5.10	8604	14
47	8506	4.00	6804	1.10	1702	5.12	8298	13
48	8746	4.00	6738	1.12	2009	5.10	7991	12
49	8986	3.98	6671	1.12	2315	5.10	7685	11
50	9.66 9225	3.98	9.94 6604	1.10	9.72 2621	5.10	10.27 7379	10
51	9464	3.98	6538	1.12	2927	5.08	7073	9
52	9703	3.98	6471	1.12	3232	5.10	6768	8
53	.66 9942	3.98	6404	1.12	3538	5.10	6462	7
54	.67 0181	3.97	6337	1.12	3844	5.08	6156	6
55	9.67 0419	3.98	9.94 6270	1.12	9.72 4149	5.08	10.27 5851	5
56	0658	3.97	6203	1.12	4454	5.10	5546	4
57	0896	3.97	6136	1.12	4760	5.08	5240	3
58	1134	3.97	6069	1.12	5065	5.08	4935	2
59	1372	3.95	6002	1.12	5370	5.07	4630	1
60	9.67 1609		9.94 5935		9.72 5674		10.27 4326	0
′	Cosine.	D.1″.	Sine.	D.1″.	Cotang.	D.1″.	Tang.	′

117° 62°

28° **Table VIII.**—*Continued* 151°

′	Sine.	D.1″.	Cosine.	D.1″.	Tang.	D.1″.	Cotang.	′
0	**9.67 1609**	3.97	**9.94 5935**	1.12	**9.72 5674**	5.08	**10.27 4326**	**60**
1	1847	3.95	5868	1.13	5979	5.08	4021	59
2	2084	3.95	5800	1.12	6284	5.07	3716	58
3	2321	3.95	5733	1.12	6588	5.07	3412	57
4	2558	3.95	5666	1.13	6892	5.05	3108	56
5	**9.67 2795**	3.95	**9.94 5598**	1.12	**9.72 7197**	5.07	**10.27 2803**	**55**
6	3032	3.93	5531	1.12	7501	5.07	2499	54
7	3268	3.95	5464	1.13	7805	5.07	2195	53
8	3505	3.93	5396	1.13	8109	5.05	1891	52
9	3741	3.93	5328	1.12	8412	5.07	1588	51
10	**9.67 3977**	3.93	**9.94 5261**	1.13	**9.72 8716**	5.07	**10.27 1284**	**50**
11	4213	3.92	5193	1.13	9020	5.05	0980	49
12	4448	3.93	5125	1.12	9323	5.05	0677	48
13	4684	3.92	5058	1.13	9626	5.05	0374	47
14	4919	3.93	4990	1.13	.72 9929	5.07	.27 0071	46
15	**9.67 5155**	3.92	**9.94 4922**	1.13	**9.73 0233**	5.03	**10.26 9767**	**45**
16	5390	3.90	4854	1.13	0535	5.05	9465	44
17	5624	3.92	4786	1.13	0838	5.05	9162	43
18	5859	3.92	4718	1.13	1141	5.05	8859	42
19	6094	3.90	4650	1.13	1444	5.03	8556	41
20	**9.67 6328**	3.90	**9.94 4582**	1.13	**9.73 1746**	5.03	**10.26 8254**	**40**
21	6562	3.90	4514	1.13	2048	5.05	7952	39
22	6796	3.90	4446	1.15	2351	5.03	7649	38
23	7030	3.90	4377	1.13	2653	5.03	7347	37
24	7264	3.90	4309	1.13	2955	5.03	7045	36
25	**9.67 7498**	3.88	**9.94 4241**	1.15	**9.73 3257**	5.02	**10.26 6743**	**35**
26	7731	3.88	4172	1.13	3558	5.03	6442	34
27	7964	3.88	4104	1.13	3860	5.03	6140	33
28	8197	3.88	4036	1.15	4162	5.02	5838	32
29	8430	3.88	3967	1.13	4463	5.02	5537	31
30	**9.67 8663**	3.87	**9.94 3899**	1.15	**9.73 4764**	5.03	**10.26 5236**	**30**
31	8895	3.88	3830	1.15	5066	5.02	4934	29
32	9128	3.87	3761	1.13	5367	5.02	4633	28
33	9360	3.87	3693	1.15	5668	5.02	4332	27
34	9592	3.87	3624	1.15	5969	5.00	4031	26
35	**9.67 9824**	3.87	**9.94 3555**	1.15	**9.73 6269**	5.02	**10.26 3731**	**25**
36	.68 0056	3.87	3486	1.15	6570	5.00	3430	24
37	0288	3.85	3417	1.15	6870	5.02	3130	23
38	0519	3.85	3348	1.15	7171	5.00	2829	22
39	0750	3.87	3279	1.15	7471	5.00	2529	21
40	**9.68 0982**	3.85	**9.94 3210**	1.15	**9.73 7771**	5.00	**10.26 2229**	**20**
41	1213	3.83	3141	1.15	8071	5.00	1929	19
42	1443	3.85	3072	1.15	8371	5.00	1629	18
43	1674	3.85	3003	1.15	8671	5.00	1329	17
44	1905	3.83	2934	1.17	8971	5.00	1029	16
45	**9.68 2135**	3.83	**9.94 2864**	1.15	**9.73 9271**	4.98	**10.26 0729**	**15**
46	2365	3.83	2795	1.15	9570	5.00	0430	14
47	2595	3.83	2726	1.17	.73 9870	4.98	.26 0130	13
48	2825	3.83	2656	1.15	.74 0169	4.98	.25 9831	12
49	3055	3.82	2587	1.17	0468	4.98	9532	11
50	**9.68 3284**	3.83	**9.94 2517**	1.15	**9.74 0767**	4.98	**10.25 9233**	**10**
51	3514	3.82	2448	1.17	1066	4.98	8934	9
52	3743	3.82	2378	1.17	1365	4.98	8635	8
53	3972	3.82	2308	1.15	1664	4.97	8336	7
54	4201	3.82	2239	1.17	1962	4.98	8038	6
55	**9.68 4430**	3.80	**9.94 2169**	1.17	**9.74 2261**	4.97	**10.25 7739**	**5**
56	4658	3.82	2099	1.17	2559	4.98	7441	4
57	4887	3.80	2029	1.17	2858	4.97	7142	3
58	5115	3.80	1959	1.17	3156	4.97	6844	2
59	5343	3.80	1889	1.17	3454	4.97	6546	1
60	**9.68 5571**		**9.94 1819**		**9.74 3752**		**10.25 6248**	**0**
′	Cosine.	D.1″.	Sine.	D.1″.	Cotang.	D.1″.	Tang.	′

118° 61°

29° TABLE VIII.—*Continued* 150°

′	Sine.	D.1″.	Cosine.	D.1″.	Tang.	D.1″.	Cotang.	′
0	**9.68 5571**	3.80	**9.94 1819**	1.17	**9.74 3752**	4.97	**10.25 6248**	60
1	5799	3.80	1749	1.17	4050	4.97	5950	59
2	6027	3.78	1679	1.17	4348	4.95	5652	58
3	6254	3.80	1609	1.17	4645	4.97	5355	57
4	6482	3.78	1539	1.17	4943	4.95	5057	56
5	**9.68 6709**	3.78	**9.94 1469**	1.18	**9.74 5240**	4.97	**10.25 4760**	55
6	6936	3.78	1398	1.17	5538	4.95	4462	54
7	7163	3.77	1328	1.17	5835	4.95	4165	53
8	7389	3.78	1258	1.18	6132	4.95	3868	52
9	7616	3.78	1187	1.17	6429	4.95	3571	51
10	**9.68 7843**	3.77	**9.94 1117**	1.18	**9.74 6726**	4.95	**10.25 3274**	50
11	8069	3.77	1046	1.18	7023	4.93	2977	49
12	8295	3.77	0975	1.17	7319	4.95	2681	48
13	8521	3.77	0905	1.18	7616	4.95	2384	47
14	8747	3.75	0834	1.18	7913	4.93	2087	46
15	**9.68 8972**	3.77	**9.94 0763**	1.17	**9.74 8209**	4.93	**10.25 1791**	45
16	9198	3.75	0693	1.18	8505	4.93	1495	44
17	9423	3.75	0622	1.18	8801	4.93	1199	43
18	9648	3.75	0551	1.18	9097	4.93	0903	42
19	.68 9873	3.75	0480	1.18	9393	4.93	0607	41
20	**9.69 0098**	3.75	**9.94 0409**	1.18	**9.74 9689**	4.93	**10.25 0311**	40
21	0323	3.75	0338	1.18	.74 9985	4.93	.25 0015	39
22	0548	3.73	0267	1.18	.75 0281	4.92	.24 9719	38
23	0772	3.73	0196	1.18	0576	4.93	9424	37
24	0996	3.73	0125	1.18	0872	4.92	9128	36
25	**9.69 1220**	3.73	**9.94 0054**	1.20	**9.75 1167**	4.92	**10.24 8833**	35
26	1444	3.73	.93 9982	1.18	1462	4.92	8538	34
27	1668	3.73	9911	1.18	1757	4.92	8243	33
28	1892	3.72	9840	1.20	2052	4.92	7948	32
29	2115	3.73	9768	1.18	2347	4.92	7653	31
30	**9.69 2339**	3.72	**9.93 9697**	1.20	**9.75 2642**	4.92	**10.24 7358**	30
31	2562	3.72	9625	1.18	2937	4.90	7063	29
32	2785	3.72	9554	1.20	3231	4.92	6769	28
33	3008	3.72	9482	1.20	3526	4.90	6474	27
34	3231	3.70	9410	1.18	3820	4.92	6180	26
35	**9.69 3453**	3.72	**9.93 9339**	1.20	**9.75 4115**	4.90	**10.24 5885**	25
36	3676	3.70	9267	1.20	4409	4.90	5591	24
37	3898	3.70	9195	1.20	4703	4.90	5297	23
38	4120	3.70	9123	1.18	4997	4.90	5003	22
39	4342	3.70	9052	1.20	5291	4.90	4709	21
40	**9.69 4564**	3.70	**9.93 8980**	1.20	**9.75 5585**	4.88	**10.24 4415**	20
41	4786	3.68	8908	1.20	5878	4.90	4122	19
42	5007	3.70	8836	1.22	6172	4.88	3828	18
43	5229	3.68	8763	1.20	6465	4.90	3535	17
44	5450	3.68	8691	1.20	6759	4.88	3241	16
45	**9.69 5671**	3.68	**9.93 8619**	1.20	**9.75 7052**	4.88	**10.24 2948**	15
46	5892	3.68	8547	1.20	7345	4.88	2655	14
47	6113	3.68	8475	1.22	7638	4.88	2362	13
48	6334	3.67	8402	1.20	7931	4.88	2069	12
49	6554	3.68	8330	1.20	8224	4.88	1776	11
50	**9.69 6775**	3.67	**9.93 8258**	1.22	**9.75 8517**	4.88	**10.24 1483**	10
51	6995	3.67	8185	1.20	8810	4.87	1190	9
52	7215	3.67	8113	1.22	9102	4.88	0898	8
53	7435	3.65	8040	1.22	9395	4.87	0605	7
54	7654	3.67	7967	1.20	9687	4.87	0313	6
55	**9.69 7874**	3.67	**9.93 7895**	1.22	**9.75 9979**	4.88	**10.24 0021**	5
56	8094	3.65	7822	1.22	.76 0272	4.87	.23 9728	4
57	8313	3.65	7749	1.22	0564	4.87	9436	3
58	8532	3.65	7676	1.20	0856	4.87	9144	2
59	8751	3.65	7604	1.22	1148	4.85	8852	1
60	**9.69 8970**		**9.93 7531**		**9.76 1439**		**10.23 8561**	0
′	Cosine.	D.1″.	Sine.	D.1″.	Cotang.	D.1″.	Tang.	′

119° 60°

30° TABLE VIII.—*Continued* 149°

′	Sine.	D.1″.	Cosine.	D.1″.	Tang.	D.1″.	Cotang.	′
0	9.69 8970	3.65	9.93 7531	1.22	9.76 1439	4.87	10.23 8561	60
1	9189	3.63	7458	1.22	1731	4.87	8269	59
2	9407	3.65	7385	1.22	2023	4.85	7977	58
3	9626	3.63	7312	1.23	2314	4.87	7686	57
4	.69 9844	3.63	7238	1.22	2606	4.85	7394	56
5	9.70 0062	3.63	9.93 7165	1.22	9.76 2897	4.85	10.23 7103	55
6	0280	3.63	7092	1.22	3188	4.85	6812	54
7	0498	3.63	7019	1.22	3479	4.85	6521	53
8	0716	3.62	6946	1.23	3770	4.85	6230	52
9	0933	3.63	6872	1.22	4061	4.85	5939	51
10	9.70 1151	3.62	9.93 6799	1.23	9.76 4352	4.85	10.23 5648	50
11	1368	3.62	6725	1.22	4643	4.83	5357	49
12	1585	3.62	6652	1.23	4933	4.85	5067	48
13	1802	3.62	6578	1.22	5224	4.83	4776	47
14	2019	3.62	6505	1.23	5514	4.85	4486	46
15	9.70 2236	3.60	9.93 6431	1.23	9.76 5805	4.83	10.23 4195	45
16	2452	3.62	6357	1.22	6095	4.83	3905	44
17	2669	3.60	6284	1.23	6385	4.83	3615	43
18	2885	3.60	6210	1.23	6675	4.83	3325	42
19	3101	3.60	6136	1.23	6965	4.83	3035	41
20	9.70 3317	3.60	9.93 6062	1.23	9.76 7255	4.83	10.23 2745	40
21	3533	3.60	5988	1.23	7545	4.82	2455	39
22	3749	3.58	5914	1.23	7834	4.83	2166	38
23	3964	3.58	5840	1.23	8124	4.83	1876	37
24	4179	3.60	5766	1.23	8414	4.82	1586	36
25	9.70 4395	3.58	9.93 5692	1.23	9.76 8703	4.82	10.23 1297	35
26	4610	3.58	5618	1.25	8992	4.82	1008	34
27	4825	3.58	5543	1.23	9281	4.83	0719	33
28	5040	3.57	5469	1.23	9571	4.82	0429	32
29	5254	3.58	5395	1.25	.76 9860	4.80	.23 0140	31
30	9.70 5469	3.57	9.93 5320	1.23	9.77 0148	4.82	10.22 9852	30
31	5683	3.58	5246	1.25	0437	4.82	9563	29
32	5898	3.57	5171	1.23	0726	4.82	9274	28
33	6112	3.57	5097	1.25	1015	4.80	8985	27
34	6326	3.55	5022	1.23	1303	4.82	8697	26
35	9.70 6539	3.57	9.93 4948	1.25	9.77 1592	4.80	10.22 8408	25
36	6753	3.57	4873	1.25	1880	4.80	8120	24
37	6967	3.55	4798	1.25	2168	4.82	7832	23
38	7180	3.55	4723	1.23	2457	4.80	7543	22
39	7393	3.55	4649	1.25	2745	4.80	7255	21
40	9.70 7606	3.55	9.93 4574	1.25	9.77 3033	4.80	10.22 6967	20
41	7819	3.55	4499	1.25	3321	4.78	6679	19
42	8032	3.55	4424	1.25	3608	4.80	6392	18
43	8245	3.55	4349	1.25	3896	4.80	6104	17
44	8458	3.53	4274	1.25	4184	4.78	5816	16
45	9.70 8670	3.53	9.93 4199	1.27	9.77 4471	4.80	10.22 5529	15
46	8882	3.53	4123	1.25	4759	4.78	5241	14
47	9094	3.53	4048	1.25	5046	4.78	4954	13
48	9306	3.53	3973	1.25	5333	4.80	4667	12
49	9518	3.53	3898	1.27	5621	4.78	4379	11
50	9.70 9730	3.52	9.93 3822	1.25	9.77 5908	4.78	10.22 4092	10
51	.70 9941	3.53	3747	1.27	6195	4.78	3805	9
52	.71 0153	3.52	3671	1.25	6482	4.77	3518	8
53	0364	3.52	3596	1.27	6768	4.78	3232	7
54	0575	3.52	3520	1.25	7055	4.78	2945	6
55	9.71 0786	3.52	9.93 3445	1.27	9.77 7342	4.77	10.22 2658	5
56	0997	3.52	3369	1.27	7628	4.78	2372	4
57	1208	3.52	3293	1.27	7915	4.77	2085	3
58	1419	3.50	3217	1.27	8201	4.78	1799	2
59	1629	3.50	3141	1.25	8488	4.77	1512	1
60	9.71 1839		9.93 3066		9.77 8774		10.22 1226	0
′	Cosine.	D.1″.	Sine.	D.1″.	Cotang.	D.1″.	Tang.	′

120° 59°

31° TABLE VIII.—*Continued* 148°

′	Sine.	D.1″.	Cosine.	D.1″.	Tang.	D.1″.	Cotang.	′
0	9.71 1839	3.52	9.93 3066	1.27	9.77 8774	4.77	10.22 1226	60
1	2050	3.50	2990	1.27	9060	4.77	0940	59
2	2260	3.48	2914	1.27	9346	4.77	0654	58
3	2469	3.50	2838	1.27	9632	4.77	0368	57
4	2679	3.50	2762	1.28	.77 9918	4.75	.22 0082	56
5	9.71 2889	3.48	9.93 2685	1.27	9.78 0203	4.77	10.21 9797	55
6	3098	3.50	2609	1.27	0489	4.77	9511	54
7	3308	3.48	2533	1.27	0775	4.75	9225	53
8	3517	3.48	2457	1.28	1060	4.77	8940	52
9	3726	3.48	2380	1.27	1346	4.75	8654	51
10	9.71 3935	3.48	9.93 2304	1.27	9.78 1631	4.75	10.21 8369	50
11	4144	3.47	2228	1.28	1916	4.75	8084	49
12	4352	3.48	2151	1.27	2201	4.75	7799	48
13	4561	3.47	2075	1.28	2486	4.75	7514	47
14	4769	3.48	1998	1.28	2771	4.75	7229	46
15	9.71 4978	3.47	9.93 1921	1.27	9.78 3056	4.75	10.21 6944	45
16	5186	3.47	1845	1.28	3341	4.75	6659	44
17	5394	3.47	1768	1.28	3626	4.73	6374	43
18	5602	3.45	1691	1.28	3910	4.75	6090	42
19	5809	3.47	1614	1.28	4195	4.73	5805	41
20	9.71 6017	3.45	9.93 1537	1.28	9.78 4479	4.75	10.21 5521	40
21	6224	3.47	1460	1.28	4764	4.73	5236	39
22	6432	3.45	1383	1.28	5048	4.73	4952	38
23	6639	3.45	1306	1.28	5332	4.73	4668	37
24	6846	3.45	1229	1.28	5616	4.73	4384	36
25	9.71 7053	3.43	9.93 1152	1.28	9.78 5900	4.73	10.21 4100	35
26	7259	3.45	1075	1.28	6184	4.73	3816	34
27	7466	3.45	0998	1.28	6468	4.73	3532	33
28	7673	3.43	0921	1.30	6752	4.73	3248	32
29	7879	3.43	0843	1.28	7036	4.72	2964	31
30	9.71 8085	3.43	9.93 0766	1.30	9.78 7319	4.73	10.21 2681	30
31	8291	3.43	0688	1.28	7603	4.72	2397	29
32	8497	3.43	0611	1.30	7886	4.73	2114	28
33	8703	3.43	0533	1.28	8170	4.72	1830	27
34	8909	3.42	0456	1.30	8453	4.72	1547	26
35	9.71 9114	3.43	9.93 0378	1.30	9.78 8736	4.72	10.21 1264	25
36	9320	3.42	0300	1.28	9019	4.72	0981	24
37	9525	3.42	0223	1.30	9302	4.72	0698	23
38	9730	3.42	0145	1.30	9585	4.72	0415	22
39	.71 9935	3.42	.93 0067	1.30	.78 9868	4.72	.21 0132	21
40	9.72 0140	3.42	9.92 9989	1.30	9.79 0151	4.72	10.20 9849	20
41	0345	3.40	9911	1.30	0434	4.70	9566	19
42	0549	3.42	9833	1.30	0716	4.72	9284	18
43	0754	3.40	9755	1.30	0999	4.70	9001	17
44	0958	3.40	9677	1.30	1281	4.70	8719	16
45	9.72 1162	3.40	9.92 9599	1.30	9.79 1563	4.72	10.20 8437	15
46	1366	3.40	9521	1.32	1846	4.70	8154	14
47	1570	3.40	9442	1.30	2128	4.70	7872	13
48	1774	3.40	9364	1.30	2410	4.70	7590	12
49	1978	3.38	9286	1.32	2692	4.70	7308	11
50	9.72 2181	3.40	9.92 9207	1.30	9.79 2974	4.70	10.20 7026	10
51	2385	3.38	9129	1.32	3256	4.70	6744	9
52	2588	3.38	9050	1.30	3538	4.68	6462	8
53	2791	3.38	8972	1.32	3819	4.70	6181	7
54	2994	3.38	8893	1.30	4101	4.70	5899	6
55	9.72 3197	3.38	9.92 8815	1.32	9.79 4383	4.68	10.20 5617	5
56	3400	3.38	8736	1.32	4664	4.70	5336	4
57	3603	3.37	8657	1.32	4946	4.68	5054	3
58	3805	3.37	8578	1.32	5227	4.68	4773	2
59	4007	3.38	8499	1.32	5508	4.68	4492	1
60	9.72 4210		9.92 8420		9.79 5789		10.20 4211	0
′	Cosine.	D.1″.	Sine.	D.1″.	Cotang.	D.1″.	Tang.	′

121° 58°

32° TABLE VIII.—*Continued* 147°

′	Sine.	D.1″.	Cosine.	D.1″.	Tang.	D.1″.	Cotang.	′
0	**9.72 4210**	3.37	**9.92 8420**	1.30	**9.79 5789**	4.68	**10.20 4211**	60
1	4412	3.37	8342	1.32	6070	4.68	3930	59
2	4614	3.37	8263	1.33	6351	4.68	3649	58
3	4816	3.35	8183	1.32	6632	4.68	3368	57
4	5017	3.37	8104	1.32	6913	4.68	3087	56
5	**9.72 5219**	3.35	**9.92 8025**	1.32	**9.79 7194**	4.67	**10.20 2806**	55
6	5420	3.37	7946	1.32	7474	4.68	2526	54
7	5622	3.35	7867	1.33	7755	4.68	2245	53
8	5823	3.35	7787	1.32	8036	4.67	1964	52
9	6024	3.35	7708	1.32	8316	4.67	1684	51
10	**9.72 6225**	3.35	**9.92 7629**	1.33	**9.79 8596**	4.68	**10.20 1404**	50
11	6426	3.33	7549	1.32	8877	4.67	1123	49
12	6626	3.35	7470	1.33	9157	4.67	0843	48
13	6827	3.33	7390	1.33	9437	4.67	0563	47
14	7027	3.35	7310	1.32	9717	4.67	0283	46
15	**9.72 7228**	3.33	**9.92 7231**	1.33	**9.79 9997**	4.67	**10.20 0003**	45
16	7428	3.33	7151	1.33	.80 0277	4.67	.19 9723	44
17	7628	3.33	7071	1.33	0557	4.65	9443	43
18	7828	3.32	6991	1.33	0836	4.67	9164	42
19	8027	3.33	6911	1.33	1116	4.67	8884	41
20	**9.72 8227**	3.33	**9.92 6831**	1.33	**9.80 1396**	4.65	**10.19 8604**	40
21	8427	3.32	6751	1.33	1675	4.67	8325	39
22	8626	3.32	6671	1.33	1955	4.65	8045	38
23	8825	3.32	6591	1.33	2234	4.65	7766	37
24	9024	3.32	6511	1.33	2513	4.65	7487	36
25	**9.72 9223**	3.32	**9.92 6431**	1.33	**9.80 2792**	4.67	**10.19 7208**	35
26	9422	3.32	6351	1.35	3072	4.65	6928	34
27	9621	3.32	6270	1.33	3351	4.65	6649	33
28	.72 9820	3.30	6190	1.33	3630	4.65	6370	32
29	.73 0018	3.32	6110	1.35	3909	4.63	6091	31
30	**9.73 0217**	3.30	**9.92 6029**	1.33	**9.80 4187**	4.65	**10.19 5813**	30
31	0415	3.30	5949	1.35	4466	4.65	5534	29
32	0613	3.30	5868	1.33	4745	4.63	5255	28
33	0811	3.30	5788	1.35	5023	4.65	4977	27
34	1009	3.28	5707	1.35	5302	4.63	4698	26
35	**9.73 1206**	3.30	**9.92 5626**	1.35	**9.80 5580**	4.65	**10.19 4420**	25
36	1404	3.30	5545	1.33	5859	4.63	4141	24
37	1602	3.28	5465	1.35	6137	4.63	3863	23
38	1799	3.28	5384	1.35	6415	4.63	3585	22
39	1996	3.28	5303	1.35	6693	4.63	3307	21
40	**9.73 2193**	3.28	**9.92 5222**	1.35	**9.80 6971**	4.63	**10.19 3029**	20
41	2390	3.28	5141	1.35	7249	4.63	2751	10
42	2587	3.28	5060	1.35	7527	4.63	2473	18
43	2784	3.27	4979	1.37	7805	4.63	2195	17
44	2980	3.28	4897	1.35	8083	4.63	1917	16
45	**9.73 3177**	3.27	**9.92 4816**	1.35	**9.80 8361**	4.62	**10.19 1639**	15
46	3373	3.27	4735	1.35	8638	4.63	1362	14
47	3569	3.27	4654	1.37	8916	4.62	1084	13
48	3765	3.27	4572	1.35	9193	4.63	0807	12
49	3961	3.27	4491	1.37	9471	4.62	0529	11
50	**9.73 4157**	3.27	**9.92 4409**	1.35	**9.80 9748**	4.62	**10.19 0252**	10
51	4353	3.27	4328	1.37	.81 0025	4.62	.18 9975	9
52	4549	3.25	4246	1.37	0302	4.63	9698	8
53	4744	3.25	4164	1.35	0580	4.62	9420	7
54	4939	3.27	4083	1.37	0857	4.62	9143	6
55	**9.73 5135**	3.25	**9.92 4001**	1.37	**9.81 1134**	4.60	**10.18 8866**	5
56	5330	3.25	3919	1.37	1410	4.62	8590	4
57	5525	3.23	3837	1.37	1687	4.62	8313	3
58	5719	3.25	3755	1.37	1964	4.62	8036	2
59	5914	3.25	3673	1.37	2241	4.60	7759	1
60	**9.73 6109**		**9.92 3591**		**9.81 2517**		**10.18 7483**	0
′	Cosine.	D.1″.	Sine.	D.1″.	Cotang.	D.1″.	Tang.	′

122° 57°

33° TABLE VIII.—*Continued* 146°

′	Sine.	D.1″.	Cosine.	D.1″.	Tang.	D.1″.	Cotang.	′
0	**9.73 6109**	3.23	**9.92 3591**	1.37	**9.81 2517**	4.62	**10.18 7483**	**60**
1	6303	3.25	3509	1.37	2794	4.60	7206	59
2	6498	3.23	3427	1.37	3070	4.62	6930	58
3	6692	3.23	3345	1.37	3347	4.60	6653	57
4	6886	3.23	3263	1.37	3623	4.60	6377	56
5	**9.73 7080**	3.23	**9.92 3181**	1.38	**9.81 3899**	4.62	**10.18 6101**	**55**
6	7274	3.22	3098	1.37	4176	4.60	5824	54
7	7467	3.23	3016	1.38	4452	4.60	5548	53
8	7661	3.23	2933	1.37	4728	4.60	5272	52
9	7855	3.22	2851	1.38	5004	4.60	4996	51
10	**9.73 8048**	3.22	**9.92 2768**	1.37	**9.81 5280**	4.58	**10.18 4720**	**50**
11	8241	3.22	2686	1.38	5555	4.60	4445	49
12	8434	3.22	2603	1.38	5831	4.60	4169	48
13	8627	3.22	2520	1.37	6107	4.58	3893	47
14	8820	3.22	2438	1.38	6382	4.60	3618	46
15	**9.73 9013**	3.22	**9.92 2355**	1.38	**9.81 6658**	4.58	**10.18 3342**	**45**
16	9206	3.20	2272	1.38	6933	4.60	3067	44
17	9398	3.20	2189	1.38	7209	4.58	2791	43
18	9590	3.22	2106	1.38	7484	4.58	2516	42
19	9783	3.20	2023	1.38	7759	4.60	2241	41
20	**9.73 9975**	3.20	**9.92 1940**	1.38	**9.81 8035**	4.58	**10.18 1965**	**40**
21	.74 0167	3.20	1857	1.38	8310	4.58	1690	39
22	0359	3.18	1774	1.38	8585	4.58	1415	38
23	0550	3.20	1691	1.40	8860	4.58	1140	37
24	0742	3.20	1607	1.38	9135	4.58	0865	36
25	**9.74 0934**	3.18	**9.92 1524**	1.38	**9.81 9410**	4.57	**10.18 0590**	**35**
26	1125	3.18	1441	1.40	9684	4.58	0316	34
27	1316	3.20	1357	1.38	.81 9959	4.58	.18 0041	33
28	1508	3.18	1274	1.40	.82 0234	4.57	.17 9766	32
29	1699	3.17	1190	1.38	0508	4.58	9492	31
30	**9.74 1889**	3.18	**9.92 1107**	1.40	**9.82 0783**	4.57	**10.17 9217**	**30**
31	2080	3.18	1023	1.40	1057	4.58	8943	29
32	2271	3.18	0939	1.38	1332	4.57	8668	28
33	2462	3.17	0856	1.40	1606	4.57	8394	27
34	2652	3.17	0772	1.40	1880	4.57	8120	26
35	**9.74 2842**	3.18	**9.92 0688**	1.40	**9.82 2154**	4.58	**10.17 7846**	**25**
36	3033	3.17	0604	1.40	2429	4.57	7571	24
37	3223	3.17	0520	1.40	2703	4.57	7297	23
38	3413	3.15	0436	1.40	2977	4.57	7023	22
39	3602	3.17	0352	1.40	3251	4.55	6749	21
40	**9.74 3792**	3.17	**9.92 0268**	1.40	**9.82 3524**	4.57	**10.17 6476**	**20**
41	3982	3.15	0184	1.42	3798	4.57	6202	19
42	4171	3.17	0099	1.40	4072	4.55	5928	18
43	4361	3.15	.92 0015	1.40	4345	4.57	5655	17
44	4550	3.15	.91 9931	1.42	4619	4.57	5381	16
45	**9.74 4739**	3.15	**9.91 9846**	1.40	**9.82 4893**	4.55	**10.17 5107**	**15**
46	4928	3.15	9762	1.42	5166	4.55	4834	14
47	5117	3.15	9677	1.40	5439	4.57	4561	13
48	5306	3.13	9593	1.42	5713	4.55	4287	12
49	5494	3.15	9508	1.40	5986	4.55	4014	11
50	**9.74 5683**	3.13	**9.91 9424**	1.42	**9.82 6254**	4.55	**10.17 3741**	**10**
51	5871	3.15	9339	1.42	6532	4.55	3468	9
52	6060	3.13	9254	1.42	6805	4.55	3195	8
53	6248	3.13	9169	1.40	7078	4.55	2922	7
54	6436	3.13	9085	1.42	7351	4.55	2649	6
55	**9.74 6624**	3.13	**9.91 9000**	1.42	**9.82 7624**	4.55	**10.17 2376**	**5**
56	6812	3.12	8915	1.42	7897	4.55	2103	4
57	6999	3.13	8830	1.42	8170	4.53	1830	3
58	7187	3.12	8745	1.43	8442	4.55	1558	2
59	7374	3.13	8659	1.42	8715	4.53	1285	1
60	**9.74 7562**		**9.91 8574**		**9.82 8987**		**10.17 1013**	**0**
′	Cosine.	D.1″.	Sine.	D.1″.	Cotang.	D.1″.	Tang.	′

123° 56°

34° TABLE VIII.—*Continued* 145°

′	Sine.	D.1″	Cosine.	D.1″	Tang.	D.1″	Cotang.	′
0	**9.74 7562**	3.12	**9.91 8574**	1.42	**9.82 8987**	4.55	**10.17 1013**	**60**
1	7749	3.12	8489	1.42	9260	4.53	0740	59
2	7936	3.12	8404	1.43	9532	4.55	0468	58
3	8123	3.12	8318	1.42	.82 9805	4.53	.17 0195	57
4	8310	3.12	8233	1.43	.83 0077	4.53	.16 9923	56
5	**9.74 8497**	3.10	**9.91 8147**	1.42	**9.83 0349**	4.53	**10.16 9651**	**55**
6	8683	3.12	8062	1.43	0621	4.53	9379	54
7	8870	3.10	7976	1.42	0893	4.53	9107	53
8	9056	3.12	7891	1.43	1165	4.53	8835	52
9	9243	3.10	7805	1.43	1437	4.53	8563	51
10	**9.74 9429**	3.10	**9.91 7719**	1.42	**9.83 1709**	4.53	**10.16 8291**	**50**
11	9615	3.10	7634	1.43	1981	4.53	8019	49
12	9801	3.10	7548	1.43	2253	4.53	7747	48
13	.74 9987	3.08	7462	1.43	2525	4.52	7475	47
14	.75 0172	3.10	7376	1.43	2796	4.53	7204	46
15	**9.75 0358**	3.08	**9.91 7290**	1.43	**9.83 3068**	4.52	**10.16 6932**	**45**
16	0543	3.10	7204	1.43	3339	4.53	6661	44
17	0729	3.08	7118	1.43	3611	4.52	6389	43
18	0914	3.08	7032	1.43	3882	4.53	6118	42
19	1099	3.08	6946	1.45	4154	4.52	5846	41
20	**9.75 1284**	3.08	**9.91 6859**	1.43	**9.83 4425**	4.52	**10.16 5575**	**40**
21	1469	3.08	6773	1.43	4696	4.52	5304	39
22	1654	3.08	6687	1.45	4967	4.52	5033	38
23	1839	3.07	6600	1.43	5238	4.52	4762	37
24	2023	3.08	6514	1.45	5509	4.52	4491	36
25	**9.75 2208**	3.07	**9.91 6427**	1.43	**9.83 5780**	4.52	**10.16 4220**	**35**
26	2392	3.07	6341	1.45	6051	4.52	3949	34
27	2576	3.07	6254	1.45	6322	4.52	3678	33
28	2760	3.07	6167	1.43	6593	4.52	3407	32
29	2944	3.07	6081	1.45	6864	4.50	3136	31
30	**9.75 3128**	3.07	**9.91 5994**	1.45	**9.83 7134**	4.52	**10.16 2866**	**30**
31	3312	3.05	5907	1.45	7405	4.50	2595	29
32	3495	3.07	5820	1.45	7675	4.52	2325	28
33	3679	3.07	5733	1.45	7946	4.50	2054	27
34	3862	3.07	5646	1.45	8216	4.52	1784	26
35	**9.75 4046**	3.05	**9.91 5559**	1.45	**9.83 8487**	4.50	**10.16 1513**	**25**
36	4229	3.05	5472	1.45	8757	4.50	1243	24
37	4412	3.05	5385	1.47	9027	4.50	0973	23
38	4595	3.05	5297	1.45	9297	4.52	0703	22
39	4778	3.03	5210	1.45	9568	4.50	0432	21
40	**9.75 4960**	3.05	**9.91 5123**	1.47	**9.83 9838**	4.50	**10.16 0162**	**20**
41	5143	3.05	5035	1.45	.84 0108	4.50	.15 9892	19
42	5326	3.03	4948	1.47	0378	4.50	9622	18
43	5508	3.03	4860	1.45	0648	4.48	9352	17
44	5690	3.03	4773	1.47	0917	4.50	9083	16
45	**9.75 5872**	3.03	**9.91 4685**	1.45	**9.84 1187**	4.50	**10.15 8813**	**15**
46	6054	3.03	4598	1.47	1457	4.50	8543	14
47	6236	3.03	4510	1.47	1727	4.48	8273	13
48	6418	3.03	4422	1.47	1996	4.50	8004	12
49	6600	3.03	4334	1.47	2266	4.48	7734	11
50	**9.75 6782**	3.02	**9.91 4246**	1.47	**9.84 2535**	4.50	**10.15 7465**	**10**
51	6963	3.02	4158	1.47	2805	4.48	7195	9
52	7144	3.03	4070	1.47	3074	4.48	6926	8
53	7326	3.02	3982	1.47	3343	4.48	6657	7
54	7507	3.02	3894	1.47	3612	4.50	6388	6
55	**9.75 7688**	3.02	**9.91 3806**	1.47	**9.84 3882**	4.48	**10.15 6118**	**5**
56	7869	3.02	3718	1.47	4151	4.48	5849	4
57	8050	3.00	3630	1.48	4420	4.48	5580	3
58	8230	3.02	3541	1.47	4689	4.48	5311	2
59	8411	3.00	3453	1.47	4958	4.48	5042	1
60	**9.75 8591**		**9.91 3365**		**9.84 5227**		**10.15 4773**	**0**
′	Cosine.	D.1″.	Sine.	D.1″.	Cotang.	D.1″.	Tang.	′

124° 55°

35° TABLE VIII.—*Continued* 144°

′	Sine.	D.1″.	Cosine.	D.1″.	Tang.	D.1″.	Cotang.	′
0	**9.75 8591**	3.02	**9.91 3365**	1.48	**9.84 5227**	4.48	**10 15 4773**	**60**
1	8772	3.00	3276	1.48	5496	4.47	4504	59
2	8952	3.00	3187	1.47	5764	4.48	4236	58
3	9132	3.00	3099	1.48	6033	4.48	3967	57
4	9312	3.00	3010	1.47	6302	4.47	3698	56
5	**9.75 9492**	3.00	**9.91 2922**	1.48	**9.84 6570**	4.48	**10.15 3430**	**55**
6	9672	3.00	2833	1.48	6839	4.48	3161	54
7	.75 9852	2.98	2744	1.48	7108	4.47	2892	53
8	.76 0031	3.00	2655	1.48	7376	4.47	2624	52
9	0211	2.98	2566	1.48	7644	4.48	2356	51
10	**9.76 0390**	2.98	**9.91 2477**	1.48	**9.84 7913**	4.47	**10.15 2087**	**50**
11	0569	2.98	2388	1.48	8181	4.47	1819	49
12	0748	2.98	2299	1.48	8449	4.47	1551	48
13	0927	2.98	2210	1.48	8717	4.48	1283	47
14	1106	2.98	2121	1.50	8986	4.47	1014	46
15	**9.76 1285**	2.98	**9.91 2031**	1.48	**9.84 9254**	4.47	**10.15 0746**	**45**
16	1464	2.97	1942	1.48	9522	4.47	0478	44
17	1642	2.98	1853	1.50	.84 9790	4.45	.15 0210	43
18	1821	2.97	1763	1.48	.85 0057	4.47	.14 9943	42
19	1999	2.97	1674	1.50	0325	4.47	9675	41
20	**9.76 2177**	2.98	**9.91 1584**	1.48	**9.85 0593**	4.47	**10.14 9407**	**40**
21	2356	2.97	1495	1.50	0861	4.47	9139	39
22	2534	2.97	1405	1.50	1129	4.45	8871	38
23	2712	2.95	1315	1.48	1396	4.47	8604	37
24	2889	2.97	1226	1.50	1664	4.45	8336	36
25	**9.76 3067**	2.97	**9.91 1136**	1.50	**9.85 1931**	4.47	**10.14 8069**	**35**
26	3245	2.95	1046	1.50	2199	4.45	7801	34
27	3422	2.97	0956	1.50	2466	4.45	7534	33
28	3600	2.95	0866	1.50	2733	4.47	7267	32
29	3777	2.95	0776	1.50	3001	4.45	6999	31
30	**9.76 3954**	2.95	**9.91 0686**	1.50	**9.85 3268**	4.45	**10.14 6732**	**30**
31	4131	2.95	0596	1.50	3535	4.45	6465	29
32	4308	2.95	0506	1.52	3802	4.45	6198	28
33	4485	2.95	0415	1.50	4069	4.45	5931	27
34	4662	2.93	0325	1.50	4336	4.45	5664	26
35	**9.76 4838**	2.95	**9.91 0235**	1.52	**9.85 4603**	4.45	**10.14 5397**	**25**
36	5015	2.93	0144	1.50	4870	4.45	5130	24
37	5191	2.93	.91 0054	1.52	5137	4.45	4863	23
38	5367	2.95	.90 9963	1.50	5404	4.45	4596	22
39	5544	2.93	9873	1.52	5671	4.45	4329	21
40	**9.76 5720**	2.93	**9.90 9782**	1.52	**9.85 5938**	4.43	**10.14 4062**	**20**
41	5896	2.93	9691	1.50	6204	4.45	3796	19
42	6072	2.92	9601	1.52	6471	4.43	3529	18
43	6247	2.93	9510	1.52	6737	4.45	3263	17
44	6423	2.92	9419	1.52	7004	4.43	2996	16
45	**9.76 6598**	2.93	**9.90 9328**	1.52	**9.85 7270**	4.45	**10.14 2730**	**15**
46	6774	2.92	9237	1.52	7537	4.43	2463	14
47	6949	2.92	9146	1.52	7803	4.43	2197	13
48	7124	2.93	9055	1.52	8069	4.45	1931	12
49	7300	2.92	8964	1.52	8336	4.43	1664	11
50	**9.76 7475**	2.90	**9.90 8873**	1.53	**9.85 8602**	4.43	**10.14 1398**	**10**
51	7649	2.92	8781	1.52	8868	4.43	1132	9
52	7824	2.92	8690	1.52	9134	4.43	0866	8
53	7999	2.90	8599	1.53	9400	4.43	0600	7
54	8173	2.92	8507	1.52	9666	4.43	0334	6
55	**9.76 8348**	2.90	**9.90 8416**	1.53	**9.85 9932**	4.43	**10.14 0068**	**5**
56	8522	2.92	8324	1.52	.86 0198	4.43	.13 9802	4
57	8697	2.90	8233	1.53	0464	4.43	9536	3
58	8871	2.90	8141	1.53	0730	4.42	9270	2
59	9045	2.90	8049	1.52	0995	4.43	9005	1
60	**9.76 9219**		**9.90 7958**		**9.86 1261**		**10.13 8739**	**0**
′	Cosine.	D.1″.	Sine.	D.1″.	Cotang.	D.1″.	Tang.	′

125° 54°

36° **Table VIII.—*Continued*** 143°

′	Sine.	D.1″.	Cosine.	D.1″.	Tang.	D.1″.	Cotang.	′
0	**9.76 9219**	2.90	**9.90 7958**	1.53	**9.86 1261**	4.43	**10.13 8739**	60
1	9393	2.88	7866	1.53	1527	4.42	8473	59
2	9566	2.90	7774	1.53	1792	4.43	8208	58
3	9740	2.88	7682	1.53	2058	4.42	7942	57
4	.76 9913	2.90	7590	1.53	2323	4.43	7677	56
5	**9.77 0087**	2.88	**9.90 7498**	1.53	**9.86 2589**	4.42	**10.13 7411**	55
6	0260	2.88	7406	1.53	2854	4.42	7146	54
7	0433	2.88	7314	1.53	3119	4.43	6881	53
8	0606	2.88	7222	1.55	3385	4.42	6615	52
9	0779	2.88	7129	1.53	3650	4.42	6350	51
10	**9.77 0952**	2.88	**9.90 7037**	1.53	**9.86 3915**	4.42	**10.13 6085**	50
11	1125	2.88	6945	1.55	4180	4.42	5820	49
12	1298	2.87	6852	1.53	4445	4.42	5555	48
13	1470	2.88	6760	1.55	4710	4.42	5290	47
14	1643	2.87	6667	1.53	4975	4.42	5025	46
15	**9.77 1815**	2.87	**9.90 6575**	1.55	**9.86 5240**	4.42	**10.13 4760**	45
16	1987	2.87	6482	1.55	5505	4.42	4495	44
17	2159	2.87	6389	1.55	5770	4.42	4230	43
18	2331	2.87	6296	1.53	6035	4.42	3965	42
19	2503	2.87	6204	1.55	6300	4.40	3700	41
20	**9.77 2675**	2.87	**9.90 6111**	1.55	**9.86 6564**	4.42	**10.13 3436**	40
21	2847	2.85	6018	1.55	6829	4.42	3171	39
22	3018	2.87	5925	1.55	7094	4.40	2906	38
23	3190	2.85	5832	1.55	7358	4.42	2642	37
24	3361	2.87	5739	1.57	7623	4.40	2377	36
25	**9.77 3533**	2.85	**9.90 5645**	1.55	**9.86 7887**	4.42	**10.13 2113**	35
26	3704	2.85	5552	1.55	8152	4.40	1848	34
27	3875	2.85	5459	1.55	8416	4.40	1584	33
28	4046	2.85	5366	1.57	8680	4.42	1320	32
29	4217	2.85	5272	1.55	8945	4.40	1055	31
30	**9.77 4388**	2.83	**9.90 5179**	1.57	**9.86 9209**	4.40	**10.13 0791**	30
31	4558	2.85	5085	1.55	9473	4.40	0527	29
32	4729	2.83	4992	1.57	.86 9737	4.40	.13 0263	28
33	4899	2.85	4898	1.57	.87 0001	4.40	.12 9999	27
34	5070	2.83	4804	1.55	0265	4.40	9735	26
35	**9.77 5240**	2.83	**9.90 4711**	1.57	**9.87 0529**	4.40	**10.12 9471**	25
36	5410	2.83	4617	1.57	0793	4.40	9207	24
37	5580	2.83	4523	1.57	1057	4.40	8943	23
38	5750	2.83	4429	1.57	1321	4.40	8679	22
39	5920	2.83	4335	1.57	1585	4.40	8415	21
40	**9.77 6090**	2.82	**9.90 4241**	1.57	**9.87 1849**	4.38	**10.12 8151**	20
41	6259	2.83	4147	1.57	2112	4.40	7888	19
42	6429	2.82	4053	1.57	2376	4.40	7624	18
43	6598	2.83	3959	1.58	2640	4.38	7360	17
44	6768	2.82	3864	1.57	2903	4.40	7097	16
45	**9.77 6937**	2.82	**9.90 3770**	1.57	**9.87 3167**	4.38	**10.12 6833**	15
46	7106	2.82	3676	1.58	3430	4.40	6570	14
47	7275	2.82	3581	1.57	3694	4.38	6306	13
48	7444	2.82	3487	1.58	3957	4.38	6043	12
49	7613	2.80	3392	1.57	4220	4.40	5780	11
50	**9.77 7781**	2.82	**9.90 3298**	1.58	**9.87 4484**	4.38	**10.12 5516**	10
51	7950	2.82	3203	1.58	4747	4.38	5253	9
52	8119	2.80	3108	1.57	5010	4.38	4990	8
53	8287	2.80	3014	1.58	5273	4.40	4727	7
54	8455	2.82	2919	1.58	5537	4.38	4463	6
55	**9.77 8624**	2.80	**9.90 2824**	1.58	**9.87 5800**	4.38	**10.12 4200**	5
56	8792	2.80	2729	1.58	6063	4.38	3937	4
57	8960	2.80	2634	1.58	6326	4.38	3674	3
58	9128	2.78	2539	1.58	6589	4.38	3411	2
59	9295	2.80	2444	1.58	6852	4.37	3148	1
60	**9.77 9463**		**9.90 2349**		**9.87 7114**		**10.12 2886**	0
′	Cosine.	D.1″.	Sine.	D.1″.	Cotang.	D.1″.	Tang.	′

126° 53°

37° TABLE VIII.—*Continued* 142°

′	Sine.	D.1″.	Cosine.	D.1″.	Tang.	D.1″.	Cotang.	′
0	**9.77 9463**	2.80	**9.90 2349**	1.60	**9.87 7114**	4.38	**10.12 2886**	**60**
1	9631	2.78	2253	1.58	7377	4.38	2623	59
2	9798	2.80	2158	1.58	7640	4.38	2360	58
3	.77 9966	2.78	2063	1.60	7903	4.37	2097	57
4	.78 0133	2.78	1967	1.58	8165	4.38	1835	56
5	**9.78 0300**	2.78	**9.90 1872**	1.60	**9.87 8428**	4.38	**10.12 1572**	**55**
6	0467	2.78	1776	1.58	8691	4.37	1309	54
7	0634	2.78	1681	1.60	8953	4.38	1047	53
8	0801	2.78	1585	1.58	9216	4.37	0784	52
9	0968	2.77	1490	1.60	9478	4.38	0522	51
10	**9.78 1134**	2.78	**9.90 1394**	1.60	**9.87 9741**	4.37	**10.12 0259**	**50**
11	1301	2.78	1298	1.60	.88 0003	4.37	.11 9997	49
12	1468	2.77	1202	1.60	0265	4.38	9735	48
13	1634	2.77	1106	1.60	0528	4.37	9472	47
14	1800	2.77	1010	1.60	0790	4.37	9210	46
15	**9.78 1966**	2.77	**9.90 0914**	1.60	**9.88 1052**	4.37	**10.11 8948**	**45**
16	2132	2.77	0818	1.60	1314	4.38	8686	44
17	2298	2.77	0722	1.60	1577	4.37	8423	43
18	2464	2.77	0626	1.62	1839	4.37	8161	42
19	2630	2.77	0529	1.60	2101	4.37	7899	41
20	**9.78 2796**	2.75	**9.90 0433**	1.60	**9.88 2363**	4.37	**10.11 7637**	**40**
21	2961	2.77	0337	1.62	2625	4.37	7375	39
22	3127	2.75	0240	1.60	2887	4.35	7113	38
23	3292	2.77	0144	1.62	3148	4.37	6852	37
24	3458	2.75	.90 0047	1.60	3410	4.37	6590	36
25	**9.78 3623**	2.75	**9.89 9951**	1.63	**9.88 3672**	4.37	**10.11 6328**	**35**
26	3788	2.75	9854	1.62	3934	4.37	6066	34
27	3953	2.75	9757	1.62	4196	4.35	5804	33
28	4118	2.73	9660	1.60	4457	4.37	5543	32
29	4282	2.75	9564	1.62	4719	4.35	5281	31
30	**9.78 4447**	2.75	**9.89 9467**	1.62	**9.88 4980**	4.37	**10.11 5020**	**30**
31	4612	2.73	9370	1.62	5242	4.37	4758	29
32	4776	2.75	9273	1.62	5504	4.35	4496	28
33	4941	2.73	9176	1.63	5765	4.35	4235	27
34	5105	2.73	9078	1.62	6026	4.37	3974	26
35	**9.78 5269**	2.73	**9.89 8981**	1.62	**9.88 6288**	4.35	**10.11 3712**	**25**
36	5433	2.73	8884	1.62	6549	4.37	3451	24
37	5597	2.73	8787	1.63	6811	4.35	3189	23
38	5761	2.73	8689	1.62	7072	4.35	2928	22
39	5925	2.73	8592	1.63	7333	4.35	2667	21
40	**9.78 6089**	2.72	**9.89 8494**	1.62	**9.88 7594**	4.35	**10.11 2406**	**20**
41	6252	2.73	8397	1.63	7855	4.35	2145	19
42	6416	2.72	8299	1.62	8116	4.37	1884	18
43	6579	2.72	8202	1.63	8378	4.35	1622	17
44	6742	2.73	8104	1.63	8639	4.35	1361	16
45	**9.78 6906**	2.72	**9.89 8006**	1.63	**9 88 8900**	4.35	**10.11 1100**	**15**
46	7069	2.72	7908	1.63	9161	4.33	0839	14
47	7232	2.72	7810	1.63	9421	4.35	0579	13
48	7395	2.70	7712	1.63	9682	4.35	0318	12
49	7557	2.72	7614	1.63	.88 9943	4.35	.11 0057	11
50	**9.78 7720**	2.72	**9.89 7516**	1.63	**9.89 0204**	4.35	**10.10 9796**	**10**
51	7883	2.70	7418	1.63	0465	4.33	9535	9
52	8045	2.72	7320	1.63	0725	4.35	9275	8
53	8208	2.70	7222	1.65	0986	4.35	9014	7
54	8370	2.70	7123	1.63	1247	4.33	8753	6
55	**9.78 8532**	2.70	**9.89 7025**	1.65	**9.89 1507**	4.35	**10.10 8493**	**5**
56	8694	2.70	6926	1.63	1768	4.33	8232	4
57	8856	2.70	6828	1.65	2028	4.35	7972	3
58	9018	2.70	6729	1.63	2289	4.33	7711	2
59	9180	2.70	6631	1.65	2549	4.35	7451	1
60	**9.78 9342**		**9.89 6532**		**9.89 2810**		**10.10 7190**	**0**
′	Cosine.	D.1″.	Sine.	D.1″.	Cotang.	D.1″.	Tang.	′

127° 52°

38° TABLE VIII.—*Continued* 141°

′	Sine.	D.1″.	Cosine.	D.1″.	Tang.	D.1″.	Cotang.	′
0	**9.78 9342**	2.70	**9.89 6532**	1.65	**9.89 2810**	4.33	**10.10 7190**	**60**
1	9504	2.68	6433	1.63	3070	4.35	6930	59
2	9665	2.70	6335	1.65	3331	4.33	6669	58
3	9827	2.68	6236	1.65	3591	4.33	6409	57
4	.78 9988	2.68	6137	1.65	3851	4.33	6149	56
5	**9.79 0149**	2.68	**9.89 6038**	1.65	**9.89 4111**	4.35	**10.10 5889**	**55**
6	0310	2.68	5939	1.65	4372	4.33	5628	54
7	0471	2.68	5840	1.65	4632	4.33	5368	53
8	0632	2.68	5741	1.67	4892	4.33	5108	52
9	0793	2.68	5641	1.65	5152	4.33	4848	51
10	**9.79 0954**	2.68	**9.89 5542**	1.65	**9.89 5412**	4.33	**10.10 4588**	**50**
11	1115	2.67	5443	1.67	5672	4.33	4328	49
12	1275	2.68	5343	1.65	5932	4.33	4068	48
13	1436	2.67	5244	1.65	6192	4.33	3808	47
14	1596	2.68	5145	1.67	6452	4.33	3548	46
15	**9.79 1757**	2.67	**9.89 5045**	1.67	**9.89 6712**	4.32	**10.10 3288**	**45**
16	1917	2.67	4945	1.65	6971	4.33	3029	44
17	2077	2.67	4846	1.67	7231	4.33	2769	43
18	2237	2.67	4746	1.67	7491	4.33	2509	42
19	2397	2.67	4646	1.67	7751	4.32	2249	41
20	**9.79 2557**	2.65	**9.89 4546**	1.67	**9.89 8010**	4.33	**10.10 1990**	**40**
21	2716	2.67	4446	1.67	8270	4.33	1730	39
22	2876	2.65	4346	1.67	8530	4.32	1470	38
23	3035	2.67	4246	1.67	8789	4.33	1211	37
24	3195	2.65	4146	1.67	9049	4.32	0951	36
25	**9.79 3354**	2.67	**9.89 4046**	1.67	**9.89 9308**	4.33	**10.10 0692**	**35**
26	3514	2.65	3946	1.67	9568	4.32	0432	34
27	3673	2.65	3846	1.68	.89 9827	4.33	.10 0173	33
28	3832	2.65	3745	1.67	.90 0087	4.32	.09 9913	32
29	3991	2.65	3645	1.68	0346	4.32	9654	31
30	**9.79 4150**	2.63	**9.89 3544**	1.67	**9.90 0605**	4.32	**10.09 9395**	**30**
31	4308	2.65	3444	1.68	0864	4.33	9136	29
32	4467	2.65	3343	1.67	1124	4.32	8876	28
33	4626	2.63	3243	1.68	1383	4.32	8617	27
34	4784	2.63	3142	1.68	1642	4.32	8358	26
35	**9.79 4942**	2.65	**9.89 3041**	1.68	**9.90 1901**	4.32	**10.09 8099**	**25**
36	5101	2.63	2940	1.68	2160	4.33	7840	24
37	5259	2.63	2839	1.67	2420	4.32	7580	23
38	5417	2.63	2739	1.68	2679	4.32	7321	22
39	5575	2.63	2638	1.70	2938	4.32	7062	21
40	**9.79 5733**	2.63	**9.89 2536**	1.68	**9.90 3197**	4.32	**10.09 6803**	**20**
41	5891	2.63	2435	1.68	3456	4.30	6544	19
42	6049	2.62	2334	1.68	3714	4.32	6286	18
43	6206	2.63	2233	1.68	3973	4.32	6027	17
44	6364	2.62	2132	1.70	4232	4.32	5768	16
45	**9.79 6521**	2.63	**9.89 2030**	1.68	**9.90 4491**	4.32	**10.09 5509**	**15**
46	6679	2.62	1929	1.70	4750	4.30	5250	14
47	6836	2.62	1827	1.68	5008	4.32	4992	13
48	6993	2.62	1726	1.70	5267	4.32	4733	12
49	7150	2.62	1624	1.68	5526	4.32	4474	11
50	**9.79 7307**	2.62	**9.89 1523**	1.70	**9.90 5785**	4.30	**10.09 4215**	**10**
51	7464	2.62	1421	1.70	6043	4.32	3957	9
52	7621	2.60	1319	1.70	6302	4.30	3698	8
53	7777	2.62	1217	1.70	6560	4.32	3440	7
54	7934	2.62	1115	1.70	6819	4.30	3181	6
55	**9.79 8091**	2.60	**9.89 1013**	1.70	**9.90 7077**	4.32	**10.09 2923**	**5**
56	8247	2.60	0911	1.70	7336	4.30	2664	4
57	8403	2.62	0809	1.70	7594	4.32	2406	3
58	8560	2.60	0707	1.70	7853	4.30	2147	2
59	8716	2.60	0605	1.70	8111	4.30	1889	1
60	**9.79 8872**		**9.89 0503**		**9.90 8369**		**10.09 1631**	**0**
′	Cosine.	D.1″.	Sine.	D.1″.	Cotang.	D.1″.	Tang.	′

128° 51°

39° TABLE VIII.—*Continued* 140°

′	Sine.	D.1″.	Cosine.	D.1″.	Tang.	D.1″.	Cotang.	′
0	9.79 8872	2.60	9.89 0503	1.72	9.90 8369	4.32	10.09 1631	60
1	9028	2.60	0400	1.70	8628	4.30	1372	59
2	9184	2.58	0298	1.72	8886	4.30	1114	58
3	9339	2.60	0195	1.70	9144	4.30	0856	57
4	9495	2.60	.89 0093	1.72	9402	4.30	0598	56
5	9.79 9651	2.58	9.88 9990	1.70	9.90 9660	4.30	10.09 0340	55
6	9806	2.60	9888	1.72	.90 9918	4.32	.09 0082	54
7	.79 9962	2.58	9785	1.72	.91 0177	4.30	.08 9823	53
8	.80 0117	2.58	9682	1.72	0435	4.30	9565	52
9	0272	2.58	9579	1.70	0693	4.30	9307	51
10	9.80 0427	2.58	9.88 9477	1.72	9.91 0951	4.30	10.08 9049	50
11	0582	2.58	9374	1.72	1209	4.30	8791	49
12	0737	2.58	9271	1.72	1467	4.30	8533	48
13	0892	2.58	9168	1.73	1725	4.28	8275	47
14	1047	2.57	9064	1.72	1982	4.30	8018	46
15	9.80 1201	2.58	9.88 8961	1.72	9.91 2240	4.30	10.08 7760	45
16	1356	2.58	8858	1.72	2498	4.30	7502	44
17	1511	2.57	8755	1.73	2756	4.30	7244	43
18	1665	2.57	8651	1.72	3014	4.28	6986	42
19	1819	2.57	8548	1.73	3271	4.30	6729	41
20	9.80 1973	2.58	9.88 8444	1.72	9.91 3529	4.30	10.08 6471	40
21	2128	2.57	8341	1.73	3787	4.28	6213	39
22	2282	2.57	8237	1.72	4044	4.30	5956	38
23	2436	2.55	8134	1.73	4302	4.30	5698	37
24	2589	2.57	8030	1.73	4560	4.28	5440	36
25	9.80 2743	2.57	9.88 7926	1.73	9.91 4817	4.30	10.08 5183	35
26	2897	2.55	7822	1.73	5075	4.28	4925	34
27	3050	2.57	7718	1.73	5332	4.30	4668	33
28	3204	2.55	7614	1.73	5590	4.28	4410	32
29	3357	2.57	7510	1.73	5847	4.28	4153	31
30	9.80 3511	2.55	9.88 7406	1.73	9.91 6104	4.30	10.08 3896	30
31	3664	2.55	7302	1.73	6362	4.28	3638	29
32	3817	2.55	7198	1.75	6619	4.30	3381	28
33	3970	2.55	7093	1.73	6877	4.28	3123	27
34	4123	2.55	6989	1.73	7134	4.28	2866	26
35	9.80 4276	2.55	9.88 6885	1.75	9.91 7391	4.28	10.08 2609	25
36	4428	2.55	6780	1.73	7648	4.30	2352	24
37	4581	2.55	6676	1.75	7906	4.28	2094	23
38	4734	2.53	6571	1.75	8163	4.28	1837	22
39	4886	2.55	6466	1.73	8420	4.28	1580	21
40	9.80 5039	2.53	9.88 6362	1.75	9.91 8677	4.28	10.08 1323	20
41	5191	2.53	6257	1.75	8934	4.28	1066	19
42	5343	2.53	6152	1.75	9191	4.28	0809	18
43	5495	2.53	6047	1.75	9448	4.28	0552	17
44	5647	2.53	5942	1.75	9705	4.28	0295	16
45	9.80 5799	2.53	9.88 5837	1.75	9.91 9962	4.28	10.08 0038	15
46	5951	2.53	5732	1.75	.92 0219	4.28	.07 9781	14
47	6103	2.52	5627	1.75	0476	4.28	9524	13
48	6254	2.53	5522	1.77	0733	4.28	9267	12
49	6406	2.52	5416	1.75	0990	4.28	9010	11
50	9.80 6557	2.53	9.88 5311	1.77	9.92 1247	4.27	10.07 8753	10
51	6709	2.52	5205	1.75	1503	4.28	8497	9
52	6860	2.52	5100	1.77	1760	4.28	8240	8
53	7011	2.53	4994	1.75	2017	4.28	7983	7
54	7163	2.52	4889	1.77	2274	4.27	7726	6
55	9.80 7314	2.52	9.88 4783	1.77	9.92 2530	4.28	10.07 7470	5
56	7465	2.50	4677	1.75	2787	4.28	7213	4
57	7615	2.52	4572	1.77	3044	4.27	6956	3
58	7766	2.52	4466	1.77	3300	4.28	6700	2
59	7917	2.50	4360	1.77	3557	4.28	6443	1
60	9.80 8067		9.88 4254		9.92 3814		10.07 6186	0
′	Cosine.	D.1″.	Sine.	D.1″.	Cotang.	D.1″.	Tang.	′

129° 50°

40° **TABLE VIII.**—*Continued* 139°

′	Sine.	D.1″.	Cosine.	D.1″.	Tang.	D.1″.	Cotang.	′
0	**9.80 8067**	2.52	**9.88 4254**	1.77	**9.92 3814**	4.27	**10.07 6186**	**60**
1	8218	2.50	4148	1.77	4070	4.28	5930	59
2	8368	2.52	4042	1.77	4327	4.27	5673	58
3	8519	2.50	3936	1.78	4583	4.28	5417	57
4	8669	2.50	3829	1.77	4840	4.27	5160	56
5	**9.80 8819**	2.50	**9.88 3723**	1.77	**9.92 5096**	4.27	**10.07 4904**	**55**
6	8969	2.50	3617	1.78	5352	4.28	4648	54
7	9119	2.50	3510	1.77	5609	4.27	4391	53
8	9269	2.50	3404	1.78	5865	4.28	4135	52
9	9419	2.50	3297	1.77	6122	4.27	3878	51
10	**9.80 9569**	2.48	**9.88 3191**	1.78	**9.92 6378**	4.27	**10.07 3622**	**50**
11	9718	2.50	3084	1.78	6634	4.27	3366	49
12	.80 9868	2.48	2977	1.77	6890	4.28	3110	48
13	.81 0017	2.50	2871	1.78	7147	4.27	2853	47
14	0167	2.48	2764	1.78	7403	4.27	2597	46
15	**9.81 0316**	2.48	**9.88 2657**	1.78	**9.92 7659**	4.27	**10.07 2341**	**45**
16	0465	2.48	2550	1.78	7915	4.27	2085	44
17	0614	2.48	2443	1.78	8171	4.27	1829	43
18	0763	2.48	2336	1.78	8427	4.28	1573	42
19	0912	2.48	2229	1.80	8684	4.27	1316	41
20	**9.81 1061**	2.48	**9.88 2121**	1.78	**9.92 8940**	4.27	**10.07 1060**	**40**
21	1210	2.47	2014	1.78	9196	4.27	0804	39
22	1358	2.48	1907	1.80	9452	4.27	0548	38
23	1507	2.47	1799	1.78	9708	4.27	0292	37
24	1655	2.48	1692	1.80	.92 9964	4.27	.07 0036	36
25	**9.81 1804**	2.47	**9.88 1584**	1.78	**9.93 0220**	4.25	**10.06 9780**	**35**
26	1952	2.47	1477	1.80	0475	4.27	9525	34
27	2100	2.47	1369	1.80	0731	4.27	9269	33
28	2248	2.47	1261	1.80	0987	4.27	9013	32
29	2396	2.47	1153	1.78	1243	4.27	8757	31
30	**9.81 2544**	2.47	**9.88 1046**	1.80	**9.93 1499**	4.27	**10.06 8501**	**30**
31	2692	2.47	0938	1.80	1755	4.25	8245	29
32	2840	2.47	0830	1.80	2010	4.27	7990	28
33	2988	2.45	0722	1.82	2266	4.27	7734	27
34	3135	2.47	0613	1.80	2522	4.27	7478	26
35	**9.81 3283**	2.45	**9.88 0505**	1.80	**9.93 2778**	4.25	**10.06 7222**	**25**
36	3430	2.47	0397	1.80	3033	4.27	6967	24
37	3578	2.45	0289	1.82	3289	4.27	6711	23
38	3725	2.45	0180	1.80	3545	4.25	6455	22
39	3872	2.45	.88 0072	1.82	3800	4.27	6200	21
40	**9.81 4019**	2.45	**9.87 9963**	1.80	**9.93 4056**	4.25	**10.06 5944**	**20**
41	4166	2.45	9855	1.82	4311	4.27	5689	19
42	4313	2.45	9746	1.82	4567	4.25	5433	18
43	4460	2.45	9637	1.80	4822	4.27	5178	17
44	4607	2.43	9529	1.82	5078	4.25	4922	16
45	**9.81 4753**	2.45	**9.87 9420**	1.82	**9.93 5333**	4.27	**10.06 4667**	**15**
46	4900	2.43	9311	1.82	5589	4.25	4411	14
47	5046	2.45	9202	1.82	5844	4.27	4156	13
48	5193	2.43	9093	1.82	6100	4.25	3900	12
49	5339	2.43	8984	1.82	6355	4.27	3645	11
50	**9.81 5485**	2.45	**9.87 8875**	1.82	**9.93 6611**	4.25	**10.06 3389**	**10**
51	5632	2.43	8766	1.83	6866	4.25	3134	9
52	5778	2.43	8656	1.82	7121	4.27	2879	8
53	5924	2.42	8547	1.82	7377	4.25	2623	7
54	6069	2.43	8438	1.83	7632	4.25	2368	6
55	**9.81 6215**	2.43	**9.87 8328**	1.82	**9.93 7887**	4.25	**10.06 2113**	**5**
56	6361	2.43	8219	1.83	8142	4.27	1858	4
57	6507	2.42	8109	1.83	8398	4.25	1602	3
58	6652	2.43	7999	1.82	8653	4.25	1347	2
59	6798	2.42	7890	1.83	8908	4.25	1092	1
60	**9.81 6943**		**9.87 7780**		**9.93 9163**		**10.06 0837**	**0**
′	Cosine.	D.1″.	Sine.	D.1″.	Cotang.	D.1″.	Tang.	′

130° 49°

41° TABLE VIII.—*Continued* 138°

′	Sine.	D.1″.	Cosine.	D.1″.	Tang.	D.1″.	Cotang.	′
0	**9.81 6943**	2.42	**9.87 7780**	1.83	**9.93 9163**	4.25	**10.06 0837**	**60**
1	7088	2.42	7670	1.83	9418	4.25	0582	59
2	7233	2.43	7560	1.83	9673	4.25	0327	58
3	7379	2.42	7450	1.83	.93 9928	4.25	.06 0072	57
4	7524	2.40	7340	1.83	.94 0183	4.27	.05 9817	56
5	**9.81 7668**	2.42	**9.87 7230**	1.83	**9.94 0439**	4.25	**10.05 9561**	**55**
6	7813	2.42	7120	1.83	0694	4.25	9306	54
7	7958	2.42	7010	1.85	0949	4.25	9051	53
8	8103	2.40	6899	1.83	1204	4.25	8796	52
9	8247	2.42	6789	1.85	1459	4.23	8541	51
10	**9.81 8392**	2.40	**9.87 6678**	1.83	**9.94 1713**	4.25	**10.05 8287**	**50**
11	8536	2.42	6568	1.85	1968	4.25	8032	49
12	8681	2.40	6457	1.83	2223	4.25	7777	48
13	8825	2.40	6347	1.85	2478	4.25	7522	47
14	8969	2.40	6236	1.85	2733	4.25	7267	46
15	**9.81 9113**	2.40	**9.87 6125**	1.85	**9.94 2988**	4.25	**10.05 7012**	**45**
16	9257	2.40	6014	1.83	3243	4.25	6757	44
17	9401	2.40	5904	1.85	3498	4.23	6502	43
18	9545	2.40	5793	1.85	3752	4.25	6248	42
19	9689	2.38	5682	1.85	4007	4.25	5993	41
20	**9.81 9832**	2.40	**9.87 5571**	1.87	**9.94 4262**	4.25	**10.05 5738**	**40**
21	.81 9976	2.40	5459	1.85	4517	4.23	5483	39
22	.82 0120	2.38	5348	1.85	4771	4.25	5229	38
23	0263	2.38	5237	1.85	5026	4.25	4974	37
24	0406	2.40	5126	1.87	5281	4.23	4719	36
25	**9.82 0550**	2.38	**9.87 5014**	1.85	**9.94 5535**	4.25	**10.05 4465**	**35**
26	0693	2.38	4903	1.87	5790	4.25	4210	34
27	0836	2.38	4791	1.85	6045	4.23	3955	33
28	0979	2.38	4680	1.87	6299	4.25	3701	32
29	1122	2.38	4568	1.87	6554	4.23	3446	31
30	**9.82 1265**	2.37	**9.87 4456**	1.87	**9 94 6808**	4.25	**10.05 3192**	**30**
31	1407	2.38	4344	1.87	7063	4.25	2937	29
32	1550	2.38	4232	1.85	7318	4.23	2682	28
33	1693	2.37	4121	1.87	7572	4.25	2428	27
34	1835	2.37	4009	1.88	7827	4.23	2173	26
35	**9.82 1977**	2.38	**9.87 3896**	1.87	**9.94 8081**	4.23	**10.05 1919**	**25**
36	2120	2.37	3784	1.87	8335	4.25	1665	24
37	2262	2.37	3672	1.87	8590	4.23	1410	23
38	2404	2.37	3560	1.87	8844	4.25	1156	22
39	2546	2.37	3448	1.88	9099	4.23	0901	21
40	**9.82 2688**	2.37	**9.87 3335**	1.87	**9.94 9353**	4.25	**10.05 0647**	**20**
41	2830	2.37	3223	1.88	9608	4.23	0392	19
42	2972	2.37	3110	1.87	.94 9862	4.23	.05 0138	18
43	3114	2.35	2998	1.88	.95 0116	4.25	.04 9884	17
44	3255	2.37	2885	1.88	0371	4.23	9629	16
45	**9.82 3397**	2.37	**9.87 2772**	1.88	**9.95 0625**	4.23	**10.04 9375**	**15**
46	3539	2.35	2659	1.87	0879	4.23	9121	14
47	3680	2.35	2547	1.88	1133	4.25	8867	13
48	3821	2.37	2434	1.88	1388	4.23	8612	12
49	3963	2.35	2321	1.88	1642	4.23	8358	11
50	**9.82 4104**	2.35	**9.87 2208**	1.88	**9.95 1896**	4.23	**10.04 8104**	**10**
51	4245	2.35	2095	1.90	2150	4.25	7850	9
52	4386	2.35	1981	1.88	2405	4.23	7595	8
53	4527	2.35	1868	1.88	2659	4.23	7341	7
54	4668	2.33	1755	1.90	2913	4.23	7087	6
55	**9.82 4808**	2.35	**9.87 1641**	1.88	**9.95 3167**	4.23	**10.04 6833**	**5**
56	4949	2.35	1528	1.90	3421	4.23	6579	4
57	5090	2.33	1414	1.88	3675	4.23	6325	3
58	5230	2.35	1301	1.90	3929	4.23	6071	2
59	5371	2.33	1187	1.90	4183	4.23	5817	1
60	**9.82 5511**		**9.87 1073**		**9.95 4437**		**10.04 5563**	**0**
′	Cosine.	D.1″.	Sine.	D.1″.	Cotang.	D.1″.	Tang.	′

131° 48°

42° TABLE VIII.—*Continued* 137°

′	Sine.	D.1″.	Cosine.	D.1″.	Tang.	D.1″.	Cotang.	′
0	9.82 5511	2.33	9.87 1073	1.88	9.95 4437	4.23	10.04 5563	60
1	5651	2.33	0960	1.90	4691	4.25	5309	59
2	5791	2.33	0846	1.90	4946	4.23	5054	58
3	5931	2.33	0732	1.90	5200	4.23	4800	57
4	6071	2.33	0618	1.90	5454	4.23	4546	56
5	9.82 6211	2.33	9.87 0504	1.90	9.95 5708	4.22	10.04 4292	55
6	6351	2.33	0390	1.90	5961	4.23	4039	54
7	6491	2.33	0276	1.92	6215	4.23	3785	53
8	6631	2.32	0161	1.90	6469	4.23	3531	52
9	6770	2.33	.87 0047	1.90	6723	4.23	3277	51
10	9.82 6910	2.32	9.86 9933	1.92	9.95 6977	4.23	10.04 3023	50
11	7049	2.33	9818	1.90	7231	4.23	2769	49
12	7189	2.32	9704	1.92	7485	4.23	2515	48
13	7328	2.32	9589	1.92	7739	4.23	2261	47
14	7467	2.32	9474	1.90	7993	4.23	2007	46
15	9.82 7606	2.32	9.86 9360	1.92	9.95 8247	4.22	10.04 1753	45
16	7745	2.32	9245	1.92	8500	4.23	1500	44
17	7884	2.32	9130	1.92	8754	4.23	1246	43
18	8023	2.32	9015	1.92	9008	4.23	0992	42
19	8162	2.32	8900	1.92	9262	4.23	0738	41
20	9.82 8301	2.30	9.86 8785	1.92	9.95 9516	4.22	10.04 0484	40
21	8439	2.32	8670	1.92	.95 9769	4.23	.04 0231	39
22	8578	2.30	8555	1.92	.96 0023	4.23	.03 9977	38
23	8716	2.32	8440	1.93	0277	4.22	9723	37
24	8855	2.30	8324	1.92	0530	4.23	9470	36
25	9.82 8993	2.30	9.86 8209	1.93	9.96 0784	4.23	10.03 9216	35
26	9131	2.30	8093	1.92	1038	4.23	8962	34
27	9269	2.30	7978	1.93	1292	4.22	8708	33
28	9407	2.30	7862	1.92	1545	4.23	8455	32
29	9545	2.30	7747	1.93	1799	4.22	8201	31
30	9.82 9683	2.30	9.86 7631	1.93	9.96 2052	4.23	10.03 7948	30
31	9821	2.30	7515	1.93	2306	4.23	7694	29
32	.82 9959	2.30	7399	1.93	2560	4.22	7440	28
33	.83 0097	2.28	7283	1.93	2813	4.23	7187	27
34	0234	2.30	7167	1.93	3067	4.22	6933	26
35	9.83 0372	2.28	9.86 7051	1.93	9.96 3320	4.23	10.03 6680	25
36	0509	2.28	6935	1.93	3574	4.23	6426	24
37	0646	2.30	6819	1.93	3828	4.22	6172	23
38	0784	2.28	6703	1.95	4081	4.23	5919	22
39	0921	2.28	6586	1.93	4335	4.22	5665	21
40	9.83 1058	2.28	9.86 6470	1.95	9.96 4588	4.23	10.03 5412	20
41	1195	2.28	6353	1.93	4842	4.22	5158	19
42	1332	2.28	6237	1.95	5095	4.23	4905	18
43	1469	2.28	6120	1.93	5349	4.22	4651	17
44	1606	2.27	6004	1.95	5602	4.22	4398	16
45	9.83 1742	2.28	9.86 5887	1.95	9.96 5855	4.23	10.03 4145	15
46	1879	2.27	5770	1.95	6109	4.22	3891	14
47	2015	2.28	5653	1.95	6362	4.23	3638	13
48	2152	2.27	5536	1.95	6616	4.22	3384	12
49	2288	2.28	5419	1.95	6869	4.23	3131	11
50	9.83 2425	2.27	9.86 5302	1.95	9.96 7123	4.22	10.03 2877	10
51	2561	2.27	5185	1.95	7376	4.22	2624	9
52	2697	2.27	5068	1.97	7629	4.23	2371	8
53	2833	2.27	4950	1.95	7883	4.22	2117	7
54	2969	2.27	4833	1.95	8136	4.22	1864	6
55	9.83 3105	2.27	9.86 4716	1.97	9.96 8389	4.23	10.03 1611	5
56	3241	2.27	4598	1.95	8643	4.22	1357	4
57	3377	2.25	4481	1.97	8896	4.22	1104	3
58	3512	2.27	4363	1.97	9149	4.23	0851	2
59	3648	2.25	4245	1.97	9403	4.22	0597	1
60	9.83 3783		9.86 4127		9.96 9656		10.03 0344	0
′	Cosine.	D.1″.	Sine.	D.1″.	Cotang.	D.1″.	Tang.	′

132° 47°

43° TABLE VIII.—*Continued* 136°

′	Sine.	D.1″.	Cosine.	D.1″.	Tang.	D.1″.	Cotang.	′
0	**9.83 3783**	2.27	**9.86 4127**	1.95	**9.96 9656**	4.22	**10.03 0344**	**60**
1	3919	2.25	4010	1.97	.96 9909	4.22	.03 0091	59
2	4054	2.25	3892	1.97	.97 0162	4.23	.02 9838	58
3	4189	2.27	3774	1.97	0416	4.22	9584	57
4	4325	2.25	3656	1.97	0669	4.22	9331	56
5	**9.83 4460**	2.25	**9.86 3538**	1.98	**9.97 0922**	4.22	**10.02 9078**	**55**
6	4595	2.25	3419	1.97	1175	4.23	8825	54
7	4730	2.25	3301	1.97	1429	4.22	8571	53
8	4865	2.23	3183	1.98	1682	4.22	8318	52
9	4999	2.25	3064	1.97	1935	4.22	8065	51
10	**9.83 5134**	2.25	**9.86 2946**	1.98	**9.97 2188**	4.22	**10.02 7812**	**50**
11	5269	2.23	2827	1.97	2441	4.23	7559	49
12	5403	2.25	2709	1.98	2695	4.22	7305	48
13	5538	2.23	2590	1.98	2948	4.22	7052	47
14	5672	2.25	2471	1.97	3201	4.22	6799	46
15	**9.83 5807**	2.23	**9.86 2353**	1.98	**9.97 3454**	4.22	**10.02 6546**	**45**
16	5941	2.23	2234	1.98	3707	4.22	6293	44
17	6075	2.23	2115	1.98	3960	4.22	6040	43
18	6209	2.23	1996	1.98	4213	4.22	5787	42
19	6343	2.23	1877	1.98	4466	4.23	5534	41
20	**9.83 6477**	2.23	**9.86 1758**	2.00	**9.97 4720**	4.22	**10.02 5280**	**40**
21	6611	2.23	1638	1.98	4973	4.22	5027	39
22	6745	2.22	1519	1.98	5226	4.22	4774	38
23	6878	2.23	1400	2.00	5479	4.22	4521	37
24	7012	2.23	1280	1.98	5732	4.22	4268	36
25	**9.83 7146**	2.22	**9.86 1161**	2.00	**9.97 5985**	4.22	**10.02 4015**	**35**
26	7279	2.22	1041	1.98	6238	4.22	3762	34
27	7412	2.23	0922	2.00	6491	4.22	3509	33
28	7546	2.22	0802	2.00	6744	4.22	3256	32
29	7679	2.22	0682	2.00	6997	4.22	3003	31
30	**9.83 7812**	2.22	**9.86 0562**	2.00	**9.97 7250**	4.22	**10.02 2750**	**30**
31	7945	2.22	0442	2.00	7503	4.22	2497	29
32	8078	2.22	0322	2.00	7756	4.22	2244	28
33	8211	2.22	0202	2.00	8009	4.22	1991	27
34	8344	2.22	.86 0082	2.00	8262	4.22	1738	26
35	**9.83 8477**	2.22	**9.85 9962**	2.00	**9.97 8515**	4.22	**10.02 1485**	**25**
36	8610	2.20	9842	2.02	8768	4.22	1232	24
37	8742	2.22	9721	2.00	9021	4.22	0979	23
38	8875	2.20	9601	2.02	9274	4.22	0726	22
39	9007	2.22	9480	2.00	9527	4.22	0473	21
40	**9.83 9140**	2.20	**9.85 9360**	2.02	**9.97 9780**	4.22	**10.02 0220**	**20**
41	9272	2.20	9239	2.00	.98 0033	4.22	.01 9967	19
42	9404	2.20	9119	2.02	0286	4.20	9714	18
43	9536	2.20	8998	2.02	0538	4.22	9462	17
44	9668	2.20	8877	2.02	0791	4.22	9209	16
45	**9.83 9800**	2.20	**9.85 8756**	2.02	**9.98 1044**	4.22	**10.01 8956**	**15**
46	.83 9932	2.20	8635	2.02	1297	4.22	8703	14
47	.84 0064	2.20	8514	2.02	1550	4.22	8450	13
48	0196	2.20	8393	2.02	1803	4.22	8197	12
49	0328	2.18	8272	2.02	2056	4.22	7944	11
50	**9.84 0459**	2.20	**9.85 8151**	2.03	**9.98 2309**	4.22	**10.01 7691**	**10**
51	0591	2.18	8029	2.02	2562	4.20	7438	9
52	0722	2.20	7908	2.03	2814	4.22	7186	8
53	0854	2.18	7786	2.02	3067	4.22	6933	7
54	0985	2.18	7665	2.03	3320	4.22	6680	6
55	**9.84 1116**	2.18	**9.85 7543**	2.02	**9.98 3573**	4.22	**10.01 6427**	**5**
56	1247	2.18	7422	2.03	3826	4.22	6174	4
57	1378	2.18	7300	2.03	4079	4.22	5921	3
58	1509	2.18	7178	2.03	4332	4.20	5668	2
59	1640	2.18	7056	2.03	4584	4.22	5416	1
60	**9.84 1771**		**9.85 6934**		**9.98 4837**		**10.01 5163**	**0**
′	Cosine.	D.1″.	Sine.	D.1″.	Cotang.	D.1″.	Tang.	′

133° 46°

44° **Table VIII.**—*Concluded* 135°

′	Sine.	D.1″.	Cosine.	D.1″.	Tang.	D.1″.	Cotang.	′
0	9.84 1771	2.18	9.85 6934	2.03	9.98 4837	4.22	10.01 5163	60
1	1902	2.18	6812	2.03	5090	4.22	4910	59
2	2033	2.17	6690	2.03	5343	4.22	4657	58
3	2163	2.18	6568	2.03	5596	4.20	4404	57
4	2294	2.17	6446	2.05	5848	4.22	4152	56
5	9.84 2424	2.18	9.85 6323	2.03	9.98 6101	4.22	10.01 3899	55
6	2555	2.17	6201	2.05	6354	4.22	3646	54
7	2685	2.17	6078	2.03	6607	4.22	3393	53
8	2815	2.18	5956	2.05	6860	4.20	3140	52
9	2940	2.17	5833	2.03	7112	4.22	2888	51
10	9.84 3076	2.17	9.85 5711	2.05	9.98 7365	4.22	10.01 2635	50
11	3206	2.17	5588	2.05	7618	4.22	2382	49
12	3336	2.17	5465	2.05	7871	4.20	2129	48
13	3466	2.15	5342	2.05	8123	4.22	1877	47
14	3595	2.17	5219	2.05	8376	4.22	1624	46
15	9.84 3725	2.17	9.85 5096	2.05	9.98 8629	4.22	10.01 1371	45
16	3855	2.15	4973	2.05	8882	4.20	1118	44
17	3984	2.17	4850	2.05	9134	4.22	0866	43
18	4114	2.15	4727	2.07	9387	4.22	0613	42
19	4243	2.15	4603	2.05	9640	4.22	0360	41
20	9.84 4372	2.17	9.85 4480	2.07	9.98 9893	4.20	10.01 0107	40
21	4502	2.15	4356	2.05	.99 0145	4.22	.00 9855	39
22	4631	2.15	4233	2.07	0398	4.22	9602	38
23	4760	2.15	4109	2.05	0651	4.20	9349	37
24	4889	2.15	3986	2.07	0903	4.22	9097	36
25	9.84 5018	2.15	9.85 3862	2.07	9.99 1156	4.22	10.00 8844	35
26	5127	2.15	3738	2.07	1409	4.22	8591	34
27	5276	2.15	3614	2.07	1662	4.20	8338	33
28	5405	2.13	3490	2.07	1914	4.22	8086	32
29	5533	2.15	3366	2.07	2167	4.22	7833	31
30	9.84 5662	2.13	9.85 3242	2.07	9.99 2420	4.20	10.00 7580	30
31	5790	2.15	3118	2.07	2672	4.22	7328	29
32	5919	2.13	2994	2.08	2925	4.22	7075	28
33	6047	2.13	2869	2.07	3178	4.22	6822	27
34	6175	2.15	2745	2.08	3431	4.20	6569	26
35	9.84 6304	2.13	9.85 2620	2.07	9.99 3683	4.22	10.00 6317	25
36	6432	2.13	2496	2.08	3936	4.22	6064	24
37	6560	2.13	2371	2.07	4189	4.20	5811	23
38	6688	2.13	2247	2.08	4441	4.22	5559	22
39	6816	2.13	2122	2.08	4694	4.22	5306	21
40	9.84 6944	2.12	9.85 1997	2.08	9.99 4947	4.20	10.00 5053	20
41	7071	2.13	1872	2.08	5199	4.22	4801	19
42	7199	2.13	1747	2.08	5452	4.22	4548	18
43	7327	2.12	1622	2.08	5705	4.20	4295	17
44	7454	2.13	1497	2.08	5957	4.22	4043	16
45	9.84 7582	2.12	9.85 1372	2.10	9.99 6210	4.22	10.00 3790	15
46	7709	2.12	1246	2.08	6463	4.20	3537	14
47	7836	2.13	1121	2.08	6715	4.22	3285	13
48	7964	2.12	0996	2.10	6968	4.22	3032	12
49	8091	2.12	0870	2.08	7221	4.20	2779	11
50	9.84 8218	2.12	9.85 0745	2.10	9.99 7473	4.22	10.00 2527	10
51	8345	2.12	0619	2.10	7726	4.22	2274	9
52	8472	2.12	0493	2.08	7979	4.20	2021	8
53	8599	2.12	0368	2.10	8231	4.22	1769	7
54	8726	2.10	0242	2.10	8484	4.22	1516	6
55	9.84 8852	2.12	9.85 0116	2.10	9.99 8737	4.20	10.00 1263	5
56	8979	2.12	84 9990	2.10	8989	4.22	1011	4
57	9106	2.10	9864	2.10	9242	4.22	0758	3
58	9232	2.12	9738	2.12	9495	4.20	0505	2
59	9359	2.10	9611	2.10	9.99 9747	4.22	0253	1
60	9.84 9485		9.84 9485		10.00 0000		10.00 0000	0
′	Cosine.	D.1″.	Sine.	D.1″.	Cotang.	D.1″.	Tang.	′

134° 45°

TABLE IX.—NATURAL SINES AND COSINES

	0°				0°				0°		
′	Sine	Cosine	′	′	Sine	Cosine	′	′	Sine	Cosine	′
0	.00000	1	60	21	.00611	.99998	39	41	.01193	.99993	19
1	.00029	1	59	22	.00640	.99998	38	42	.01222	.99993	18
2	.00058	1	58	23	.00669	.99998	37	43	.01251	.99992	17
3	.00087	1	57	24	.00698	.99998	36	44	.01280	.99992	16
4	.00116	1	56	25	.00727	.99997	35	45	.01309	.99991	15
5	.00145	1	55	26	.00756	.99997	34	46	.01338	.99991	14
6	.00175	1	54	27	.00785	.99997	33	47	.01367	.99991	13
7	.00204	1	53	28	.00814	.99997	32	48	.01396	.99990	12
8	.00233	1	52	29	.00844	.99996	31	49	.01425	.99990	11
9	.00262	1	51	30	.00873	.99996	30	50	.01454	.99989	10
10	.00291	1	50	31	.00902	.99996	29	51	.01483	.99989	9
11	.00320	.99999	49	32	.00931	.99996	28	52	.01513	.99989	8
12	.00349	.99999	48	33	.00960	.99995	27	53	.01542	.99988	7
13	.00378	.99999	47	34	.00989	.99995	26	54	.01571	.99988	6
14	.00407	.99999	46	35	.01018	.99995	25	55	.01600	.99987	5
15	.00436	.99999	45	36	.01047	.99995	24	56	.01629	.99987	4
16	.00465	.99999	44	37	.01076	.99994	23	57	.01658	.99986	3
17	.00495	.99999	43	38	.01105	.99994	22	58	.01687	.99986	2
18	.00524	.99999	42	39	.01134	.99994	21	59	.01716	.99985	1
19	.00553	.99998	41	40	.01164	.99993	20	60	.01745	.99985	0
20	.00582	.99998	40								
′	Cosine	Sine	′	′	Cosine	Sine	′	′	Cosine	Sine	′
	89°				89°				89°		

Table IX.—*Continued*

′	1° Sine	1° Cosine	2° Sine	2° Cosine	3° Sine	3° Cosine	4° Sine	4° Cosine	′
0	.01745	.99985	.03490	.99939	.05234	.99863	.06976	.99756	60
1	.01774	.99984	.03519	.99938	.05263	.99861	.07005	.99754	59
2	.01803	.99984	.03548	.99937	.05292	.99860	.07034	.99752	58
3	.01832	.99983	.03577	.99936	.05321	.99858	.07063	.99750	57
4	.01862	.99983	.03606	.99935	.05350	.99857	.07092	.99748	56
5	.01891	.99982	.03635	.99934	.05379	.99855	.07121	.99746	55
6	.01920	.99982	.03664	.99933	.05408	.99854	.07150	.99744	54
7	.01949	.99981	.03693	.99932	.05437	.99852	.07179	.99742	53
8	.01978	.99980	.03723	.99931	.05466	.99851	.07208	.99740	52
9	.02007	.99980	.03752	.99930	.05495	.99849	.07237	.99738	51
10	.02036	.99979	.03781	.99929	.05524	.99847	.07266	.99736	50
11	.02065	.99979	.03810	.99927	.05553	.99846	.07295	.99734	49
12	.02094	.99978	.03839	.99926	.05582	.99844	.07324	.99731	48
13	.02123	.99977	.03868	.99925	.05611	.99842	.07353	.99729	47
14	.02152	.99977	.03897	.99924	.05640	.99841	.07382	.99727	46
15	.02181	.99976	.03926	.99923	.05669	.99839	.07411	.99725	45
16	.02211	.99976	.03955	.99922	.05698	.99838	.07440	.99723	44
17	.02240	.99975	.03984	.99921	.05727	.99836	.07469	.99721	43
18	.02269	.99974	.04013	.99919	.05756	.99834	.07498	.99719	42
19	.02298	.99974	.04042	.99918	.05785	.99833	.07527	.99716	41
20	.02327	.99973	.04071	.99917	.05814	.99831	.07556	.99714	40
21	.02356	.99972	.04100	.99916	.05844	.99829	.07585	.99712	39
22	.02385	.99972	.04129	.99915	.05873	.99827	.07614	.99710	38
23	.02414	.99971	.04159	.99913	.05902	.99826	.07643	.99708	37
24	.02443	.99970	.04188	.99912	.05931	.99824	.07672	.99705	36
25	.02472	.99969	.04217	.99911	.05960	.99822	.07701	.99703	35
26	.02501	.99969	.04246	.99910	.05989	.99821	.07730	.99701	34
27	.02530	.99968	.04275	.99909	.06018	.99819	.07759	.99699	33
28	.02560	.99967	.04304	.99907	.06047	.99817	.07788	.99696	32
29	.02589	.99966	.04333	.99906	.06076	.99815	.07817	.99694	31
30	.02618	.99966	.04362	.99905	.06105	.99813	.07846	.99692	30
31	.02647	.99965	.04391	.99904	.06134	.99812	.07875	.99689	29
32	.02676	.99964	.04420	.99902	.06163	.99810	.07904	.99687	28
33	.02705	.99963	.04449	.99901	.06192	.99808	.07933	.99685	27
34	.02734	.99963	.04478	.99900	.06221	.99806	.07962	.99683	26
35	.02763	.99962	.04507	.99898	.06250	.99804	.07991	.99680	25
36	.02792	.99961	.04536	.99897	.06279	.99803	.08020	.99678	24
37	.02821	.99960	.04565	.99896	.06308	.99801	.08049	.99676	23
38	.02850	.99959	.04594	.99894	.06337	.99799	.08078	.99673	22
39	.02879	.99959	.04623	.99893	.06366	.99797	.08107	.99671	21
40	.02908	.99958	.04653	.99892	.06395	.99795	.08136	.99668	20
41	.02938	.99957	.04682	.99890	.06424	.99793	.08165	.99666	19
42	.02967	.99956	.04711	.99889	.06453	.99792	.08194	.99664	18
43	.02996	.99955	.04740	.99888	.06482	.99790	.08223	.99661	17
44	.03025	.99954	.04769	.99886	.06511	.99788	.08252	.99659	16
45	.03054	.99953	.04798	.99885	.06540	.99786	.08281	.99657	15
46	.03083	.99952	.04827	.99883	.06569	.99784	.08310	.99654	14
47	.03112	.99952	.04856	.99882	.06598	.99782	.08339	.99652	13
48	.03141	.99951	.04885	.99881	.06627	.99780	.08368	.99649	12
49	.03170	.99950	.04914	.99879	.06656	.99778	.08397	.99647	11
50	.03199	.99949	.04943	.99878	.06685	.99776	.08426	.99644	10
51	.03228	.99948	.04972	.99876	.06714	.99774	.08455	.99642	9
52	.03257	.99947	.05001	.99875	.06743	.99772	.08484	.99639	8
53	.03286	.99946	.05030	.99873	.06773	.99770	.08513	.99637	7
54	.03316	.99945	.05059	.99872	.06802	.99768	.08542	.99635	6
55	.03345	.99944	.05088	.99870	.06831	.99766	.08571	.99632	5
56	.03374	.99943	.05117	.99869	.06860	.99764	.08600	.99630	4
57	.03403	.99942	.05146	.99867	.06889	.99762	.08629	.99627	3
58	.03432	.99941	.05175	.99866	.06918	.99760	.08658	.99625	2
59	.03461	.99940	.05205	.99864	.06947	.99758	.08687	.99622	1
60	.03490	.99939	.05234	.99863	.06976	.99756	.08716	.99619	0
′	Cosine 88°	Sine 88°	Cosine 87°	Sine 87°	Cosine 86°	Sine 86°	Cosine 85°	Sine 85°	′

TABLE IX.—*Continued*

′	5° Sine	5° Cosine	6° Sine	6° Cosine	7° Sine	7° Cosine	8° Sine	8° Cosine	′
0	.08716	.99619	.10453	.99452	.12187	.99255	.13917	.99027	60
1	.08745	.99617	.10482	.99449	.12216	.99251	.13946	.99023	59
2	.08774	.99614	.10511	.99446	.12245	.99248	.13975	.99019	58
3	.08803	.99612	.10540	.99443	.12274	.99244	.14004	.99015	57
4	.08831	.99609	.10569	.99440	.12302	.99240	.14033	.99011	56
5	.08860	.99607	.10597	.99437	.12331	.99237	.14061	.99006	55
6	.08889	.99604	.10626	.99434	.12360	.99233	.14090	.99002	54
7	.08918	.99602	.10655	.99431	.12389	.99230	.14119	.98998	53
8	.08947	.99599	.10684	.99428	.12418	.99226	.14148	.98994	52
9	.08976	.99596	.10713	.99424	.12447	.99222	.14177	.98990	51
10	.09005	.99594	.10742	.99421	.12476	.99219	.14205	.98986	50
11	.09034	.99591	.10771	.99418	.12504	.99215	.14234	.98982	49
12	.09063	.99588	.10800	.99415	.12533	.99211	.14263	.98978	48
13	.09092	.99586	.10829	.99412	.12562	.99208	.14292	.98973	47
14	.09121	.99583	.10858	.99409	.12591	.99204	.14320	.98969	46
15	.09150	.99580	.10887	.99406	.12620	.99200	.14349	.98965	45
16	.09179	.99578	.10916	.99402	.12649	.99197	.14378	.98961	44
17	.09208	.99575	.10945	.99399	.12678	.99193	.14407	.98957	43
18	.09237	.99572	.10973	.99396	.12706	.99189	.14436	.98953	42
19	.09266	.99570	.11002	.99393	.12735	.99186	.14464	.98948	41
20	.09295	.99567	.11031	.99390	.12764	.99182	.14493	.98944	40
21	.09324	.99564	.11060	.99386	.12793	.99178	.14522	.98940	39
22	.09353	.99562	.11089	.99383	.12822	.99175	.14551	.98936	38
23	.09382	.99559	.11118	.99380	.12851	.99171	.14580	.98931	37
24	.09411	.99556	.11147	.99377	.12880	.99167	.14608	.98927	36
25	.09440	.99553	.11176	.99374	.12908	.99163	.14637	.98923	35
26	.09469	.99551	.11205	.99370	.12937	.99160	.14666	.98919	34
27	.09498	.99548	.11234	.99367	.12966	.99156	.14695	.98914	33
28	.09527	.99545	.11263	.99364	.12995	.99152	.14723	.98910	32
29	.09556	.99542	.11291	.99360	.13024	.99148	.14752	.98906	31
30	.09585	.99540	.11320	.99357	.13053	.99144	.14781	.98902	30
31	.09614	.99537	.11349	.99354	.13081	.99141	.14810	.98897	29
32	.09642	.99534	.11378	.99351	.13110	.99137	.14838	.98893	28
33	.09671	.99531	.11407	.99347	.13139	.99133	.14867	.98889	27
34	.09700	.99528	.11436	.99344	.13168	.99129	.14896	.98884	26
35	.09729	.99526	.11465	.99341	.13197	.99125	.14925	.98880	25
36	.09758	.99523	.11494	.99337	.13226	.99122	.14954	.98876	24
37	.09787	.99520	.11523	.99334	.13254	.99118	.14982	.98871	23
38	.09816	.99517	.11552	.99331	.13283	.99114	.15011	.98867	22
39	.09845	.99514	.11580	.99327	.13312	.99110	.15040	.98863	21
40	.09874	.99511	.11609	.99324	.13341	.99106	.15069	.98858	20
41	.09903	.99508	.11638	.99320	.13370	.99102	.15097	.98854	19
42	.09932	.99506	.11667	.99317	.13399	.99098	.15126	.98849	18
43	.09961	.99503	.11696	.99314	.13427	.99094	.15155	.98845	17
44	.09990	.99500	.11725	.99310	.13456	.99091	.15184	.98841	16
45	.10019	.99497	.11754	.99307	.13485	.99087	.15212	.98836	15
46	.10048	.99494	.11783	.99303	.13514	.99083	.15241	.98832	14
47	.10077	.99491	.11812	.99300	.13543	.99079	.15270	.98827	13
48	.10106	.99488	.11840	.99297	.13572	.99075	.15299	.98823	12
49	.10135	.99485	.11869	.99293	.13600	.99071	.15327	.98818	11
50	.10164	.99482	.11898	.99290	.13629	.99067	.15356	.98814	10
51	.10192	.99479	.11927	.99286	.13658	.99063	.15385	.98809	9
52	.10221	.99476	.11956	.99283	.13687	.99059	.15414	.98805	8
53	.10250	.99473	.11985	.99279	.13716	.99055	.15442	.98800	7
54	.10279	.99470	.12014	.99276	.13744	.99051	.15471	.98796	6
55	.10308	.99467	.12043	.99272	.13773	.99047	.15500	.98791	5
56	.10337	.99464	.12071	.99269	.13802	.99043	.15529	.98787	4
57	.10366	.99461	.12100	.99265	.13831	.99039	.15557	.98782	3
58	.10395	.99458	.12129	.99262	.13860	.99035	.15586	.98778	2
59	.10424	.99455	.12158	.99258	.13889	.99031	.15615	.98773	1
60	.10453	.99452	.12187	.99255	.13917	.99027	.15643	.98769	0
′	84° Cosine	84° Sine	83° Cosine	83° Sine	82° Cosine	82° Sine	81° Cosine	81° Sine	′

TABLE IX.—*Continued*

′	9° Sine	9° Cosine	10° Sine	10° Cosine	11° Sine	11° Cosine	12° Sine	12° Cosine	′
0	.15643	.98769	.17365	.98481	.19081	.98163	.20791	.97815	60
1	.15672	.98764	.17393	.98476	.19109	.98157	.20820	.97809	59
2	.15701	.98760	.17422	.98471	.19138	.98152	.20848	.97803	58
3	.15730	.98755	.17451	.98466	.19167	.98146	.20877	.97797	57
4	.15758	.98751	.17479	.98461	.19195	.98140	.20905	.97791	56
5	.15787	.98746	.17508	.98455	.19224	.98135	.20933	.97784	55
6	.15816	.98741	.17537	.98450	.19252	.98129	.20962	.97778	54
7	.15845	.98737	.17565	.98445	.19281	.98124	.20990	.97772	53
8	.15873	.98732	.17594	.98440	.19309	.98118	.21019	.97766	52
9	.15902	.98728	.17623	.98435	.19338	.98112	.21047	.97760	51
10	.15931	.98723	.17651	.98430	.19366	.98107	.21076	.97754	50
11	.15959	.98718	.17680	.98425	.19395	.98101	.21104	.97748	49
12	.15988	.98714	.17708	.98420	.19423	.98096	.21132	.97742	48
13	.16017	.98709	.17737	.98414	.19452	.98090	.21161	.97735	47
14	.16046	.98704	.17766	.98409	.19481	.98084	.21189	.97729	46
15	.16074	.98700	.17794	.98404	.19509	.98079	.21218	.97723	45
16	.16103	.98695	.17823	.98399	.19538	.98073	.21246	.97717	44
17	.16132	.98690	.17852	.98394	.19566	.98067	.21275	.97711	43
18	.16160	.98686	.17880	.98389	.19595	.98061	.21303	.97705	42
19	.16189	.98681	.17909	.98383	.19623	.98056	.21331	.97698	41
20	.16218	.98676	.17937	.98378	.19652	.98050	.21360	.97692	40
21	.16246	.98671	.17966	.98373	.19680	.98044	.21388	.97686	39
22	.16275	.98667	.17995	.98368	.19709	.98039	.21417	.97680	38
23	.16304	.98662	.18023	.98362	.19737	.98033	.21445	.97673	37
24	.16333	.98657	.18052	.98357	.19766	.98027	.21474	.97667	36
25	.16361	.98652	.18081	.98352	.19794	.98021	.21502	.97661	35
26	.16390	.98648	.18109	.98347	.19823	.98016	.21530	.97655	34
27	.16419	.98643	.18138	.98341	.19851	.98010	.21559	.97648	33
28	.16447	.98638	.18166	.98336	.19880	.98004	.21587	.97642	32
29	.16476	.98633	.18195	.98331	.19908	.97998	.21616	.97636	31
30	.16505	.98629	.18224	.98325	.19937	.97992	.21644	.97630	30
31	.16533	.98624	.18252	.98320	.19965	.97987	.21672	.97623	29
32	.16562	.98619	.18281	.98315	.19994	.97981	.21701	.97617	28
33	.16591	.98614	.18309	.98310	.20022	.97975	.21729	.97611	27
34	.16620	.98609	.18338	.98304	.20051	.97969	.21758	.97604	26
35	.16648	.98604	.18367	.98299	.20079	.97963	.21786	.97598	25
36	.16677	.98600	.18395	.98294	.20108	.97958	.21814	.97592	24
37	.16706	.98595	.18424	.98288	.20136	.97952	.21843	.97585	23
38	.16734	.98590	.18452	.98283	.20165	.97946	.21871	.97579	22
39	.16763	.98585	.18481	.98277	.20193	.97940	.21899	.97573	21
40	.16792	.98580	.18509	.98272	.20222	.97934	.21928	.97566	20
41	.16820	.98575	.18538	.98267	.20250	.97928	.21956	.97560	19
42	.16849	.98570	.18567	.98261	.20279	.97922	.21985	.97553	18
43	.16878	.98565	.18595	.98256	.20307	.97916	.22013	.97547	17
44	.16906	.98561	.18624	.98250	.20336	.97910	.22041	.97541	16
45	.16935	.98556	.18652	.98245	.20364	.97905	.22070	.97534	15
46	.16964	.98551	.18681	.98240	.20393	.97899	.22098	.97528	14
47	.16992	.98546	.18710	.98234	.20421	.97893	.22126	.97521	13
48	.17021	.98541	.18738	.98229	.20450	.97887	.22155	.97515	12
49	.17050	.98536	.18767	.98223	.20478	.97881	.22183	.97508	11
50	.17078	.98531	.18795	.98218	.20507	.97875	.22212	.97502	10
51	.17107	.98526	.18824	.98212	.20535	.97869	.22240	.97496	9
52	.17136	.98521	.18852	.98207	.20563	.97863	.22268	.97489	8
53	.17164	.98516	.18881	.98201	.20592	.97857	.22297	.97483	7
54	.17193	.98511	.18910	.98196	.20620	.97851	.22325	.97476	6
55	.17222	.98506	.18938	.98190	.20649	.97845	.22353	.97470	5
56	.17250	.98501	.18967	.98185	.20677	.97839	.22382	.97463	4
57	.17279	.98496	.18995	.98179	.20706	.97833	.22410	.97457	3
58	.17308	.98491	.19024	.98174	.20734	.97827	.22438	.97450	2
59	.17336	.98486	.19052	.98168	.20763	.97821	.22467	.97444	1
60	.17365	.98481	.19081	.98163	.20791	.97815	.22495	.97437	0
′	80° Cosine	80° Sine	79° Cosine	79° Sine	78° Cosine	78° Sine	77° Cosine	77° Sine	′

TABLE IX.—*Continued*

′	13° Sine	13° Cosine	14° Sine	14° Cosine	15° Sine	15° Cosine	16° Sine	16° Cosine	′
0	.22495	.97437	.24192	.97030	.25882	.96593	.27564	.96126	60
1	.22523	.97430	.24220	.97023	.25910	.96585	.27592	.96118	59
2	.22552	.97424	.24249	.97015	.25938	.96578	.27620	.96110	58
3	.22580	.97417	.24277	.97008	.25966	.96570	.27648	.96102	57
4	.22608	.97411	.24305	.97001	.25994	.96562	.27676	.96094	56
5	.22637	.97404	.24333	.96994	.26022	.96555	.27704	.96086	55
6	.22665	.97398	.24362	.96987	.26050	.96547	.27731	.96078	54
7	.22693	.97391	.24390	.96980	.26079	.96540	.27759	.96070	53
8	.22722	.97384	.24418	.96973	.26107	.96532	.27787	.96062	52
9	.22750	.97378	.24446	.96966	.26135	.96524	.27815	.96054	51
10	.22778	.97371	.24474	.96959	.26163	.96517	.27843	.96046	50
11	.22807	.97365	.24503	.96952	.26191	.96509	.27871	.96037	49
12	.22835	.97358	.24531	.96945	.26219	.96502	.27899	.96029	48
13	.22863	.97351	.24559	.96937	.26247	.96494	.27927	.96021	47
14	.22892	.97345	.24587	.96930	.26275	.96486	.27955	.96013	46
15	.22920	.97338	.24615	.96923	.26303	.96479	.27983	.96005	45
16	.22948	.97331	.24644	.96916	.26331	.96471	.28011	.95997	44
17	.22977	.97325	.24672	.96909	.26359	.96463	.28039	.95989	43
18	.23005	.97318	.24700	.96902	.26387	.96456	.28067	.95981	42
19	.23033	.97311	.24728	.96894	.26415	.96448	.28095	.95972	41
20	.23062	.97304	.24756	.96887	.26443	.96440	.28123	.95964	40
21	.23090	.97298	.24784	.96880	.26471	.96433	.28150	.95956	39
22	.23118	.97291	.24813	.96873	.26500	.96425	.28178	.95948	38
23	.23146	.97284	.24841	.96866	.26528	.96417	.28206	.95940	37
24	.23175	.97278	.24869	.96858	.26556	.96410	.28234	.95931	36
25	.23203	.97271	.24897	.96851	.26584	.96402	.28262	.95923	35
26	.23231	.97264	.24925	.96844	.26612	.96394	.28290	.95915	34
27	.23260	.97257	.24954	.96837	.26640	.96386	.28318	.95907	33
28	.23288	.97251	.24982	.96829	.26668	.96379	.28346	.95898	32
29	.23316	.97244	.25010	.96822	.26696	.96371	.28374	.95890	31
30	.23345	.97237	.25038	.96815	.26724	.96363	.28402	.95882	30
31	.23373	.97230	.25066	.96807	.26752	.96355	.28429	.95874	29
32	.23401	.97223	.25094	.96800	.26780	.96347	.28457	.95865	28
33	.23429	.97217	.25122	.96793	.26808	.96340	.28485	.95857	27
34	.23458	.97210	.25151	.96786	.26836	.96332	.28513	.95849	26
35	.23486	.97203	.25179	.96778	.26864	.96324	.28541	.95841	25
36	.23514	.97196	.25207	.96771	.26892	.96316	.28569	.95832	24
37	.23542	.97189	.25235	.96764	.26920	.96308	.28597	.95824	23
38	.23571	.97182	.25263	.96756	.26948	.96301	.28625	.95816	22
39	.23599	.97176	.25291	.96749	.26976	.96293	.28652	.95807	21
40	.23627	.97169	.25320	.96742	.27004	.96285	.28680	.95799	20
41	.23656	.97162	.25348	.96734	.27032	.96277	.28708	.95791	19
42	.23684	.97155	.25376	.96727	.27060	.96269	.28736	.95782	18
43	.23712	.97148	.25404	.96719	.27088	.96261	.28764	.95774	17
44	.23740	.97141	.25432	.96712	.27116	.96253	.28792	.95766	16
45	.23769	.97134	.25460	.96705	.27144	.96246	.28820	.95757	15
46	.23797	.97127	.25488	.96697	.27172	.96238	.28847	.95749	14
47	.23825	.97120	.25516	.96690	.27200	.96230	.28875	.95740	13
48	.23853	.97113	.25545	.96682	.27228	.96222	.28903	.95732	12
49	.23882	.97106	.25573	.96675	.27256	.96214	.28931	.95724	11
50	.23910	.97100	.25601	.96667	.27284	.96206	.28959	.95715	10
51	.23938	.97093	.25629	.96660	.27312	.96198	.28987	.95707	9
52	.23966	.97086	.25657	.96653	.27340	.96190	.29015	.95698	8
53	.23995	.97079	.25685	.96645	.27368	.96182	.29042	.95690	7
54	.24023	.97072	.25713	.96638	.27396	.96174	.29070	.95681	6
55	.24051	.97065	.25741	.96630	.27424	.96166	.29098	.95673	5
56	.24079	.97058	.25769	.96623	.27452	.96158	.29126	.95664	4
57	.24108	.97051	.25798	.96615	.27480	.96150	.29154	.95656	3
58	.24136	.97044	.25826	.96608	.27508	.96142	.29182	.95647	2
59	.24164	.97037	.25854	.96600	.27536	.96134	.29209	.95639	1
60	.24192	.97030	.25882	.96593	.27564	.96126	.29237	.95630	0
′	76° Cosine	76° Sine	75° Cosine	75° Sine	74° Cosine	74° Sine	73° Cosine	73° Sine	′

TABLE IX.—*Continued*

′	17° Sine	17° Cosine	18° Sine	18° Cosine	19° Sine	19° Cosine	20° Sine	20° Cosine	′
0	.29237	.95630	.30902	.95106	.32557	.94552	.34202	.93969	60
1	.29265	.95622	.30929	.95097	.32584	.94542	.34229	.93959	59
2	.29293	.95613	.30957	.95088	.32612	.94533	.34257	.93949	58
3	.29321	.95605	.30985	.95079	.32639	.94523	.34284	.93939	57
4	.29348	.95596	.31012	.95070	.32667	.94514	.34311	.93929	56
5	.29376	.95588	.31040	.95061	.32694	.94504	.34339	.93919	55
6	.29404	.95579	.31068	.95052	.32722	.94495	.34366	.93909	54
7	.29432	.95571	.31095	.95043	.32749	.94485	.34393	.93899	53
8	.29460	.95562	.31123	.95033	.32777	.94476	.34421	.93889	52
9	.29487	.95554	.31151	.95024	.32804	.94466	.34448	.93879	51
10	.29515	.95545	.31178	.95015	.32832	.94457	.34475	.93869	50
11	.29543	.95536	.31206	.95006	.32859	.94447	.34503	.93859	49
12	.29571	.95528	.31233	.94997	.32887	.94438	.34530	.93849	48
13	.29599	.95519	.31261	.94988	.32914	.94428	.34557	.93839	47
14	.29626	.95511	.31289	.94979	.32942	.94418	.34584	.93829	46
15	.29654	.95502	.31316	.94970	.32969	.94409	.34612	.93819	45
16	.29682	.95493	.31344	.94961	.32997	.94399	.34639	.93809	44
17	.29710	.95485	.31372	.94952	.33024	.94390	.34666	.93799	43
18	.29737	.95476	.31399	.94943	.33051	.94380	.34694	.93789	42
19	.29765	.95467	.31427	.94933	.33079	.94370	.34721	.93779	41
20	.29793	.95459	.31454	.94924	.33106	.94361	.34748	.93769	40
21	.29821	.95450	.31482	.94915	.33134	.94351	.34775	.93759	39
22	.29849	.95441	.31510	.94906	.33161	.94342	.34803	.93748	38
23	.29876	.95433	.31537	.94897	.33189	.94332	.34830	.93738	37
24	.29904	.95424	.31565	.94888	.33216	.94322	.34857	.93728	36
25	.29932	.95415	.31593	.94878	.33244	.94313	.34884	.93718	35
26	.29960	.95407	.31620	.94869	.33271	.94303	.34912	.93708	34
27	.29987	.95398	.31648	.94860	.33298	.94293	.34939	.93698	33
28	.30015	.95389	.31675	.94851	.33326	.94284	.34966	.93688	32
29	.30043	.95380	.31703	.94842	.33353	.94274	.34993	.93677	31
30	.30071	.95372	.31730	.94832	.33381	.94264	.35021	.93667	30
31	.30098	.95363	.31758	.94823	.33408	.94254	.35048	.93657	29
32	.30126	.95354	.31786	.94814	.33436	.94245	.35075	.93647	28
33	.30154	.95345	.31813	.94805	.33463	.94235	.35102	.93637	27
34	.30182	.95337	.31841	.94795	.33490	.94225	.35130	.93626	26
35	.30209	.95328	.31868	.94786	.33518	.94215	.35157	.93616	25
36	.30237	.95319	.31896	.94777	.33545	.94206	.35184	.93606	24
37	.30265	.95310	.31923	.94768	.33573	.94196	.35211	.93596	23
38	.30292	.95301	.31951	.94758	.33600	.94186	.35239	.93585	22
39	.30320	.95293	.31979	.94749	.33627	.94176	.35266	.93575	21
40	.30348	.95284	.32006	.94740	.33655	.94167	.35293	.93565	20
41	.30376	.95275	.32034	.94730	.33682	.94157	.35320	.93555	19
42	.30403	.95266	.32061	.94721	.33710	.94147	.35347	.93544	18
43	.30431	.95257	.32089	.94712	.33737	.94137	.35375	.93534	17
44	.30459	.95248	.32116	.94702	.33764	.94127	.35402	.93524	16
45	.30486	.95240	.32144	.94693	.33792	.94118	.35429	.93514	15
46	.30514	.95231	.32171	.94684	.33819	.94108	.35456	.93503	14
47	.30542	.95222	.32199	.94674	.33846	.94098	.35484	.93493	13
48	.30570	.95213	.32227	.94665	.33874	.94088	.35511	.93483	12
49	.30597	.95204	.32254	.94656	.33901	.94078	.35538	.93472	11
50	.30625	.95195	.32282	.94646	.33929	.94068	.35565	.93462	10
51	.30653	.95186	.32309	.94637	.33956	.94058	.35592	.93452	9
52	.30680	.95177	.32337	.94627	.33983	.94049	.35619	.93441	8
53	.30708	.95168	.32364	.94618	.34011	.94039	.35647	.93431	7
54	.30736	.95159	.32392	.94609	.34038	.94029	.35674	.93420	6
55	.30763	.95150	.32419	.94599	.34065	.94019	.35701	.93410	5
56	.30791	.95142	.32447	.94590	.34093	.94009	.35728	.93400	4
57	.30819	.95133	.32474	.94580	.34120	.93999	.35755	.93389	3
58	.30846	.95124	.32502	.94571	.34147	.93989	.35782	.93379	2
59	.30874	.95115	.32529	.94561	.34175	.93979	.35810	.93368	1
60	.30902	.95106	.32557	.94552	.34202	.93969	.35837	.93358	0
′	Cosine	Sine	Cosine	Sine	Cosine	Sine	Cosine	Sine	′
	72°		71°		70°		69°		

TABLE IX.—*Continued*

′	21° Sine	21° Cosine	22° Sine	22° Cosine	23° Sine	23° Cosine	24° Sine	24° Cosine	′
0	.35837	.93358	.37461	.92718	.39073	.92050	.40674	.91355	60
1	.35864	.93348	.37488	.92707	.39100	.92039	.40700	.91343	59
2	.35891	.93337	.37515	.92697	.39127	.92028	.40727	.91331	58
3	.35918	.93327	.37542	.92686	.39153	.92016	.40753	.91319	57
4	.35945	.93316	.37569	.92675	.39180	.92005	.40780	.91307	56
5	.35973	.93306	.37595	.92664	.39207	.91994	.40806	.91295	55
6	.36000	.93295	.37622	.92653	.39234	.91982	.40833	.91283	54
7	.36027	.93285	.37649	.92642	.39260	.91971	.40860	.91272	53
8	.36054	.93274	.37676	.92631	.39287	.91959	.40886	.91260	52
9	.36081	.93264	.37703	.92620	.39314	.91948	.40913	.91248	51
10	.36108	.93253	.37730	.92609	.39341	.91936	.40939	.91236	50
11	.36135	.93243	.37757	.92598	.39367	.91925	.40966	.91224	49
12	.36162	.93232	.37784	.92587	.39394	.91914	.40992	.91212	48
13	.36190	.93222	.37811	.92576	.39421	.91902	.41019	.91200	47
14	.36217	.93211	.37838	.92565	.39448	.91891	.41045	.91188	46
15	.36244	.93201	.37865	.92554	.39474	.91879	.41072	.91176	45
16	.36271	.93190	.37892	.92543	.39501	.91868	.41098	.91164	44
17	.36298	.93180	.37919	.92532	.39528	.91856	.41125	.91152	43
18	.36325	.93169	.37946	.92521	.39555	.91845	.41151	.91140	42
19	.36352	.93159	.37973	.92510	.39581	.91833	.41178	.91128	41
20	.36379	.93148	.37999	.92499	.39608	.91822	.41204	.91116	40
21	.36406	.93137	.38026	.92488	.39635	.91810	.41231	.91104	39
22	.36434	.93127	.38053	.92477	.39661	.91799	.41257	.91092	38
23	.36461	.93116	.38080	.92466	.39688	.91787	.41284	.91080	37
24	.36488	.93106	.38107	.92455	.39715	.91775	.41310	.91068	36
25	.36515	.93095	.38134	.92444	.39741	.91764	.41337	.91056	35
26	.36542	.93084	.38161	.92432	.39768	.91752	.41363	.91044	34
27	.36569	.93074	.38188	.92421	.39795	.91741	.41390	.91032	33
28	.36596	.93063	.38215	.92410	.39822	.91729	.41416	.91020	32
29	.36623	.93052	.38241	.92399	.39848	.91718	.41443	.91008	31
30	.36650	.93042	.38268	.92388	.39875	.91706	.41469	.90996	30
31	.36677	.93031	.38295	.92377	.39902	.91694	.41496	.90984	29
32	.36704	.93020	.38322	.92366	.39928	.91683	.41522	.90972	28
33	.36731	.93010	.38349	.92355	.39955	.91671	.41549	.90960	27
34	.36758	.92999	.38376	.92343	.39982	.91660	.41575	.90948	26
35	.36785	.92988	.38403	.92332	.40008	.91648	.41602	.90936	25
36	.36812	.92978	.38430	.92321	.40035	.91636	.41628	.90924	24
37	.36839	.92967	.38456	.92310	.40062	.91625	.41655	.90911	23
38	.36867	.92956	.38483	.92299	.40088	.91613	.41681	.90899	22
39	.36894	.92945	.38510	.92287	.40115	.91601	.41707	.90887	21
40	.36921	.92935	.38537	.92276	.40141	.91590	.41734	.90875	20
41	.36948	.92924	.38564	.92265	.40168	.91578	.41760	.90863	19
42	.36975	.92913	.38591	.92254	.40195	.91566	.41787	.90851	18
43	.37002	.92902	.38617	.92243	.40221	.91555	.41813	.90839	17
44	.37029	.92892	.38644	.92231	.40248	.91543	.41840	.90826	16
45	.37056	.92881	.38671	.92220	.40275	.91531	.41866	.90814	15
46	.37083	.92870	.38698	.92209	.40301	.91519	.41892	.90802	14
47	.37110	.92859	.38725	.92198	.40328	.91508	.41919	.90790	13
48	.37137	.92849	.38752	.92186	.40355	.91496	.41945	.90778	12
49	.37164	.92838	.38778	.92175	.40381	.91484	.41972	.90766	11
50	.37191	.92827	.38805	.92164	.40408	.91472	.41998	.90753	10
51	.37218	.92816	.38832	.92152	.40434	.91461	.42024	.90741	9
52	.37245	.92805	.38859	.92141	.40461	.91449	.42051	.90729	8
53	.37272	.92794	.38886	.92130	.40488	.91437	.42077	.90717	7
54	.37299	.92784	.38912	.92119	.40514	.91425	.42104	.90704	6
55	.37326	.92773	.38939	.92107	.40541	.91414	.42130	.90692	5
56	.37353	.92762	.38966	.92096	.40567	.91402	.42156	.90680	4
57	.37380	.92751	.38993	.92085	.40594	.91390	.42183	.90668	3
58	.37407	.92740	.39020	.92073	.40621	.91378	.42209	.90655	2
59	.37434	.92729	.39046	.92062	.40647	.91366	.42235	.90643	1
60	.37461	.92718	.39073	.92050	.40674	.91355	.42262	.90631	0
′	Cosine 68°	Sine 68°	Cosine 67°	Sine 67°	Cosine 66°	Sine 66°	Cosine 65°	Sine 65°	′

TABLE IX.—*Continued*

′	25° Sine	25° Cosine	26° Sine	26° Cosine	27° Sine	27° Cosine	28° Sine	28° Cosine	′
0	.42262	.90631	.43837	.89879	.45399	.89101	.46947	.88295	60
1	.42288	.90618	.43863	.89867	.45425	.89087	.46973	.88281	59
2	.42315	.90606	.43889	.89854	.45451	.89074	.46999	.88267	58
3	.42341	.90594	.43916	.89841	.45477	.89061	.47024	.88254	57
4	.42367	.90582	.43942	.89828	.45503	.89048	.47050	.88240	56
5	.42394	.90569	.43968	.89816	.45529	.89035	.47076	.88226	55
6	.42420	.90557	.43994	.89803	.45554	.89021	.47101	.88213	54
7	.42446	.90545	.44020	.89790	.45580	.89008	.47127	.88199	53
8	.42473	.90532	.44046	.89777	.45606	.88995	.47153	.88185	52
9	.42499	.90520	.44072	.89764	.45632	.88981	.47178	.88172	51
10	.42525	.90507	.44098	.89752	.45658	.88968	.47204	.88158	50
11	.42552	.90495	.44124	.89739	.45684	.88955	.47229	.88144	49
12	.42578	.90483	.44151	.89726	.45710	.88942	.47255	.88130	48
13	.42604	.90470	.44177	.89713	.45736	.88928	.47281	.88117	47
14	.42631	.90458	.44203	.89700	.45762	.88915	.47306	.88103	46
15	.42657	.90446	.44229	.89687	.45787	.88902	.47332	.88089	45
16	.42683	.90433	.44255	.89674	.45813	.88888	.47358	.88075	44
17	.42709	.90421	.44281	.89662	.45839	.88875	.47383	.88062	43
18	.42736	.90408	.44307	.89649	.45865	.88862	.47409	.88048	42
19	.42762	.90396	.44333	.89636	.45891	.88848	.47434	.88034	41
20	.42788	.90383	.44359	.89623	.45917	.88835	.47460	.88020	40
21	.42815	.90371	.44385	.89610	.45942	.88822	.47486	.88006	39
22	.42841	.90358	.44411	.89597	.45968	.88808	.47511	.87993	38
23	.42867	.90346	.44437	.89584	.45994	.88795	.47537	.87979	37
24	.42894	.90334	.44464	.89571	.46020	.88782	.47562	.87965	36
25	.42920	.90321	.44490	.89558	.46046	.88768	.47588	.87951	35
26	.42946	.90309	.44516	.89545	.46072	.88755	.47614	.87937	34
27	.42972	.90296	.44542	.89532	.46097	.88741	.47639	.87923	33
28	.42999	.90284	.44568	.89519	.46123	.88728	.47665	.87909	32
29	.43025	.90271	.44594	.89506	.46149	.88715	.47690	.87896	31
30	.43051	.90259	.44620	.89493	.46175	.88701	.47716	.87882	30
31	.43077	.90246	.44646	.89480	.46201	.88688	.47741	.87868	29
32	.43104	.90233	.44672	.89467	.46226	.88674	.47767	.87854	28
33	.43130	.90221	.44698	.89454	.46252	.88661	.47793	.87840	27
34	.43156	.90208	.44724	.89441	.46278	.88647	.47818	.87826	26
35	.43182	.90196	.44750	.89428	.46304	.88634	.47844	.87812	25
36	.43209	.90183	.44776	.89415	.46330	.88620	.47869	.87798	24
37	.43235	.90171	.44802	.89402	.46355	.88607	.47895	.87784	23
38	.43261	.90158	.44828	.89389	.46381	.88593	.47920	.87770	22
39	.43287	.90146	.44854	.89376	.46407	.88580	.47946	.87756	21
40	.43313	.90133	.44880	.89363	.46433	.88566	.47971	.87743	20
41	.43340	.90120	.44906	.89350	.46458	.88553	.47997	.87729	19
42	.43366	.90108	.44932	.89337	.46484	.88539	.48022	.87715	18
43	.43392	.90095	.44958	.89324	.46510	.88526	.48048	.87701	17
44	.43418	.90082	.44984	.89311	.46536	.88512	.48073	.87687	16
45	.43445	.90070	.45010	.89298	.46561	.88499	.48099	.87673	15
46	.43471	.90057	.45036	.89285	.46587	.88485	.48124	.87659	14
47	.43497	.90045	.45062	.89272	.46613	.88472	.48150	.87645	13
48	.43523	.90032	.45088	.89259	.46639	.88458	.48175	.87631	12
49	.43549	.90019	.45114	.89245	.46664	.88445	.48201	.87617	11
50	.43575	.90007	.45140	.89232	.46690	.88431	.48226	.87603	10
51	.43602	.89994	.45166	.89219	.46716	.88417	.48252	.87589	9
52	.43628	.89981	.45192	.89206	.46742	.88404	.48277	.87575	8
53	.43654	.89968	.45218	.89193	.46767	.88390	.48303	.87561	7
54	.43680	.89956	.45243	.89180	.46793	.88377	.48328	.87546	6
55	.43706	.89943	.45269	.89167	.46819	.88363	.48354	.87532	5
56	.43733	.89930	.45295	.89153	.46844	.88349	.48379	.87518	4
57	.43759	.89918	.45321	.89140	.46870	.88336	.48405	.87504	3
58	.43785	.89905	.45347	.89127	.46896	.88322	.48430	.87490	2
59	.43811	.89892	.45373	.89114	.46921	.88308	.48456	.87476	1
60	.43837	.89879	.45399	.89101	.46947	.88295	.48481	.87462	0
′	64° Cosine	64° Sine	63° Cosine	63° Sine	62° Cosine	62° Sine	61° Cosine	61° Sine	′

TABLE IX.—*Continued*

′	29°		30°		31°		32°		′
	Sine	Cosine	Sine	Cosine	Sine	Cosine	Sine	Cosine	
0	.48481	.87462	.50000	.86603	.51504	.85717	.52992	.84805	60
1	.48506	.87448	.50025	.86588	.51529	.85702	.53017	.84789	59
2	.48532	.87434	.50050	.86573	.51554	.85687	.53041	.84774	58
3	.48557	.87420	.50076	.86559	.51579	.85672	.53066	.84759	57
4	.48583	.87406	.50101	.86544	.51604	.85657	.53091	.84743	56
5	.48608	.87391	.50126	.86530	.51628	.85642	.53115	.84728	55
6	.48634	.87377	.50151	.86515	.51653	.85627	.53140	.84712	54
7	.48659	.87363	.50176	.86501	.51678	.85612	.53164	.84697	53
8	.48684	.87349	.50201	.86486	.51703	.85597	.53189	.84681	52
9	.48710	.87335	.50227	.86471	.51728	.85582	.53214	.84666	51
10	.48735	.87321	.50252	.86457	.51753	.85567	.53238	.84650	50
11	.48761	.87306	.50277	.86442	.51778	.85551	.53263	.84635	49
12	.48786	.87292	.50302	.86427	.51803	.85536	.53288	.84619	48
13	.48811	.87278	.50327	.86413	.51828	.85521	.53312	.84604	47
14	.48837	.87264	.50352	.86398	.51852	.85506	.53337	.84588	46
15	.48862	.87250	.50377	.86384	.51877	.85491	.53361	.84573	45
16	.48888	.87235	.50403	.86369	.51902	.85476	.53386	.84557	44
17	.48913	.87221	.50428	.86354	.51927	.85461	.53411	.84542	43
18	.48938	.87207	.50453	.86340	.51952	.85446	.53435	.84526	42
19	.48964	.87193	.50478	.86325	.51977	.85431	.53460	.84511	41
20	.48989	.87178	.50503	.86310	.52002	.85416	.53484	.84495	40
21	.49014	.87164	.50528	.86295	.52026	.85401	.53509	.84480	39
22	.49040	.87150	.50553	.86281	.52051	.85385	.53534	.84464	38
23	.49065	.87136	.50578	.86266	.52076	.85370	.53558	.84448	37
24	.49090	.87121	.50603	.86251	.52101	.85355	.53583	.84433	36
25	.49116	.87107	.50628	.86237	.52126	.85340	.53607	.84417	35
26	.49141	.87093	.50654	.86222	.52151	.85325	.53632	.84402	34
27	.49166	.87079	.50679	.86207	.52175	.85310	.53656	.84386	33
28	.49192	.87064	.50704	.86192	.52200	.85294	.53681	.84370	32
29	.49217	.87050	.50729	.86178	.52225	.85279	.53705	.84355	31
30	.49242	.87036	.50754	.86163	.52250	.85264	.53730	.84339	30
31	.49268	.87021	.50779	.86148	.52275	.85249	.53754	.84324	29
32	.49293	.87007	.50804	.86133	.52299	.85234	.53779	.84308	28
33	.49318	.86993	.50829	.86119	.52324	.85218	.53804	.84292	27
34	.49344	.86978	.50854	.86104	.52349	.85203	.53828	.84277	26
35	.49369	.86964	.50879	.86089	.52374	.85188	.53853	.84261	25
36	.49394	.86949	.50904	.86074	.52399	.85173	.53877	.84245	24
37	.49419	.86935	.50929	.86059	.52423	.85157	.53902	.84230	23
38	.49445	.86921	.50954	.86045	.52448	.85142	.53926	.84214	22
39	.49470	.86906	.50979	.86030	.52473	.85127	.53951	.84198	21
40	.49495	.86892	.51004	.86015	.52498	.85112	.53975	.84182	20
41	.49521	.86878	.51029	.86000	.52522	.85096	.54000	.84167	19
42	.49546	.86863	.51054	.85985	.52547	.85081	.54024	.84151	18
43	.49571	.86849	.51079	.85970	.52572	.85066	.54049	.84135	17
44	.49596	.86834	.51104	.85956	.52597	.85051	.54073	.84120	16
45	.49622	.86820	.51129	.85941	.52621	.85035	.54097	.84104	15
46	.49647	.86805	.51154	.85926	.52646	.85020	.54122	.84088	14
47	.49672	.86791	.51179	.85911	.52671	.85005	.54146	.84072	13
48	.49697	.86777	.51204	.85896	.52696	.84989	.54171	.84057	12
49	.49723	.86762	.51229	.85881	.52720	.84974	.54195	.84041	11
50	.49748	.86748	.51254	.85866	.52745	.84959	.54220	.84025	10
51	.49773	.86733	.51279	.85851	.52770	.84943	.54244	.84009	9
52	.49798	.86719	.51304	.85836	.52794	.84928	.54269	.83994	8
53	.49824	.86704	.51329	.85821	.52819	.84913	.54293	.83978	7
54	.49849	.86690	.51354	.85806	.52844	.84897	.54317	.83962	6
55	.49874	.86675	.51379	.85792	.52869	.84882	.54342	.83946	5
56	.49899	.86661	.51404	.85777	.52893	.84866	.54366	.83930	4
57	.49924	.86646	.51429	.85762	.52918	.84851	.54391	.83915	3
58	.49950	.86632	.51454	.85747	.52943	.84836	.54415	.83899	2
59	.49975	.86617	.51479	.85732	.52967	.84820	.54440	.83883	1
60	.50000	.86603	.51504	.85717	.52992	.84805	.54464	.83867	0
′	Cosine	Sine	Cosine	Sine	Cosine	Sine	Cosine	Sine	′
	60°		59°		58°		57°		

TABLE IX.—*Continued*

′	33° Sine	33° Cosine	34° Sine	34° Cosine	35° Sine	35° Cosine	36° Sine	36° Cosine	′
0	.54464	.83867	.55919	.82904	.57358	.81915	.58779	.80902	60
1	.54488	.83851	.55943	.82887	.57381	.81899	.58802	.80885	59
2	.54513	.83835	.55968	.82871	.57405	.81882	.58826	.80867	58
3	.54537	.83819	.55992	.82855	.57429	.81865	.58849	.80850	57
4	.54561	.83804	.56016	.82839	.57453	.81848	.58873	.80833	56
5	.54586	.83788	.56040	.82822	.57477	.81832	.58896	.80816	55
6	.54610	.83772	.56064	.82806	.57501	.81815	.58920	.80799	54
7	.54635	.83756	.56088	.82790	.57524	.81798	.58943	.80782	53
8	.54659	.83740	.56112	.82773	.57548	.81782	.58967	.80765	52
9	.54683	.83724	.56136	.82757	.57572	.81765	.58990	.80748	51
10	.54708	.83708	.56160	.82741	.57596	.81748	.59014	.80730	50
11	.54732	.83692	.56184	.82724	.57619	.81731	.59037	.80713	49
12	.54756	.83676	.56208	.82708	.57643	.81714	.59061	.80696	48
13	.54781	.83660	.56232	.82692	.57667	.81698	.59084	.80679	47
14	.54805	.83645	.56256	.82675	.57691	.81681	.59108	.80662	46
15	.54829	.83629	.56280	.82659	.57715	.81664	.59131	.80644	45
16	.54854	.83613	.56305	.82643	.57738	.81647	.59154	.80627	44
17	.54878	.83597	.56329	.82626	.57762	.81631	.59178	.80610	43
18	.54902	.83581	.56353	.82610	.57786	.81614	.59201	.80593	42
19	.54927	.83565	.56377	.82593	.57810	.81597	.59225	.80576	41
20	.54951	.83549	.56401	.82577	.57833	.81580	.59248	.80558	40
21	.54975	.83533	.56425	.82561	.57857	.81563	.59272	.80541	39
22	.54999	.83517	.56449	.82544	.57881	.81546	.59295	.80524	38
23	.55024	.83501	.56473	.82528	.57904	.81530	.59318	.80507	37
24	.55048	.83485	.56497	.82511	.57928	.81513	.59342	.80489	36
25	.55072	.83469	.56521	.82495	.57952	.81496	.59365	.80472	35
26	.55097	.83453	.56545	.82478	.57976	.81479	.59389	.80455	34
27	.55121	.83437	.56569	.82462	.57999	.81462	.59412	.80438	33
28	.55145	.83421	.56593	.82446	.58023	.81445	.59436	.80420	32
29	.55169	.83405	.56617	.82429	.58047	.81428	.59459	.80403	31
30	.55194	.83389	.56641	.82413	.58070	.81412	.59482	.80386	30
31	.55218	.83373	.56565	.82396	.58094	.81395	.59506	.80368	29
32	.55242	.83356	.56689	.82380	.58118	.81378	.59529	.80351	28
33	.55266	.83340	.56713	.82363	.58141	.81361	.59552	.80334	27
34	.55291	.83324	.56736	.82347	.58165	.81344	.59576	.80316	26
35	.55315	.83308	.56760	.82330	.58189	.81327	.59599	.80299	25
36	.55339	.83292	.56784	.82314	.58212	.81310	.59622	.80282	24
37	.55363	.83276	.56808	.82297	.58236	.81293	.59646	.80264	23
38	.55388	.83260	.56832	.82281	.58260	.81276	.59669	.80247	22
39	.55412	.83244	.56856	.82264	.58283	.81259	.59693	.80230	21
40	.55436	.83228	.56880	.82248	.58307	.81242	.59716	.80212	20
41	.55460	.83212	.56904	.82231	.58330	.81225	.59739	.80195	19
42	.55484	.83195	.56928	.82214	.58354	.81208	.59763	.80178	18
43	.55509	.83179	.56952	.82198	.58378	.81191	.59786	.80160	17
44	.55533	.83163	.56976	.82181	.58401	.81174	.59809	.80143	16
45	.55557	.83147	.57000	.82165	.58425	.81157	.59832	.80125	15
46	.55581	.83131	.57024	.82148	.58449	.81140	.59856	.80108	14
47	.55605	.83115	.57047	.82132	.58472	.81123	.59879	.80091	13
48	.55630	.83098	.57071	.82115	.58496	.81106	.59902	.80073	12
49	.55654	.83082	.57095	.82098	.58519	.81089	.59926	.80056	11
50	.55678	.83066	.57119	.82082	.58543	.81072	.59949	.80038	10
51	.55702	.83050	.57143	.82065	.58567	.81055	.59972	.80021	9
52	.55726	.83034	.57167	.82048	.58590	.81038	.59995	.80003	8
53	.55750	.83017	.57191	.82032	.58614	.81021	.60019	.79986	7
54	.55775	.83001	.57215	.82015	.58637	.81004	.60042	.79968	6
55	.55799	.82985	.57238	.81999	.58661	.80987	.60065	.79951	5
56	.55823	.82969	.57262	.81982	.58684	.80970	.60089	.79934	4
57	.55847	.82953	.57286	.81965	.58708	.80953	.60112	.79916	3
58	.55871	.82936	.57310	.81949	.58731	.80936	.60135	.79899	2
59	.55895	.82920	.57334	.81932	.58755	.80919	.60158	.79881	1
60	.55919	.82904	.57358	.81915	.58779	.80902	.60182	.79864	0
′	56° Cosine	56° Sine	55° Cosine	55° Sine	54° Cosine	54° Sine	53° Cosine	53° Sine	′

TABLE IX.—*Continued*

′	37° Sine	37° Cosine	38° Sine	38° Cosine	39° Sine	39° Cosine	40° Sine	40° Cosine	′
0	.60182	.79864	.61566	.78801	.62932	.77715	.64279	.76604	60
1	.60205	.79846	.61589	.78783	.62955	.77696	.64301	.76586	59
2	.60228	.79829	.61612	.78765	.62977	.77678	.64323	.76567	58
3	.60251	.79811	.61635	.78747	.63000	.77660	.64346	.76548	57
4	.60274	.79793	.61658	.78729	.63022	.77641	.64368	.76530	56
5	.60298	.79776	.61681	.78711	.63045	.77623	.64390	.76511	55
6	.60321	.79758	.61704	.78694	.63068	.77605	.64412	.76492	54
7	.60344	.79741	.61726	.78676	.63090	.77586	.64435	.76473	53
8	.60367	.79723	.61749	.78658	.63113	.77568	.64457	.76455	52
9	.60390	.79706	.61772	.78640	.63135	.77550	.64479	.76436	51
10	.60414	.79688	.61795	.78622	.63158	.77531	.64501	.76417	50
11	.60437	.79671	.61818	.78604	.63180	.77513	.64524	.76398	49
12	.60460	.79653	.61841	.78586	.63203	.77494	.64546	.76380	48
13	.60483	.79635	.61864	.78568	.63225	.77476	.64568	.76361	47
14	.60506	.79618	.61887	.78550	.63248	.77458	.64590	.76342	46
15	.60529	.79600	.61909	.78532	.63271	.77439	.64612	.76323	45
16	.60553	.79583	.61932	.78514	.63293	.77421	.64635	.76304	44
17	.60576	.79565	.61955	.78496	.63316	.77402	.64657	.76286	43
18	.60599	.79547	.61978	.78478	.63338	.77384	.64679	.76267	42
19	.60622	.79530	.62001	.78460	.63361	.77366	.64701	.76248	41
20	.60645	.79512	.62024	.78442	.63383	.77347	.64723	.76229	40
21	.60668	.79494	.62046	.78424	.63406	.77329	.64746	.76210	39
22	.60691	.79477	.62069	.78405	.63428	.77310	.64768	.76192	38
23	.60714	.79459	.62092	.78387	.63451	.77292	.64790	.76173	37
24	.60738	.79441	.62115	.78369	.63473	.77273	.64812	.76154	36
25	.60761	.79424	.62138	.78351	.63496	.77255	.64834	.76135	35
26	.60784	.79406	.62160	.78333	.63518	.77236	.64856	.76116	34
27	.60807	.79388	.62183	.78315	.63540	.77218	.64878	.76097	33
28	.60830	.79371	.62206	.78297	.63563	.77199	.64901	.76078	32
29	.60853	.79353	.62229	.78279	.63585	.77181	.64923	.76059	31
30	.60876	.79335	.62251	.78261	.63608	.77162	.64945	.76041	30
31	.60899	.79318	.62274	.78243	.63630	.77144	.64967	.76022	29
32	.60922	.79300	.62297	.78225	.63653	.77125	.64989	.76003	28
33	.60945	.79282	.62320	.78206	.63675	.77107	.65011	.75984	27
34	.60968	.79264	.62342	.78188	.63698	.77088	.65033	.75965	26
35	.60991	.79247	.62365	.78170	.63720	.77070	.65055	.75946	25
36	.61015	.79229	.62388	.78152	.63742	.77051	.65077	.75927	24
37	.61038	.79211	.62411	.78134	.63765	.77033	.65100	.75908	23
38	.61061	.79193	.62433	.78116	.63787	.77014	.65122	.75889	22
39	.61084	.79176	.62456	.78098	.63810	.76996	.65144	.75870	21
40	.61107	.79158	.62479	.78079	.63832	.76977	.65166	.75851	20
41	.61130	.79140	.62502	.78061	.63854	.76959	.65188	.75832	19
42	.61153	.79122	.62524	.78043	.63877	.76940	.65210	.75813	18
43	.61176	.79105	.62547	.78025	.63899	.76921	.65232	.75794	17
44	.61199	.79087	.62570	.78007	.63922	.76903	.65254	.75775	16
45	.61222	.79069	.62592	.77988	.63944	.76884	.65276	.75756	15
46	.61245	.79051	.62615	.77970	.63966	.76866	.65298	.75738	14
47	.61268	.79033	.62638	.77952	.63989	.76847	.65320	.75719	13
48	.61291	.79016	.62660	.77934	.64011	.76828	.65342	.75700	12
49	.61314	.78998	.62683	.77916	.64033	.76810	.65364	.75680	11
50	.61337	.78980	.62706	.77897	.64056	.76791	.65386	.75661	10
51	.61360	.78962	.62728	.77879	.64078	.76772	.65408	.75642	9
52	.61383	.78944	.62751	.77861	.64100	.76754	.65430	.75623	8
53	.61406	.78926	.62774	.77843	.64123	.76735	.65452	.75604	7
54	.61429	.78908	.62796	.77824	.64145	.76717	.65474	.75585	6
55	.61451	.78891	.62819	.77806	.64167	.76698	.65496	.75566	5
56	.61474	.78873	.62842	.77788	.64190	.76679	.65518	.75547	4
57	.61497	.78855	.62864	.77769	.64212	.76661	.65540	.75528	3
58	.61520	.78837	.62887	.77751	.64234	.76642	.65562	.75509	2
59	.61543	.78819	.62909	.77733	.64256	.76623	.65584	.75490	1
60	.61566	.78801	.62932	.77715	.64279	.76604	.65606	.75471	0
′	Cosine 52°	Sine 52°	Cosine 51°	Sine 51°	Cosine 50°	Sine 50°	Cosine 49°	Sine 49°	′

TABLE IX.—*Concluded*

′	41° Sine	41° Cosine	42° Sine	42° Cosine	43° Sine	43° Cosine	44° Sine	44° Cosine	′
0	.65606	.75471	.66913	.74314	.68200	.73135	.69466	.71934	60
1	.65628	.75452	.66935	.74295	.68221	.73116	.69487	.71914	59
2	.65650	.75433	.66956	.74276	.68242	.73096	.69508	.71894	58
3	.65672	.75414	.66978	.74256	.68264	.73076	.69529	.71873	57
4	.65694	.75395	.66999	.74237	.68285	.73056	.69549	.71853	56
5	.65716	.75375	.67021	.74217	.68306	.73036	.69570	.71833	55
6	.65738	.75356	.67043	.74198	.68327	.73016	.69591	.71813	54
7	.65759	.75337	.67064	.74178	.68349	.72996	.69612	.71792	53
8	.65781	.75318	.67086	.74159	.68370	.72976	.69633	.71772	52
9	.65803	.75299	.67107	.74139	.68391	.72957	.69654	.71752	51
10	.65825	.75280	.67129	.74120	.68412	.72937	.69675	.71732	50
11	.65847	.75261	.67151	.74100	.68434	.72917	.69696	.71711	49
12	.65869	.75241	.67172	.74080	.68455	.72897	.69717	.71691	48
13	.65891	.75222	.67194	.74061	.68476	.72877	.69737	.71671	47
14	.65913	.75203	.67215	.74041	.68497	.72857	.69758	.71650	46
15	.65935	.75184	.67237	.74022	.68518	.72837	.69779	.71630	45
16	.65956	.75165	.67258	.74002	.68539	.72817	.69800	.71610	44
17	.65978	.75146	.67280	.73983	.68561	.72797	.69821	.71590	43
18	.66000	.75126	.67301	.73963	.68582	.72777	.69842	.71569	42
19	.66022	.75107	.67323	.73944	.68603	.72757	.69862	.71549	41
20	.66044	.75088	.67344	.73924	.68624	.72737	.69883	.71529	40
21	.66066	.75069	.67366	.73904	.68645	.72717	.69904	.71508	39
22	.66088	.75050	.67387	.73885	.68666	.72697	.69925	.71488	38
23	.66109	.75030	.67409	.73865	.68688	.72677	.69946	.71468	37
24	.66131	.75011	.67430	.73846	.68709	.72657	.69966	.71447	36
25	.66153	.74992	.67452	.73826	.68730	.72637	.69987	.71427	35
26	.66175	.74973	.67473	.73806	.68751	.72617	.70008	.71407	34
27	.66197	.74953	.67495	.73787	.68772	.72597	.70029	.71386	33
28	.66218	.74934	.67516	.73767	.68793	.72577	.70049	.71366	32
29	.66240	.74915	.67538	.73747	.68814	.72557	.70070	.71345	31
30	.66262	.74896	.67559	.73728	.68835	.72537	.70091	.71325	30
31	.66284	.74876	.67580	.73708	.68857	.72517	.70112	.71305	29
32	.66306	.74857	.67602	.73688	.68878	.72497	.70132	.71284	28
33	.66327	.74838	.67623	.73669	.68899	.72477	.70153	.71264	27
34	.66349	.74818	.67645	.73649	.68920	.72457	.70174	.71243	26
35	.66371	.74799	.67666	.73629	.68941	.72437	.70195	.71223	25
36	.66393	.74780	.67688	.73610	.68962	.72417	.70215	.71203	24
37	.66414	.74760	.67709	.73590	.68983	.72397	.70236	.71182	23
38	.66436	.74741	.67730	.73570	.69004	.72377	.70257	.71162	22
39	.66458	.74722	.67752	.73551	.69025	.72357	.70277	.71141	21
40	.66480	.74703	.67773	.73531	.69046	.72337	.70298	.71121	20
41	.66501	.74683	.67795	.73511	.69067	.72317	.70319	.71100	19
42	.66523	.74664	.67816	.73491	.69088	.72297	.70339	.71080	18
43	.66545	.74644	.67837	.73472	.69109	.72277	.70360	.71059	17
44	.66566	.74625	.67859	.73452	.69130	.72257	.70381	.71039	16
45	.66588	.74606	.67880	.73432	.69151	.72236	.70401	.71019	15
46	.66610	.74586	.67901	.73413	.69172	.72216	.70422	.70998	14
47	.66632	.74567	.67923	.73393	.69193	.72196	.70443	.70978	13
48	.66653	.74548	.67944	.73373	.69214	.72176	.70463	.70957	12
49	.66675	.74528	.67965	.73353	.69235	.72156	.70484	.70937	11
50	.66697	.74509	.67987	.73333	.69256	.72136	.70505	.70916	10
51	.66718	.74489	.68008	.73314	.69277	.72116	.70525	.70896	9
52	.66740	.74470	.68029	.73294	.69298	.72095	.70546	.70875	8
53	.66762	.74451	.68051	.73274	.69319	.72075	.70567	.70855	7
54	.66783	.74431	.68072	.73254	.69340	.72055	.70587	.70834	6
55	.66805	.74412	.68093	.73234	.69361	.72035	.70608	.70813	5
56	.66827	.74392	.68115	.73215	.69382	.72015	.70628	.70793	4
57	.66848	.74373	.68136	.73195	.69403	.71995	.70649	.70772	3
58	.66870	.74353	.68157	.73175	.69424	.71974	.70670	.70752	2
59	.66891	.74334	.68179	.73155	.69445	.71954	.70690	.70731	1
60	.66913	.74314	.68200	.73135	.69466	.71934	.70711	.70711	0
′	48° Cosine	48° Sine	47° Cosine	47° Sine	46° Cosine	46° Sine	45° Cosine	45° Sine	′

Table X.—Natural Tangents and Cotangents

′	0°		1°		2°		3°		′
′	Tan.	Co-tan.	Tan.	Co-tan.	Tan.	Co-tan.	Tan.	Co-tan.	′
0	.00000	Infinite.	.01746	57.2900	.03492	28.6363	.05241	19.0811	60
1	.00029	3437.750	.01775	56.3506	.03521	28.3994	.05270	18.9755	59
2	.00058	1718.870	.01804	55.4415	.03550	28.1664	.05299	18.8711	58
3	.00087	1145.920	.01833	54.5613	.03579	27.9372	.05328	18.7678	57
4	.00116	859.436	.01862	53.7086	.03609	27.7117	.05357	18.6656	56
5	.00145	687.549	.01891	52.8821	.03638	27.4899	.05387	18.5645	55
6	.00175	572.957	.01920	52.0807	.03667	27.2715	.05416	18.4645	54
7	.00204	491.106	.01949	51.3032	.03696	27.0566	.05445	18.3655	53
8	.00233	429.718	.01978	50.5485	.03725	26.8450	.05474	18.2677	52
9	.00262	381.971	.02007	49.8157	.03754	26.6367	.05503	18.1708	51
10	.00291	343.774	.02036	49.1039	.03783	26.4316	.05533	18.0750	50
11	.00320	312.521	.02066	48.4121	.03812	26.2296	.05562	17.9802	49
12	.00349	286.478	.02095	47.7395	.03842	26.0307	.05591	17.8863	48
13	.00378	264.441	.02124	47.0853	.03871	25.8348	.05620	17.7934	47
14	.00407	245.552	.02153	46.4489	.03900	25.6418	.05649	17.7015	46
15	.00436	229.182	.02182	45.8294	.03929	25.4517	.05678	17.6106	45
16	.00465	214.858	.02211	45.2261	.03958	25.2644	.05708	17.5205	44
17	.00495	202.219	.02240	44.6386	.03987	25.0798	.05737	17.4314	43
18	.00524	190.984	.02269	44.0661	.04016	24.8978	.05766	17.3432	42
19	.00553	180.932	.02298	43.5081	.04046	24.7185	.05795	17.2558	41
20	.00582	171.885	.02328	42.9641	.04075	24.5418	.05824	17.1693	40
21	.00611	163.700	.02357	42.4335	.04104	24.3675	.05854	17.0837	39
22	.00640	156.259	.02386	41.9158	.04133	24.1957	.05883	16.9990	38
23	.00669	149.465	.02415	41.4106	.04162	24.0263	.05912	16.9150	37
24	.00698	143.237	.02444	40.9174	.04191	23.8593	.05941	16.8319	36
25	.00727	137.507	.02473	40.4358	.04220	23.6945	.05970	16.7496	35
26	.00756	132.219	.02502	39.9655	.04250	23.5321	.05999	16.6681	34
27	.00785	127.321	.02531	39.5059	.04279	23.3718	.06029	16.5874	33
28	.00814	122.774	.02560	39.0568	.04308	23.2137	.06058	16.5075	32
29	.00844	118.540	.02589	38.6177	.04337	23.0577	.06087	16.4283	31
30	.00873	114.589	.02619	38.1885	.04366	22.9038	.06116	16.3499	30
31	.00902	110.892	.02648	37.7686	.04395	22.7519	.06145	16.2722	29
32	.00931	107.426	.02677	37.3579	.04424	22.6020	.06175	16.1952	28
33	.00960	104.171	.02706	36.9560	.04454	22.4541	.06204	16.1190	27
34	.00989	101.107	.02735	36.5627	.04483	22.3081	.06233	16.0435	26
35	.01018	98.2179	.02764	36.1776	.04512	22.1640	.06262	15.9687	25
36	.01047	95.4895	.02793	35.8006	.04541	22.0217	.06291	15.8945	24
37	.01076	92.9085	.02822	35.4313	.04570	21.8813	.06321	15.8211	23
38	.01105	90.4633	.02851	35.0695	.04599	21.7426	.06350	15.7483	22
39	.01135	88.1436	.02881	34.7151	.04628	21.6056	.06379	15.6762	21
40	.01164	85.9398	.02910	34.3678	.04658	21.4704	.06408	15.6048	20
41	.01193	83.8435	.02939	34.0273	.04687	21.3369	.06437	15.5340	19
42	.01222	81.8470	.02968	33.6935	.04716	21.2049	.06467	15.4638	18
43	.01251	79.9434	.02997	33.3662	.04745	21.0747	.06496	15.3943	17
44	.01280	78.1263	.03026	33.0452	.04774	20.9460	.06525	15.3254	16
45	.01309	76.3900	.03055	32.7303	.04803	20.8188	.06554	15.2571	15
46	.01338	74.7292	.03084	32.4213	.04832	20.6932	.06584	15.1893	14
47	.01367	73.1390	.03114	32.1181	.04862	20.5691	.06613	15.1222	13
48	.01396	71.6151	.03143	31.8205	.04891	20.4465	.06642	15.0557	12
49	.01425	70.1533	.03172	31.5284	.04920	20.3253	.06671	14.9898	11
50	.01455	68.7501	.03201	31.2416	.04949	20.2056	.06700	14.9244	10
51	.01484	67.4019	.03230	30.9599	.04978	20.0872	.06730	14.8596	9
52	.01513	66.1055	.03259	30.6833	.05007	19.9702	.06759	14.7954	8
53	.01542	64.8580	.03288	30.4116	.05037	19.8546	.06788	14.7317	7
54	.01571	63.6567	.03317	30.1446	.05066	19.7403	.06817	14.6685	6
55	.01600	62.4992	.03346	29.8823	.05095	19.6273	.06847	14.6059	5
56	.01629	61.3829	.03376	29.6245	.05124	19.5156	.06876	14.5438	4
57	.01658	60.3058	.03405	29.3711	.05153	19.4051	.06905	14.4823	3
58	.01687	59.2659	.03434	29.1220	.05182	19.2959	.06934	14.4212	2
59	.01716	58.2612	.03463	28.8771	.05212	19.1879	.06963	14.3607	1
60	.01746	57.2900	.03492	28.6363	.05241	19.0811	.06993	14.3007	0
′	Co-tan.	Tan.	Co-tan.	Tan.	Co-tan.	Tan.	Co-tan.	Tan.	′
	89°		88°		87°		86°		

TABLE X.—*Continued*

′	4°		5°		6°		7°		′
	TAN.	CO-TAN.	TAN.	CO-TAN.	TAN.	CO-TAN.	TAN.	CO-TAN.	
0	.06993	14.3007	.08749	11.4301	.10510	9.51436	.12278	8.14435	60
1	.07022	14.2411	.08778	11.3919	.10540	9.48781	.12308	8.12481	59
2	.07051	14.1821	.08807	11.3540	.10569	9.46141	.12338	8.10536	58
3	.07080	14.1235	.08837	11.3163	.10599	9.43515	.12367	8.08600	57
4	.07110	14.0655	.08866	11.2789	.10628	9.40904	.12397	8.06674	56
5	.07139	14.0079	.08895	11.2417	.10657	9.38307	.12426	8.04756	55
6	.07168	13.9507	.08925	11.2048	.10687	9.35724	.12456	8.02848	54
7	.07197	13.8940	.08954	11.1681	.10716	9.33154	.12485	8.00948	53
8	.07227	13.8378	.08983	11.1316	.10746	9.30599	.12515	7.99058	52
9	.07256	13.7821	.09013	11.0954	.10775	9.28058	.12544	7.97176	51
10	.07285	13.7267	.09042	11.0594	.10805	9.25530	.12574	7.95302	50
11	.07314	13.6719	.09071	11.0237	.10834	9.23016	.12603	7.93438	49
12	.07344	13.6174	.09101	10.9882	.10863	9.20516	.12633	7.91582	48
13	.07373	13.5634	.09130	10.9529	.10893	9.18028	.12662	7.89734	47
14	.07402	13.5098	.09159	10.9178	.10922	9.15554	.12692	7.87895	46
15	.07431	13.4566	.09189	10.8829	.10952	9.13093	.12722	7.86064	45
16	.07461	13.4039	.09218	10.8483	.10981	9.10646	.12751	7.84242	44
17	.07490	13.3515	.09247	10.8139	.11011	9.08211	.12781	7.82428	43
18	.07519	13.2996	.09277	10.7797	.11040	9.05789	.12810	7.80622	42
19	.07548	13.2480	.09306	10.7457	.11070	9.03379	.12840	7.78825	41
20	.07578	13.1969	.09335	10.7119	.11099	9.00983	.12869	7.77035	40
21	.07607	13.1461	.09365	10.6783	.11128	8.98598	.12899	7.75254	39
22	.07636	13.0958	.09394	10.6450	.11158	8.96227	.12929	7.73480	38
23	.07665	13.0458	.09423	10.6118	.11187	8.93867	.12958	7.71715	37
24	.07695	12.9962	.09453	10.5789	.11217	8.91520	.12988	7.69957	36
25	.07724	12.9469	.09482	10.5462	.11246	8.89185	.13017	7.68208	35
26	.07753	12.8981	.09511	10.5136	.11276	8.86862	.13047	7.66466	34
27	.07782	12.8496	.09541	10.4813	.11305	8.84551	.13076	7.64732	33
28	.07812	12.8014	.09570	10.4491	.11335	8.82252	.13106	7.63005	32
29	.07841	12.7536	.09600	10.4172	.11364	8.79964	.13136	7.61287	31
30	.07870	12.7062	.09629	10.3854	.11394	8.77689	.13165	7.59575	30
31	.07899	12.6591	.09658	10.3538	.11423	8.75425	.13195	7.57872	29
32	.07929	12.6124	.09688	10.3224	.11452	8.73172	.13224	7.56176	28
33	.07958	12.5660	.09717	10.2913	.11482	8.70931	.13254	7.54487	27
34	.07987	12.5199	.09746	10.2602	.11511	8.68701	.13284	7.52806	26
35	.08017	12.4742	.09776	10.2204	.11541	8.66482	.13313	7.51132	25
36	.08046	12.4288	.09805	10.1988	.11570	8.64275	.13343	7.49465	24
37	.08075	12.3838	.09834	10.1683	.11600	8.62078	.13372	7.47806	23
38	.08104	12.3390	.09864	10.1381	.11629	8.59893	.13402	7.46154	22
39	.08134	12.2946	.09893	10.1080	.11659	8.57718	.13432	7.44509	21
40	.08163	12.2505	.09923	10.0780	.11688	8.55555	.13461	7.42871	20
41	.08192	12.2067	.09952	10.0483	.11718	8.53402	.13491	7.41240	19
42	.08221	12.1632	.09981	10.0187	.11747	8.51259	.13521	7.39616	18
43	.08251	12.1201	.10011	9.98931	.11777	8.49128	.13550	7.37999	17
44	.08280	12.0772	.10040	9.96007	.11806	8.47007	.13580	7.36389	16
45	.08309	12.0346	.10069	9.93101	.11836	8.44896	.13609	7.34786	15
46	.08339	11.9923	.10099	9.90211	.11865	8.42795	.13639	7.33190	14
47	.08368	11.9504	.10128	9.87338	.11895	8.40705	.13669	7.31600	13
48	.08397	11.9087	.10158	9.84482	.11924	8.38625	.13698	7.30018	12
49	.08427	11.8673	.10187	9.81641	.11954	8.36555	.13728	7.28442	11
50	.08456	11.8262	.10216	9.78817	.11983	8.34496	.13758	7.26873	10
51	.08485	11.7853	.10246	9.76009	.12013	8.32446	.13787	7.25310	9
52	.08514	11.7448	.10275	9.73217	.12042	8.30406	.13817	7.23754	8
53	.08544	11.7045	.10305	9.70441	.12072	8.28376	.13846	7.22204	7
54	.08573	11.6645	.10334	9.67680	.12101	8.26355	.13876	7.20661	6
55	.08602	11.6248	.10363	9.64935	.12131	8.24345	.13906	7.19125	5
56	.08632	11.5853	.10393	9.62205	.12160	8.22344	.13935	7.17594	4
57	.08661	11.5461	.10422	9.59490	.12190	8.20352	.13965	7.16071	3
58	.08690	11.5072	.10452	9.56791	.12219	8.18370	.13995	7.14553	2
59	.08720	11.4685	.10481	9.54106	.12249	8.16398	.14024	7.13042	1
60	.08749	11.4301	.10510	9.51436	.12278	8.14435	.14054	7.11537	0
′	CO-TAN.	TAN.	CO-TAN.	TAN.	CO-TAN.	TAN.	CO-TAN.	TAN.	′
	85°		84°		83°		82°		

TABLE X.—*Continued*

′	8°		9°		10°		11°		′
	Tan.	Co-tan.	Tan.	Co-tan.	Tan.	Co-tan.	Tan.	Co-tan.	
0	.14054	7.11537	.15838	6.31375	.17633	5.67128	.19438	5.14455	60
1	.14084	7.10038	.15868	6.30189	.17663	5.66165	.19468	5.13658	59
2	.14113	7.08546	.15898	6.29007	.17693	5.65205	.19498	5.12862	58
3	.14143	7.07059	.15928	6.27829	.17723	5.64248	.19529	5.12069	57
4	.14173	7.05579	.15958	6.26655	.17753	5.63295	.19559	5.11279	56
5	.14202	7.04105	.15988	6.25486	.17783	5.62344	.19589	5.10490	55
6	.14232	7.02637	.16017	6.24321	.17813	5.61397	.19619	5.09704	54
7	.14262	7.01174	.16047	6.23160	.17843	5.60452	.19649	5.08921	53
8	.14291	6.99718	.16077	6.22003	.17873	5.59511	.19680	5.08139	52
9	.14321	6.98268	.16107	6.20851	.17903	5.58573	.19710	5.07360	51
10	.14351	6.96823	.16137	6.19703	.17933	5.57638	.19740	5.06584	50
11	.14381	6.95385	.16167	6.18559	.17963	5.56706	.19770	5.05809	49
12	.14410	6.93952	.16196	6.17419	.17993	5.55777	.19801	5.05037	48
13	.14440	6.92525	.16226	6.16283	.18023	5.54851	.19831	5.04267	47
14	.14470	6.91104	.16256	6.15151	.18053	5.53927	.19861	5.03499	46
15	.14499	6.89688	.16286	6.14023	.18083	5.53007	.19891	5.02734	45
16	.14529	6.88278	.16316	6.12899	.18113	5.52090	.19921	5.01971	44
17	.14559	6.86874	.16346	6.11779	.18143	5.51176	.19952	5.01210	43
18	.14588	6.85475	.16376	6.10664	.18173	5.50264	.19982	5.00451	42
19	.14618	6.84082	.16405	6.09552	.18203	5.49356	.20012	4.99695	41
20	.14648	6.82694	.16435	6.08444	.18233	5.48451	.20042	4.98940	40
21	.14678	6.81312	.16465	6.07340	.18263	5.47548	.20073	4.98188	39
22	.14707	6.79936	.16495	6.06240	.18293	5.46648	.20103	4.97438	38
23	.14737	6.78564	.16525	6.05143	.18323	5.45751	.20133	4.96690	37
24	.14767	6.77199	.16555	6.04051	.18353	5.44857	.20164	4.95945	36
25	.14796	6.75838	.16585	6.02962	.18383	5.43966	.20194	4.95201	35
26	.14826	6.74483	.16615	6.01878	.18414	5.43077	.20224	4.94460	34
27	.14856	6.73133	.16645	6.00797	.18444	5.42192	.20254	4.93721	33
28	.14886	6.71789	.16674	5.99720	.18474	5.41309	.20285	4.92984	32
29	.14915	6.70450	.16704	5.98646	.18504	5.40429	.20315	4.92249	31
30	.14945	6.69116	.16734	5.97576	.18534	5.39552	.20345	4.91516	30
31	.14975	6.67787	.16764	5.96510	.18564	5.38677	.20376	4.90785	29
32	.15005	6.66463	.16794	5.95448	.18594	5.37805	.20406	4.90056	28
33	.15034	6.65144	.16824	5.94390	.18624	5.36936	.20436	4.89330	27
34	.15064	6.63831	.16854	5.93335	.18654	5.36070	.20466	4.83605	26
35	.15094	6.62523	.16884	5.92283	.18684	5.35206	.20497	4.87882	25
36	.15124	6.61219	.16914	5.91235	.18714	5.34345	.20527	4.87162	24
37	.15153	6.59921	.16944	5.90191	.18745	5.33487	.20557	4.86444	23
38	.15183	6.58627	.16974	5.89151	.18775	5.32631	.20588	4.85727	22
39	.15213	6.57339	.17004	5.88114	.18805	5.31778	.20618	4.85013	21
40	.15243	6.56055	.17033	5.87080	.18835	5.30928	.20648	4.84300	20
41	.15272	6.54777	.17063	5.86051	.18865	5.30080	.20679	4.83590	19
42	.15302	6.53503	.17093	5.85024	.18895	5.29235	.20709	4.82882	18
43	.15332	6.52234	.17123	5.84001	.18925	5.28393	.20739	4.82175	17
44	.15362	6.50970	.17153	5.82982	.18955	5.27553	.20770	4.81471	16
45	.15391	6.49710	.17183	5.81966	.18986	5.26715	.20800	4.80769	15
46	.15421	6.48456	.17213	5.80953	.19016	5.25880	.20830	4.80068	14
47	.15451	6.47206	.17243	5.79944	.19046	5.25048	.20861	4.79370	13
48	.15481	6.45961	.17273	5.78938	.19076	5.24218	.20891	4.78673	12
49	.15511	6.44720	.17303	5.77936	.19106	5.23391	.20921	4.77978	11
50	.15540	6.43484	.17333	5.76937	.19136	5.22566	.20952	4.77286	10
51	.15570	6.42253	.17363	5.75941	.19166	5.21744	.20982	4.76595	9
52	.15600	6.41026	.17393	5.74949	.19197	5.20925	.21013	4.75906	8
53	.15630	6.39804	.17423	5.73960	.19227	5.20107	.21043	4.75219	7
54	.15660	6.38587	.17453	5.72974	.19257	5.19293	.21073	4.74534	6
55	.15689	6.37374	.17483	5.71992	.19287	5.18480	.21104	4.73851	5
56	.15719	6.36165	.17513	5.71013	.19317	5.17671	.21134	4.73170	4
57	.15749	6.34961	.17543	5.70037	.19347	5.16863	.21164	4.72490	3
58	.15779	6.33761	.17573	5.69064	.19378	5.16058	.21195	4.71813	2
59	.15809	6.32566	.17603	5.68094	.19408	5.15256	.21225	4.71137	1
60	.15838	6.31375	.17633	5.67128	.19438	5.14455	.21256	4.70463	0
′	Co-tan.	Tan.	Co-tan.	Tan.	Co-tan.	Tan.	Co-tan.	Tan.	′
	81°		80°		79°		78°		

TABLE X.—*Continued*

′	12° Tan.	12° Co-tan.	13° Tan.	13° Co-tan.	14° Tan.	14° Co-tan.	15° Tan.	15° Co-tan.	′
0	.21256	4.70463	.23087	4.33148	.24933	4.01078	.26795	3.73205	60
1	.21286	4.69791	.23117	4.32573	.24964	4.00582	.26826	3.72771	59
2	.21316	4.69121	.23148	4.32001	.24995	4.00086	.26857	3.72338	58
3	.21347	4.68452	.23179	4.31430	.25026	3.99592	.26888	3.71907	57
4	.21377	4.67786	.23209	4.30860	.25056	3.99099	.26920	3.71476	56
5	.21408	4.67121	.23240	4.30291	.25087	3.98607	.26951	3.71046	55
6	.21438	4.66458	.23271	4.29724	.25118	3.98117	.26982	3.70616	54
7	.21469	4.65797	.23301	4.29159	.25149	3.97627	.27013	3.70188	53
8	.21499	4.65138	.23332	4.28595	.25180	3.97139	.27044	3.69761	52
9	.21529	4.64480	.23363	4.28032	.25211	3.96651	.27076	3.69335	51
10	.21560	4.63825	.23393	4.27471	.25242	3.96165	.27107	3.68909	50
11	.21590	4.63171	.23424	4.26911	.25273	3.95680	.27138	3.68485	49
12	.21621	4.62518	.23455	4.26352	.25304	3.95196	.27169	3.68061	48
13	.21651	4.61868	.23485	4.25795	.25335	3.94713	.27201	3.67638	47
14	.21682	4.61219	.23516	4.25239	.25366	3.94232	.27232	3.67217	46
15	.21712	4.60572	.23547	4.24685	.25397	3.93751	.27263	3.66796	45
16	.21743	4.59927	.23578	4.24132	.25428	3.93271	.27294	3.66376	44
17	.21773	4.59283	.23608	4.23580	.25459	3.92793	.27326	3.65957	43
18	.21804	4.58641	.23639	4.23030	.25490	3.92316	.27357	3.65538	42
19	.21834	4.58001	.23670	4.22481	.25521	3.91839	.27388	3.65121	41
20	.21864	4.57363	.23700	4.21933	.25552	3.91364	.27419	3.64705	40
21	.21895	4.56726	.23731	4.21387	.25583	3.90890	.27451	3.64289	39
22	.21925	4.56091	.23762	4.20842	.25614	3.90417	.27482	3.63874	38
23	.21956	4.55458	.23793	4.20298	.25645	3.89945	.27513	3.63461	37
24	.21986	4.54826	.23823	4.19756	.25676	3.89474	.27545	3.63048	36
25	.22017	4.54196	.23854	4.19215	.25707	3.89004	.27576	3.62636	35
26	.22047	4.53568	.23885	4.18675	.25738	3.88536	.27607	3.62224	34
27	.22078	4.52941	.23916	4.18137	.25769	3.88068	.27638	3.61814	33
28	.22108	4.52316	.23946	4.17600	.25800	3.87601	.27670	3.61405	32
29	.22139	4.51693	.23977	4.17064	.25831	3.87136	.27701	3.60996	31
30	.22169	4.51071	.24008	4.16530	.25862	3.86671	.27732	3.60588	30
31	.22200	4.50451	.24039	4.15997	.25893	3.86208	.27764	3.60181	29
32	.22231	4.49832	.24069	4.15465	.25924	3.85745	.27795	3.59775	28
33	.22261	4.49215	.24100	4.14934	.25955	3.85284	.27826	3.59370	27
34	.22292	4.48600	.24131	4.14405	.25986	3.84824	.27858	3.58966	26
35	.22322	4.47986	.24162	4.13877	.26017	3.84364	.27889	3.58562	25
36	.22353	4.47374	.24193	4.13350	.26048	3.83906	.27920	3.58160	24
37	.22383	4.46764	.24223	4.12825	.26079	3.83449	.27952	3.57758	23
38	.22414	4.46155	.24254	4.12301	.26110	3.82992	.27983	3.57357	22
39	.22444	4.45548	.24285	4.11778	.26141	3.82537	.28015	3.56957	21
40	.22475	4.44942	.24316	4.11256	.26172	3.82083	.28046	3.56557	20
41	.22505	4.44338	.24347	4.10736	.26203	3.81630	.28077	3.56159	19
42	.22536	4.43735	.24377	4.10216	.26235	3.81177	.28109	3.55761	18
43	.22567	4.43134	.24408	4.09699	.26266	3.80726	.28140	3.55364	17
44	.22597	4.42534	.24439	4.09182	.26297	3.80276	.28172	3.54968	16
45	.22628	4.41936	.24470	4.08666	.26328	3.79827	.28203	3.54573	15
46	.22658	4.41340	.24501	4.08152	.26359	3.79378	.28234	3.54179	14
47	.22689	4.40745	.24532	4.07639	.26390	3.78931	.28266	3.53785	13
48	.22719	4.40152	.24562	4.07127	.26421	3.78485	.28297	3.53393	12
49	.22750	4.39560	.24593	4.06616	.26452	3.78040	.28329	3.53001	11
50	.22781	4.38969	.24624	4.06107	.26483	3.77595	.28360	3.52609	10
51	.22811	4.38381	.24655	4.05599	.26515	3.77152	.28391	3.52219	9
52	.22842	4.37793	.24686	4.05092	.26546	3.76709	.28423	3.51829	8
53	.22872	4.37207	.24717	4.04586	.26577	3.76268	.28454	3.51441	7
54	.22903	4.36623	.24747	4.04081	.26608	3.75828	.28486	3.51053	6
55	.22934	4.36040	.24778	4.03578	.26639	3.75388	.28517	3.50666	5
56	.22964	4.35459	.24809	4.03075	.26670	3.74950	.28549	3.50279	4
57	.22995	4.34879	.24840	4.02574	.26701	3.74512	.28580	3.49894	3
58	.23026	4.34300	.24871	4.02074	.26733	3.74075	.28612	3.49509	2
59	.23056	4.33723	.24902	4.01576	.26764	3.73640	.28643	3.49125	1
60	.23087	4.33148	.24933	4.01078	.26795	3.73205	.28675	3.48741	0
′	Co-tan.	Tan.	Co-tan.	Tan.	Co-tan.	Tan.	Co-tan.	Tan.	′
	77°		76°		75°		74°		

TABLE X.—*Continued*

′	16°		17°		18°		19°		′
	TAN.	CO-TAN.	TAN.	CO-TAN.	TAN.	CO-TAN.	TAN.	CO-TAN.	
0	.28675	3.48741	.30573	3.27085	.32492	3.07768	.34433	2.90421	60
1	.28706	3.48359	.30605	3.26745	.32524	3.07464	.34465	2.90147	59
2	.28738	3.47977	.30637	3.26406	.32556	3.07160	.34498	2.89873	58
3	.28769	3.47596	.30669	3.26067	.32588	3.06857	.34530	2.89600	57
4	.28800	3.47216	.30700	3.25729	.32621	3.06554	.34563	2.89327	56
5	.28832	3.46837	.30732	3.25392	.32653	3.06252	.34596	2.89055	55
6	.28864	3.46458	.30764	3.25055	.32685	3.05950	.34628	2.88783	54
7	.28895	3.46080	.30796	3.24719	.32717	3.05649	.34661	2.88511	53
8	.28927	3.45703	.30828	3.24383	.32749	3.05349	.34693	2.88240	52
9	.28958	3.45327	.30860	3.24049	.32782	3.05049	.34726	2.87970	51
10	.28990	3.44951	.30891	3.23714	.32814	3.04749	.34758	2.87700	50
11	.29021	3.44576	.30923	3.23381	.32846	3.04450	.34791	2.87430	49
12	.29053	3.44202	.30955	3.23048	.32878	3.04152	.34824	2.87161	48
13	.29084	3.43829	.30987	3.22715	.32911	3.03854	.34856	2.86892	47
14	.29116	3.43456	.31019	3.22384	.32943	3.03556	.34889	2.86624	46
15	.29147	3.43084	.31051	3.22053	.32975	3.03260	.34922	2.86356	45
16	.29179	3.42713	.31083	3.21722	.33007	3.02963	.34954	2.86089	44
17	.29210	3.42343	.31115	3.21392	.33040	3.02667	.34987	2.85822	43
18	.29242	3.41973	.31147	3.21063	.33072	3.02372	.35019	2.85555	42
19	.29274	3.41604	.31178	3.20734	.33104	3.02077	.35052	2.85289	41
20	.29305	3.41236	.31210	3.20406	.33136	3.01783	.35085	2.85023	40
21	.29337	3.40869	.31242	3.20079	.33160	3.01489	.35117	2.84758	39
22	.29368	3.40502	.31274	3.19752	.33201	3.01196	.35150	2.84494	38
23	.29400	3.40136	.31306	3.19426	.33233	3.00903	.35183	2.84229	37
24	.29432	3.39771	.31338	3.19100	.33266	3.00611	.35216	2.83965	36
25	.29463	3.39406	.31370	3.18775	.33298	3.00319	.35248	2.83702	35
26	.29495	3.39042	.31402	3.18451	.33330	3.00028	.35281	2.83439	34
27	.29526	3.38679	.31434	3.18127	.33363	2.99738	.35314	2.83176	33
28	.29558	3.38317	.31466	3.17804	.33395	2.99447	.35346	2.82914	32
29	.29590	3.37955	.31498	3.17481	.33427	2.99158	.35379	2.82653	31
30	.29621	3.37594	.31530	3.17159	.33460	2.98868	.35412	2.82391	30
31	.29653	3.37234	.31562	3.16838	.33492	2.98580	.35445	2.82130	29
32	.29685	3.36875	.31594	3.16517	.33524	2.98292	.35477	2.81870	28
33	.29716	3.36516	.31626	3.16197	.33557	2.98004	.35510	2.81610	27
34	.29748	3.36158	.31658	3.15877	.33589	2.97717	.35543	2.81350	26
35	.29780	3.35800	.31690	3.15558	.33621	2.97430	.35576	2.81091	25
36	.29811	3.35443	.31722	3.15240	.33654	2.97144	.35608	2.80833	24
37	.29843	3.35087	.31754	3.14922	.33686	2.96858	.35641	2.80574	23
38	.29875	3.34732	.31786	3.14605	.33718	2.96573	.35674	2.80316	22
39	.29906	3.34377	.31818	3.14288	.33751	2.96288	.35707	2.80059	21
40	.29938	3.34023	.31850	3.13972	.33783	2.96004	.35740	2.79802	20
41	.29970	3.33670	.31882	3.13656	.33816	2.95721	.35772	2.79545	19
42	.30001	3.33317	.31914	3.13341	.33848	2.95437	.35805	2.79289	18
43	.30033	3.32965	.31946	3.13027	.33881	2.95155	.35838	2.79033	17
44	.30065	3.32614	.31978	3.12713	.33913	2.94872	.35871	2.78778	16
45	.30097	3.32264	.32010	3.12400	.33945	2.94590	.35904	2.78523	15
46	.30128	3.31914	.32042	3.12087	.33978	2.94309	.35937	2.78269	14
47	.30160	3.31565	.32074	3.11775	.34010	2.94028	.35969	2.78014	13
48	.30192	3.31216	.32106	3.11464	.34043	2.93748	.36002	2.77761	12
49	.30224	3.30868	.32139	3.11153	.34075	2.93468	.36035	2.77507	11
50	.30255	3.30521	.32171	3.10842	.34108	2.93189	.36068	2.77254	10
51	.30287	3.30174	.32203	3.10532	.34140	2.92910	.36101	2.77002	9
52	.30319	3.29829	.32235	3.10223	.34173	2.92632	.36134	2.76750	8
53	.30351	3.29483	.32267	3.09914	.34205	2.92354	.36167	2.76498	7
54	.30382	3.29139	.32299	3.09606	.34238	2.92076	.36199	2.76247	6
55	.30414	3.28795	.32331	3.09298	.34270	2.91799	.36232	2.75996	5
56	.30446	3.28452	.32363	3.08991	.34303	2.91523	.36265	2.75746	4
57	.30478	3.28109	.32396	3.08685	.34335	2.91246	.36298	2.75496	3
58	.30509	3.27767	.32428	3.08379	.34368	2.90971	.36331	2.75246	2
59	.30541	3.27426	.32460	3.08073	.34400	2.90696	.36364	2.74997	1
60	.30573	3.27085	.32492	3.07768	.34433	2.90421	.36397	2.74748	0
′	CO-TAN.	TAN.	CO-TAN.	TAN.	CO-TAN.	TAN.	CO-TAN.	TAN.	′
	73°		72°		71°		70°		

TABLE X.—*Continued*

′	20° Tan.	20° Co-tan.	21° Tan.	21° Co-tan.	22° Tan.	22° Co-tan.	23° Tan.	23° Co-tan.	′
0	.36397	2.74748	.38386	2.60509	.40403	2.47509	.42447	2.35585	60
1	.36430	2.74499	.38420	2.60283	.40436	2.47302	.42482	2.35395	59
2	.36463	2.74251	.38453	2.60057	.40470	2.47095	.42516	2.35205	58
3	.36496	2.74004	.38487	2.59831	.40504	2.46888	.42551	2.35015	57
4	.36529	2.73756	.38520	2.59606	.40538	2.46682	.42585	2.34825	56
5	.36562	2.73509	.38553	2.59381	.40572	2.46476	.42619	2.34636	55
6	.36595	2.73263	.38587	2.59156	.40606	2.46270	.42654	2.34447	54
7	.36628	2.73017	.38620	2.58932	.40640	2.46065	.42688	2.34258	53
8	.36661	2.72771	.38654	2.58708	.40674	2.45860	.42722	2.34069	52
9	.36694	2.72526	.38687	2.58484	.40707	2.45655	.42757	2.33881	51
10	.36727	2.72281	.38721	2.58261	.40741	2.45451	.42791	2.33693	50
11	.36760	2.72036	.38754	2.58038	.40775	2.45246	.42826	2.33505	49
12	.36793	2.71792	.38787	2.57815	.40809	2.45043	.42860	2.33317	48
13	.36826	2.71548	.38821	2.57593	.40843	2.44839	.42894	2.33130	47
14	.36859	2.71305	.38854	2.57371	.40877	2.44636	.42929	2.32943	46
15	.36892	2.71062	.38888	2.57150	.40911	2.44433	.42963	2.32756	45
16	.36925	2.70819	.38921	2.56928	.40945	2.44230	.42998	2.32570	44
17	.36958	2.70577	.38955	2.56707	.40979	2.44027	.43032	2.32383	43
18	.36991	2.70335	.38988	2.56487	.41013	2.43825	.43067	2.32197	42
19	.37024	2.70094	.39022	2.56266	.41047	2.43623	.43101	2.32012	41
20	.37057	2.69853	.39055	2.56046	.41081	2.43422	.43136	2.31826	40
21	.37090	2.69612	.39089	2.55827	.41115	2.43220	.43170	2.31641	39
22	.37124	2.69371	.39122	2.55608	.41149	2.43019	.43205	2.31456	38
23	.37157	2.69131	.39156	2.55389	.41183	2.42819	.43239	2.31271	37
24	.37190	2.68892	.39190	2.55170	.41217	2.42618	.43274	2.31086	36
25	.37223	2.68653	.39223	2.54952	.41251	2.42418	.43308	2.30902	35
26	.37256	2.68414	.39257	2.54734	.41285	2.42218	.43343	2.30718	34
27	.37289	2.68175	.39290	2.54516	.41319	2.42019	.43378	2.30534	33
28	.37322	2.67937	.39324	2.54299	.41353	2.41819	.43412	2.30351	32
29	.37355	2.67700	.39357	2.54082	.41387	2.41620	.43447	2.30167	31
30	.37388	2.67462	.39391	2.53865	.41421	2.41421	.43481	2.29984	30
31	.37422	2.67225	.39425	2.53648	.41455	2.41223	.43516	2.29801	29
32	.37455	2.66989	.39458	2.53432	.41490	2.41025	.43550	2.29619	28
33	.37488	2.66752	.39492	2.53217	.41524	2.40827	.43585	2.29437	27
34	.37521	2.66516	.39526	2.53001	.41558	2.40629	.43620	2.29254	26
35	.37554	2.66281	.39559	2.52786	.41592	2.40432	.43654	2.29073	25
36	.37588	2.66046	.39593	2.52571	.41626	2.40235	.43689	2.28891	24
37	.37621	2.65811	.39626	2.52357	.41660	2.40038	.43724	2.28710	23
38	.37654	2.65576	.39660	2.52142	.41694	2.39841	.43758	2.28528	22
39	.37687	2.65342	.39694	2.51929	.41728	2.39645	.43793	2.28348	21
40	.37720	2.65109	.39727	2.51715	.41763	2.39449	.43828	2.28167	20
41	.37754	2.64875	.39761	2.51502	.41797	2.39253	.43862	2.27987	19
42	.37787	2.64642	.39795	2.51289	.41831	2.39058	.43897	2.27806	18
43	.37820	2.64410	.39829	2.51076	.41865	2.38862	.43932	2.27626	17
44	.37853	2.64177	.39862	2.50864	.41899	2.38668	.43966	2.27447	16
45	.37887	2.63945	.39896	2.50652	.41933	2.38473	.44001	2.27267	15
46	.37920	2.63714	.39930	2.50440	.41968	2.38279	.44036	2.27088	14
47	.37953	2.63483	.39963	2.50229	.42002	2.38084	.44071	2.26909	13
48	.37986	2.63252	.39997	2.50018	.42036	2.37891	.44105	2.26730	12
49	.38020	2.63021	.40031	2.49807	.42070	2.37697	.44140	2.26552	11
50	.38053	2.62791	.40065	2.49597	.42105	2.37504	.44175	2.26374	10
51	.38086	2.62561	.40098	2.49386	.42139	2.37311	.44210	2.26196	9
52	.38120	2.62332	.40132	2.49177	.42173	2.37118	.44244	2.26018	8
53	.38153	2.62103	.40166	2.48967	.42207	2.36925	.44279	2.25840	7
54	.38186	2.61874	.40200	2.48758	.42242	2.36733	.44314	2.25663	6
55	.38220	2.61646	.40234	2.48549	.42276	2.36541	.44349	2.25486	5
56	.38253	2.61418	.40267	2.48340	.42310	2.36349	.44384	2.25309	4
57	.38286	2.61190	.40301	2.48132	.42345	2.36158	.44418	2.25132	3
58	.38320	2.60963	.40335	2.47924	.42379	2.35967	.44453	2.24956	2
59	.38353	2.60736	.40369	2.47716	.42413	2.35776	.44488	2.24780	1
60	.38386	2.60509	.40403	2.47509	.42447	2.35585	.44523	2.24604	0
′	69° Co-tan.	69° Tan.	68° Co-tan.	68° Tan.	67° Co-tan.	67° Tan.	66° Co-tan.	66° Tan.	′

TABLE X.—*Continued*

′	24° Tan.	24° Co-tan.	25° Tan.	25° Co-tan.	26° Tan.	26° Co-tan.	27° Tan.	27° Co-tan.	′
0	.44523	2.24604	.46631	2.14451	.48773	2.05030	.50953	1.96261	60
1	.44558	2.24428	.46666	2.14288	.48809	2.04879	.50989	1.96120	59
2	.44593	2.24252	.46702	2.14125	.48845	2.04728	.51026	1.95979	58
3	.44627	2.24077	.46737	2.13963	.48881	2.04577	.51063	1.95838	57
4	.44662	2.23902	.46772	2.13801	.48917	2.04426	.51099	1.95698	56
5	.44697	2.23727	.46808	2.13639	.48953	2.04276	.51136	1.95557	55
6	.44732	2.23553	.46843	2.13477	.48989	2.04125	.51173	1.95417	54
7	.44767	2.23378	.46879	2.13316	.49026	2.03975	.51209	1.95277	53
8	.44802	2.23204	.46914	2.13154	.49062	2.03825	.51246	1.95137	52
9	.44837	2.23030	.46950	2.12993	.49098	2.03675	.51283	1.94997	51
10	.44872	2.22857	.46985	2.12832	.49134	2.03526	.51319	1.94858	50
11	.44907	2.22683	.47021	2.12671	.49170	2.03376	.51356	1.94718	49
12	.44942	2.22510	.47056	2.12511	.49206	2.03227	.51393	1.94579	48
13	.44977	2.22337	.47092	2.12350	.49242	2.03078	.51430	1.94440	47
14	.45012	2.22164	.47128	2.12190	.49278	2.02929	.51467	1.94301	46
15	.45047	2.21992	.47163	2.12030	.49315	2.02780	.51503	1.94162	45
16	.45082	2.21819	.47199	2.11871	.49351	2.02631	.51540	1.94023	44
17	.45117	2.21647	.47234	2.11711	.49387	2.02483	.51577	1.93885	43
18	.45152	2.21475	.47270	2.11552	.49423	2.02335	.51614	1.93746	42
19	.45187	2.21304	.47305	2.11392	.49459	2.02187	.51651	1.93608	41
20	.45222	2.21132	.47341	2.11233	.49495	2.02039	.51688	1.98470	40
21	.45257	2.20961	.47377	2.11075	.49532	2.01891	.51724	1.93332	39
22	.45292	2.20790	.47412	2.10916	.49568	2.01743	.51761	1.93195	38
23	.45327	2.20619	.47448	2.10758	.49604	2.01596	.51798	1.93057	37
24	.45362	2.20449	.47483	2.10600	.49640	2.01449	.51835	1.92920	36
25	.45397	2.20278	.47519	2.10442	.49677	2.01302	.51872	1.92782	35
26	.45432	2.20108	.47555	2.10284	.49713	2.01155	.51909	1.92645	34
27	.45467	2.19938	.47590	2.10126	.49749	2.01008	.51946	1.92508	33
28	.45502	2.19769	.47626	2.09969	.49786	2.00862	.51983	1.92371	32
29	.45537	2.19599	.47662	2.09811	.49822	2.00715	.52020	1.92235	31
30	.45573	2.19430	.47698	2.09654	.49858	2.00569	.52057	1.92098	30
31	.45608	2.19261	.47733	2.09498	.49894	2.00423	.52094	1.91962	29
32	.45643	2.19092	.47769	2.09341	.49931	2.00277	.52131	1.91826	28
33	.45678	2.18923	.47805	2.09184	.49967	2.00131	.52168	1.91690	27
34	.45713	2.18755	.47840	2.09028	.50004	1.99986	.52205	1.91554	26
35	.45748	2.18587	.47876	2.08872	.50040	1.99841	.52242	1.91418	25
36	.45784	2.18419	.47912	2.08716	.50076	1.99695	.52279	1.91282	24
37	.45819	2.18251	.47948	2.08560	.50113	1.99550	.52316	1.91147	23
38	.45854	2.18084	.47984	2.08405	.50149	1.99406	.52353	1.91012	22
39	.45889	2.17916	.48019	2.08250	.50185	1.99261	.52390	1.90876	21
40	.45924	2.17749	.48055	2.08094	.50222	1.99116	.52427	1.90741	20
41	.45960	2.17582	.48091	2.07939	.50258	1.98972	.52464	1.90607	19
42	.45995	2.17416	.48127	2.07785	.50295	1.98828	.52501	1.90472	18
43	.46030	2.17249	.48163	2.07630	.50331	1.98684	.52538	1.90337	17
44	.46065	2.17083	.48198	2.07476	.50368	1.98540	.52575	1.90203	16
45	.46101	2.16917	.48234	2.07321	.50404	1.98396	.52613	1.90069	15
46	.46136	2.16751	.48270	2.07167	.50441	1.98253	.52650	1.89935	14
47	.46171	2.16585	.48306	2.07014	.50477	1.98110	.52687	1.89801	13
48	.46206	2.16420	.48342	2.06860	.50514	1.97966	.52724	1.89667	12
49	.46242	2.16255	.48378	2.06706	.50550	1.97823	.52761	1.89533	11
50	.46277	2.16090	.48414	2.06553	.50587	1.97680	.52798	1.89400	10
51	.46312	2.15925	.48450	2.06400	.50623	1.97538	.52836	1.89266	9
52	.46348	2.15760	.48486	2.06247	.50660	1.97395	.52873	1.89133	8
53	.46383	2.15596	.48521	2.06094	.50696	1.97253	.52910	1.89000	7
54	.46418	2.15432	.48557	2.05942	.50733	1.97111	.52947	1.88867	6
55	.46454	2.15268	.48593	2.05790	.50769	1.96969	.52984	1.88734	5
56	.46489	2.15104	.48629	2.05637	.50806	1.96827	.53022	1.88602	4
57	.46525	2.14940	.48665	2.05485	.50843	1.96685	.53059	1.88469	3
58	.46560	2.14777	.48701	2.05333	.50879	1.96544	.53096	1.88337	2
59	.46595	2.14614	.48737	2.05182	.50916	1.96402	.53134	1.88205	1
60	.46631	2.14451	.48773	2.05030	.50953	1.96261	.53171	1.88073	0
′	Co-tan. 65°	Tan. 65°	Co-tan. 64°	Tan. 64°	Co-tan. 63°	Tan. 63°	Co-tan. 62°	Tan. 62°	′

TABLE X.—*Continued*

′	28°		29°		30°		31°		′
	TAN.	CO-TAN.	TAN.	CO-TAN.	TAN.	CO-TAN.	TAN.	CO-TAN.	
0	.53171	1.88073	.55431	1.80405	.57735	1.73205	.60086	1.66428	60
1	.53208	1.87941	.55469	1.80281	.57774	1.73089	.60126	1.66318	59
2	.53246	1.87809	.55507	1.80158	.57813	1.72973	.60165	1.66209	58
3	.53283	1.87677	.55545	1.80034	.57851	1.72857	.60205	1.66099	57
4	.53320	1.87546	.55583	1.79911	.57890	1.72741	.60245	1.65990	56
5	.53358	1.87415	.55621	1.79788	.57929	1.72625	.60284	1.65881	55
6	.53395	1.87283	.55659	1.79665	.57968	1.72509	.60324	1.65772	54
7	.53432	1.87152	.55697	1.79542	.58007	1.72393	.60364	1.65663	53
8	.53470	1.87021	.55736	1.79419	.58046	1.72278	.60403	1.65534	52
9	.53507	1.86891	.55774	1.79296	.58085	1.72163	.60443	1.65445	51
10	.53545	1.86760	.55812	1.79174	.58124	1.72047	.60483	1.65337	50
11	.53582	1.86630	.55850	1.79051	.58162	1.71932	.60522	1.65228	49
12	.53620	1.86499	.55888	1.78929	.58201	1.71817	.60562	1.65120	48
13	.53657	1.86369	.55926	1.78807	.58240	1.71702	.60602	1.65011	47
14	.53694	1.86239	.55964	1.78685	.58279	1.71588	.60642	1.64903	46
15	.53732	1.86109	.56003	1.78563	.58318	1.71473	.60681	1.64795	45
16	.53769	1.85979	.56041	1.78441	.58357	1.71358	.60721	1.64687	44
17	.53807	1.85850	.56079	1.78319	.58396	1.71244	.60761	1.64579	43
18	.53844	1.85720	.56117	1.78198	.58435	1.71129	.60801	1.64471	42
19	.53882	1.85591	.56156	1.78077	.58474	1.71015	.60841	1.64363	41
20	.53920	1.85462	.56194	1.77955	.58513	1.70901	.60881	1.64256	40
21	.53957	1.85333	.56232	1.77834	.58552	1.70787	.60921	1.64148	39
22	.53995	1.85204	.56270	1.77713	.58591	1.70673	.60960	1.64041	38
23	.54032	1.85075	.56309	1.77592	.58631	1.70560	.61000	1.63934	37
24	.54070	1.84946	.56347	1.77471	.58670	1.70446	.61040	1.63826	36
25	.54107	1.84818	.56385	1.77351	.58709	1.70332	.61080	1.63719	35
26	.54145	1.84689	.56424	1.77230	.58748	1.70219	.61120	1.63612	34
27	.54183	1.84561	.56462	1.77110	.58787	1.70106	.61160	1.63505	33
28	.54220	1.84433	.56500	1.76990	.58826	1.69992	.61200	1.63398	32
29	.54258	1.84305	.56539	1.76869	.58865	1.69879	.61240	1.63292	31
30	.54296	1.84177	.56577	1.76749	.58904	1.69766	.61280	1.63185	30
31	.54333	1.84049	.56616	1.76630	.58944	1.69653	.61320	1.63079	29
32	.54371	1.83922	.56654	1.76510	.58983	1.69541	.61360	1.62972	28
33	.54409	1.83794	.56693	1.76390	.59022	1.69428	.61400	1.62866	27
34	.54446	1.83667	.56731	1.76271	.59061	1.69316	.61440	1.62760	26
35	.54484	1.83540	.56769	1.76151	.59101	1.69203	.61480	1.62654	25
36	.54522	1.83413	.56808	1·76032	.59140	1.69091	.61520	1.62548	24
37	.54560	1.83286	.56846	1.75913	.59179	1.68979	.61561	1.62442	23
38	.54597	1.83159	.56885	1.75794	.59218	1.68866	.61601	1.62336	22
39	.54635	1.83033	.56923	1.75675	.59258	1.68754	.61641	1.62230	21
40	.54673	1.82906	.56962	1.75556	.59297	1.68643	.61681	1.62125	20
41	.54711	1.82780	.57000	1.75437	.59336	1.68531	.61721	1.62019	19
42	.54748	1.82654	.57039	1.75319	.59376	1.68419	.61761	1.61914	18
43	.54786	1.82528	.57078	1.75200	.59415	1.68308	.61801	1.61808	17
44	.54824	1.82402	.57116	1.75082	.59454	1.68196	.61842	1.61703	16
45	.54862	1.82276	.57155	1.74964	.59494	1.68085	.61882	1.61598	15
46	.54900	1.82150	.57193	1.74846	.59533	1.67974	.61922	1.61493	14
47	.54938	1.82025	.57232	1.74728	.59573	1.67863	.61962	1.61388	13
48	.54975	1.81899	.57271	1.74610	.59612	1.67752	.62003	1.61283	12
49	.55013	1.81774	.57309	1.74492	.59651	1.67641	.62043	1.61179	11
50	.55051	1.81649	.57348	1.74375	.59691	1.67530	.62083	1.61074	10
51	.55089	1.81524	.57386	1.74257	.59730	1.67419	.62124	1.60970	9
52	.55127	1.81399	.57425	1.74140	.59770	1.67309	.62164	1.60865	8
53	.55165	1.81274	.57464	1.74022	.59809	1.67198	.62204	1.60761	7
54	.55203	1.81150	.57503	1.73905	.59849	1.67088	.62245	1.60657	6
55	.55241	1.81025	.57541	1.73788	.59888	1.66978	.62285	1.60553	5
56	.55279	1.80901	.57580	1.73671	.59928	1.66867	.62325	1.60449	4
57	.55317	1.80777	.57619	1.73555	.59967	1.66757	.62366	1.60345	3
58	.55355	1.80653	.57657	1.73438	.60007	1.66647	.62406	1.60241	2
59	.55393	1.80529	.57696	1.73321	.60046	1.66538	.62446	1.60137	1
60	.55431	1.80405	.57735	1.73205	.60086	1.66428	.62487	1.60033	0
′	CO-TAN.	TAN.	CO-TAN.	TAN.	CO-TAN.	TAN.	CO-TAN.	TAN.	′
	61°		60°		59°		58°		

TABLE X.—*Continued*

′	32°		33°		34°		35°		′
	TAN.	CO-TAN.	TAN.	CO-TAN.	TAN.	CO-TAN.	TAN.	CO-TAN.	
0	.62487	1.60033	.64941	1.53986	.67451	1.48256	.70021	1.42815	60
1	.62527	1.59930	.64982	1.53888	.67493	1.48163	.70064	1.42726	59
2	.62568	1.59826	.65023	1.53791	.67536	1.48070	.70107	1.42638	58
3	.62608	1.59723	.65065	1.53693	.67578	1.47977	.70151	1.42550	57
4	.62649	1.59620	.65106	1.53595	.67620	1.47885	.70194	1.42462	56
5	.62689	1.59517	.65148	1.53497	.67663	1.47792	.70238	1.42374	55
6	.62730	1.59414	.65189	1.53400	.67705	1.47699	.70281	1.42286	54
7	.62770	1.59311	.65231	1.53302	.67748	1.47607	.70325	1.42198	53
8	.62811	1.59208	.65272	1.53205	.67790	1.47514	.70368	1.42110	52
9	.62852	1.59105	.65314	1.53107	.67832	1.47422	.70412	1.42022	51
10	.62892	1.59002	.65355	1.53010	.67875	1.47330	.70455	1.41934	50
11	.62933	1.58900	.65397	1.52913	.67917	1.47238	.70499	1.41847	49
12	.62973	1.58797	.65438	1.52816	.67960	1.47146	.70542	1.41759	48
13	.63014	1.58695	.65480	1.52719	.68002	1.47053	.70586	1.41672	47
14	.63055	1.58593	.65521	1.52622	.68045	1.46962	.70629	1.41584	46
15	.63095	1.58490	.65563	1.52525	.68088	1.46870	.70673	1.41497	45
16	.63136	1.58388	.65604	1.52429	.68130	1.46778	.70717	1.41409	44
17	.63177	1.58286	.65646	1.52332	.68173	1.46686	.70760	1.41322	43
18	.63217	1.58184	.65688	1.52235	.68215	1.46595	.70804	1.41235	42
19	.63258	1.58083	.65729	1.52139	.68258	1.46503	.70848	1.41148	41
20	.63299	1.57981	.65771	1.52043	.68301	1.46411	.70891	1.41061	40
21	.63340	1.57879	.65813	1.51946	.68343	1.46320	.70935	1.40974	39
22	.63380	1.57778	.65854	1.51850	.68386	1.46229	.70979	1.40887	38
23	.63421	1.57676	.65896	1.51754	.68429	1.46137	.71023	1.40800	37
24	.63462	1.57575	.65938	1.51658	.68471	1.46046	.71066	1.40714	36
25	.63503	1.57474	.65980	1.51562	.68514	1.45955	.71110	1.40627	35
26	.63544	1.57372	.66021	1.51466	.68557	1.45864	.71154	1.40540	34
27	.63584	1.57271	.66063	1.51370	.68600	1.45773	.71198	1.40454	33
28	.63625	1.57170	.66105	1.51275	.68642	1.45682	.71242	1.40367	32
29	.63666	1.57069	.66147	1.51179	.68685	1.45592	.71285	1.40281	31
30	.63707	1.56969	.66189	1.51084	.68728	1.45501	.71329	1.40195	30
31	.63748	1.56868	.66230	1.50988	.68771	1.45410	.71373	1.40109	29
32	.63789	1.56767	.66272	1.50893	.68814	1.45320	.71417	1.40022	28
33	.63830	1.56667	.66314	1.50797	.68857	1.45229	.71461	1.39936	27
34	.63871	1.56566	.66356	1.50702	.68900	1.45139	.71505	1.39850	26
35	.63912	1.56466	.66398	1.50607	.68942	1.45049	.71549	1.39764	25
36	.63953	1.56366	.66440	1.50512	.68985	1.44958	.71593	1.39679	24
37	.63994	1.56265	.66482	1.50417	.69028	1.44868	.71637	1.39593	23
38	.64035	1.56165	.66524	1.50322	.69071	1.44778	.71681	1.39507	22
39	.64076	1.56065	.66566	1.50228	.69114	1.44688	.71725	1.39421	21
40	.64117	1.55966	.66608	1.50133	.69157	1.44598	.71769	1.39336	20
41	.64158	1.55866	.66650	1.50038	.69200	1.44508	.71813	1.39250	19
42	.64199	1.55766	.66692	1.49944	.69243	1.44418	.71857	1.39165	18
43	.64240	1.55666	.66734	1.49849	.69286	1.44329	.71901	1.39079	17
44	.64281	1.55567	.66776	1.49755	.69329	1.44239	.71946	1.38994	16
45	.64322	1.55467	.66818	1.49661	.69372	1.44149	.71990	1.38909	15
46	.64363	1.55368	.66860	1.49566	.69416	1.44060	.72034	1.38824	14
47	.64404	1.55269	.66902	1.49472	.69459	1.43970	.72078	1.38738	13
48	.64446	1.55170	.66944	1.49378	.69502	1.43881	.72122	1.38653	12
49	.64487	1.55071	.66986	1.49284	.69545	1.43792	.72166	1.38568	11
50	.64528	1.54972	.67028	1.49190	.69588	1.43703	.72211	1.38484	10
51	.64569	1.54873	.67071	1.49097	.69631	1.43614	.72255	1.38399	9
52	.64610	1.54774	.67113	1.49003	.69675	1.43525	.72299	1.38314	8
53	.64652	1.54675	.67155	1.48909	.69718	1.43436	.72344	1.38229	7
54	.64693	1.54576	.67197	1.48816	.69761	1.43347	.72388	1.38145	6
55	.64734	1.54478	.67239	1.48722	.69804	1.43258	.72432	1.38060	5
56	.64775	1.54379	.67282	1.48629	.69847	1.43169	.72477	1.37976	4
57	.64817	1.54281	.67324	1.48536	.69891	1.43080	.72521	1.37891	3
58	.64858	1.54183	.67366	1.48442	.69934	1.42992	.72565	1.37807	2
59	.64899	1.54085	.67409	1.48349	.69977	1.42903	.72610	1.37722	1
60	.64941	1.53986	.67451	1.48256	.70021	1.42815	.72654	1.37638	0
′	CO-TAN.	TAN.	CO-TAN	TAN.	CO-TAN.	TAN.	CO-TAN.	TAN.	′
	57°		56°		55°		54°		

TABLE X.—*Continued*

′	36° Tan.	36° Co-tan.	37° Tan.	37° Co-tan.	38° Tan.	38° Co-tan.	39° Tan.	39° Co-tan.	′
0	.72654	1.37638	.75355	1.32704	.78129	1.27994	.80978	1.23490	60
1	.72699	1.37554	.75401	1.32624	.78175	1.27917	.81027	1.23416	59
2	.72743	1.37470	.75447	1.32544	.78222	1.27841	.81075	1.23343	58
3	.72788	1.37386	.75492	1.32464	.78269	1.27764	.81123	1.23270	57
4	.72832	1.37302	.75538	1.32384	.78316	1.27688	.81171	1.23196	56
5	.72877	1.37218	.75584	1.32304	.78363	1.27611	.81220	1.23123	55
6	.72921	1.37134	.75629	1.32224	.78410	1.27535	.81268	1.23050	54
7	.72966	1.37050	.75675	1.32144	.78457	1.27458	.81316	1.22977	53
8	.73010	1.36967	.75721	1.32064	.78504	1.27382	.81364	1.22904	52
9	.73055	1.36883	.75767	1.31984	.78551	1.27306	.81413	1.22831	51
10	.73100	1.36800	.75812	1.31904	.78598	1.27230	.81461	1.22758	50
11	.73144	1.36716	.75858	1.31825	.78645	1.27153	.81510	1.22685	49
12	.73189	1.36633	.75904	1.31745	.78692	1.27077	.81558	1.22612	48
13	.73234	1.36549	.75950	1.31666	.78739	1.27001	.81606	1.22539	47
14	.73278	1.36466	.75996	1.31586	.78786	1.26925	.81655	1.22467	46
15	.73323	1.36383	.76042	1.31507	.78834	1.26849	.81703	1.22394	45
16	.73368	1.36300	.76088	1.31427	.78881	1.26774	.81752	1.22321	44
17	.73413	1.36217	.76134	1.31348	.78928	1.26698	.81800	1.22249	43
18	.73457	1.36133	.76180	1.31269	.78975	1.26622	.81849	1.22176	42
19	.73502	1.36051	.76226	1.31190	.79022	1.26546	.81898	1.22104	41
20	.73547	1.35968	.76272	1.31110	.79070	1.26471	.81946	1.22031	40
21	.73592	1.35885	.76318	1.31031	.79117	1.26395	.81995	1.21959	39
22	.73637	1.35802	.76364	1.30952	.79164	1.26319	.82044	1.21886	38
23	.73681	1.35719	.76410	1.30873	.79212	1.26244	.82092	1.21814	37
24	.73726	1.35637	.76456	1.30795	.79259	1.26169	.82141	1.21742	36
25	.73771	1.35554	.76502	1.30716	.79306	1.26093	.82190	1.21670	35
26	.73816	1.35472	.76548	1.30637	.79354	1.26018	.82238	1.21598	34
27	.73861	1.35389	.76594	1.30558	.79401	1.25943	.82287	1.21526	33
28	.73906	1.35307	.76640	1.30480	.79449	1.25867	.82336	1.21454	32
29	.73951	1.35224	.76686	1.30401	.79496	1.25792	.82385	1.21382	31
30	.73996	1.35142	.76733	1.30323	.79544	1.25717	.82434	1.21310	30
31	.74041	1.35060	.76779	1.30244	.79591	1.25642	.82483	1.21238	29
32	.74086	1.34978	.76825	1.30166	.79639	1.25567	.82531	1.21166	28
33	.74131	1.34896	.76871	1.30087	.79686	1.25492	.82580	1.21094	27
34	.74176	1.34814	.76918	1.30009	.79734	1.25417	.82629	1.21023	26
35	.74221	1.34732	.76964	1.29931	.79781	1.25343	.82678	1.20951	25
36	.74267	1.34650	.77010	1.29853	.79829	1.25268	.82727	1.20879	24
37	.74312	1.34568	.77057	1.29775	.79877	1.25193	.82776	1.20808	23
38	.74357	1.34487	.77103	1.29696	.79924	1.25118	.82825	1.20736	22
39	.74402	1.34405	.77149	1.29618	.79972	1.25044	.82874	1.20665	21
40	.74447	1.34323	.77196	1.29541	.80020	1.24969	.82923	1.20593	20
41	.74492	1.34242	.77242	1.29463	.80067	1.24895	.82972	1.20522	19
42	.74538	1.34160	.77289	1.29385	.80115	1.24820	.83022	1.20451	18
43	.74583	1.34079	.77335	1.29307	.80163	1.24746	.83071	1.20379	17
44	.74628	1.33998	.77382	1.29229	.80211	1.24672	.83120	1.20308	16
45	.74674	1.33916	.77428	1.29152	.80258	1.24597	.83169	1.20237	15
46	.74719	1.33835	.77475	1.29074	.80306	1.24523	.83218	1.20166	14
47	.74764	1.33754	.77521	1.28997	.80354	1.24449	.83268	1.20095	13
48	.74810	1.33673	.77568	1.28919	.80402	1.24375	.83317	1.20024	12
49	.74855	1.33592	.77615	1.28842	.80450	1.24301	.83366	1.19953	11
50	.74900	1.33511	.77661	1.28764	.80498	1.24227	.83415	1.19882	10
51	.74946	1.33430	.77708	1.28687	.80546	1.24153	.83465	1.19811	9
52	.74991	1.33349	.77754	1.28610	.80594	1.24079	.83514	1.19740	8
53	.75037	1.33268	.77801	1.28533	.80642	1.24005	.83564	1.19669	7
54	.75082	1.33187	.77848	1.28456	.80690	1.23931	.83613	1.19599	6
55	.75128	1.33107	.77895	1.28379	.80738	1.23858	.83662	1.19528	5
56	.75173	1.33026	.77941	1.28302	.80786	1.23784	.83712	1.19457	4
57	.75219	1.32946	.77988	1.28225	.80834	1.23710	.83761	1.19387	3
58	.75264	1.32865	.78035	1.28148	.80882	1.23637	.83811	1.19316	2
59	.75310	1.32785	.78082	1.28071	.80930	1.23563	.83860	1.19246	1
60	.75355	1.32704	.78129	1.27994	.80978	1.23490	.83910	1.19175	0
′	Co-tan. 53°	Tan. 53°	Co-tan. 52°	Tan. 52°	Co-tan. 51°	Tan. 51°	Co-tan. 50°	Tan. 50°	′

TABLE X.—*Continued*

′	40° Tan.	40° Co-tan.	41° Tan.	41° Co-tan.	42° Tan.	42° Co-tan.	43° Tan.	43° Co-tan.	′
0	.83910	1.19175	.86929	1.15037	.90040	1.11061	.93252	1.07237	60
1	.83960	1.19105	.86980	1.14969	.90093	1.10996	.93306	1.07174	59
2	.84009	1.19035	.87031	1.14902	.90146	1.10931	.93360	1.07112	58
3	.84059	1.18964	.87082	1.14834	.90199	1.10867	.93415	1.07049	57
4	.84108	1.18894	.87133	1.14767	.90251	1.10802	.93469	1.06987	56
5	.84158	1.18824	.87184	1.14699	.90304	1.10737	.93524	1.06925	55
6	.84208	1.18754	.87236	1.14632	.90357	1.10672	.93578	1.06862	54
7	.84258	1.18684	.87287	1.14565	.90410	1.10607	.93633	1.06800	53
8	.84307	1.18614	.87338	1.14498	.90463	1.10543	.93688	1.06738	52
9	.84357	1.18544	.87389	1.14430	.90516	1.10478	.93742	1.06676	51
10	.84407	1.18474	.87441	1.14363	.90569	1.10414	.93797	1.06613	50
11	.84457	1.18404	.87492	1.14296	.90621	1.10349	.93852	1.06551	49
12	.84507	1.18334	.87543	1.14229	.90674	1.10285	.93906	1.06489	48
13	.84556	1.18264	.87595	1.14162	.90727	1.10220	.93961	1.06427	47
14	.84606	1.18194	.87646	1.14095	.90781	1.10156	.94016	1.06365	46
15	.84656	1.18125	.87698	1.14028	.90834	1.10091	.94071	1.06303	45
16	.84706	1.18055	.87749	1.13961	.90887	1.10027	.94125	1.06241	44
17	.84756	1.17986	.87801	1.13894	.90940	1.09963	.94180	1.06179	43
18	.84806	1.17916	.87852	1.13828	.90993	1.09899	.94235	1.06117	42
19	.84856	1.17846	.87904	1.13761	.91046	1.09834	.94290	1.06056	41
20	.84906	1.17777	.87955	1.13694	.91099	1.09770	.94345	1.05994	40
21	.84956	1.17708	.88007	1.13627	.91153	1.09706	.94400	1.05932	39
22	.85006	1.17638	.88059	1.13561	.91206	1.09642	.94455	1.05870	38
23	.85057	1.17569	.88110	1.13494	.91259	1.09578	.94510	1.05809	37
24	.85107	1.17500	.88162	1.13428	.91313	1.09514	.94565	1.05747	36
25	.85157	1.17430	.88214	1.13361	.91366	1.09450	.94620	1.05685	35
26	.85207	1.17361	.88265	1.13295	.91419	1.09386	.94676	1.05624	34
27	.85257	1.17292	.88317	1.13228	.91473	1.09322	.94731	1.05562	33
28	.85307	1.17223	.88369	1.13162	.91526	1.09258	.94786	1.05501	32
29	.85358	1.17154	.88421	1.13096	.91580	1.09195	.94841	1.05439	31
30	.85408	1.17085	.88473	1.13029	.91633	1.09131	.94896	1.05378	30
31	.85458	1.17016	.88524	1.12963	.91687	1.09067	.94952	1.05317	29
32	.85509	1.16947	.88576	1.12897	.91740	1.09003	.95007	1.05255	28
33	.85559	1.16878	.88628	1.12831	.91794	1.08940	.95062	1.05194	27
34	.85609	1.16809	.88680	1.12765	.91847	1.08876	.95118	1.05133	26
35	.85660	1.16741	.88732	1.12699	.91901	1.08813	.95173	1.05072	25
36	.85710	1.16672	.88784	1.12633	.91955	1.08749	.95229	1.05010	24
37	.85761	1.16603	.88836	1.12567	.92008	1.08686	.95284	1.04949	23
38	.85811	1.16535	.88888	1.12501	.92062	1.08622	.95340	1.04888	22
39	.85862	1.16466	.88940	1.12435	.92116	1.08559	.95395	1.04827	21
40	.85912	1.16398	.88992	1.12369	.92170	1.08496	.95451	1.04766	20
41	.85963	1.16329	.89045	1.12303	.92224	1.08432	.95506	1.04705	19
42	.86014	1.16261	.89097	1.12238	.92277	1.08369	.95562	1.04644	18
43	.86064	1.16192	.89149	1.12172	.92331	1.08306	.95618	1.04583	17
44	.86115	1.16124	.89201	1.12106	.92385	1.08243	.95673	1.04522	16
45	.86166	1.16056	.89253	1.12041	.92439	1.08179	.95729	1.04461	15
46	.86216	1.15987	.89306	1.11975	.92493	1.08116	.95785	1.04401	14
47	.86267	1.15919	.89358	1.11909	.92547	1.08053	.95841	1.04340	13
48	.86318	1.15851	.89410	1.11844	.92601	1.07990	.95897	1.04279	12
49	.86368	1.15783	.89463	1.11778	.92655	1.07927	.95952	1.04218	11
50	.86419	1.15715	.89515	1.11713	.92709	1.07864	.96008	1.04158	10
51	.86470	1.15647	.89567	1.11648	.92763	1.07801	.96064	1.04097	9
52	.86521	1.15579	.89620	1.11582	.92817	1.07738	.96120	1.04036	8
53	.86572	1.15511	.89672	1.11517	.92872	1.07676	.96176	1.03976	7
54	.86623	1.15443	.89725	1.11452	.92926	1.07613	.96232	1.03915	6
55	.86674	1.15375	.89777	1.11387	.92980	1.07550	.96288	1.03855	5
56	.86725	1.15308	.89830	1.11321	.93034	1.07487	.96344	1.03794	4
57	.86776	1.15240	.89883	1.11256	.93088	1.07425	.96400	1.03734	3
58	.86827	1.15172	.89935	1.11191	.93143	1.07362	.96457	1.03674	2
59	.86878	1.15104	.89988	1.11126	.93197	1.07299	.96513	1.03613	1
60	.86929	1.15037	.90040	1.11061	.93252	1.07237	.96569	1.03553	0
′	49° Co-tan.	49° Tan.	48° Co-tan.	48° Tan.	47° Co-tan.	47° Tan.	46° Co-tan.	46° Tan.	′

TABLE X.—*Concluded*

	44°				44°				44°		
′	TAN.	CO-TAN.	′	′	TAN.	CO-TAN.	′	′	TAN.	CO-TAN.	′
0	.96569	1.03553	60	21	.97756	1.02295	39	41	.98901	1.01112	19
1	.96625	1.03493	59	22	.97813	1.02236	38	42	.98958	1.01053	18
2	.96681	1.03433	58	23	.97870	1.02176	37	43	.99016	1.00994	17
3	.96738	1.03372	57	24	.97927	1.02117	36	44	.99073	1.00935	16
4	.96794	1.03312	56	25	.97984	1.02057	35	45	.99131	1.00876	15
5	.96850	1.03252	55	26	.98041	1.01998	34	46	.99189	1.00818	14
6	.96907	1.03192	54	27	.98098	1.01939	33	47	.99247	1.00759	13
7	.96963	1.03132	53	28	.98155	1.01879	32	48	.99304	1.00701	12
8	.97020	1.03072	52	29	.98213	1.01820	31	49	.99362	1.00642	11
9	.97076	1.03012	51	30	.98270	1.01761	30	50	.99420	1.00583	10
10	.97133	1.02952	50	31	.98327	1.01702	29	51	.99478	1.00525	9
11	.97189	1.02892	49	32	.98384	1.01642	28	52	.99536	1.00467	8
12	.97246	1.02832	48	33	.98441	1.01583	27	53	.99594	1.00408	7
13	.97302	1.02772	47	34	.98499	1.01524	26	54	.99652	1.00350	6
14	.97359	1.02713	46	35	.98556	1.01465	25	55	.99710	1.00291	5
15	.97416	1.02653	45	36	.98613	1.01406	24	56	.99768	1.00233	4
16	.97472	1.02593	44	37	.98671	1.01347	23	57	.99826	1.00175	3
17	.97529	1.02533	43	38	.98728	1.01288	22	58	.99884	1.00116	2
18	.97586	1.02474	42	39	.98786	1.01229	21	59	.99942	1.00058	1
19	.97643	1.02414	41	40	.98843	1.01170	20	60	1	1	0
20	.97700	1.02355	40								
′	CO-TAN.	TAN.	′	′	CO-TAN.	TAN.	′	′	CO-TAN.	TAN.	′
	45°				45°				45°		

TABLE XI.—TRIGONOMETRIC FORMULAS

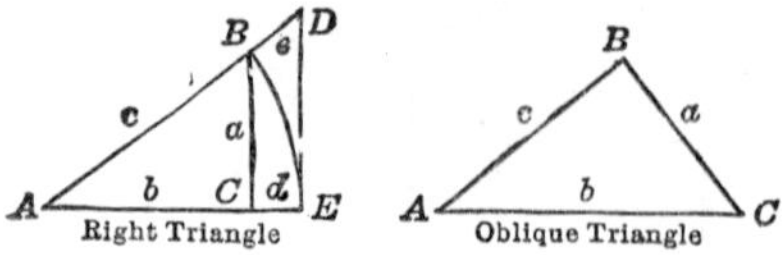

RIGHT TRIANGLES

$$\sin A = \frac{a}{c} = \cos B \qquad \sec A = \frac{c}{b} = \operatorname{cosec} B$$

$$\cos A = \frac{b}{c} = \sin B \qquad \operatorname{cosec} A = \frac{c}{a} = \sec B$$

$$\tan A = \frac{a}{b} = \cot B \qquad \operatorname{vers} A = \frac{c - b}{c} = \frac{d}{c}$$

$$\cot A = \frac{b}{a} = \tan B \qquad \operatorname{exsec} A = \frac{e}{c}$$

$$a = c \sin A = c \cos B = b \tan A = b \cot B = \sqrt{c^2 - b^2}$$

$$b = c \cos A = c \sin B = a \cot A = a \tan B = \sqrt{c^2 - a^2}$$

$$c = \frac{a}{\sin A} = \frac{a}{\cos B} = \frac{b}{\sin B} = \frac{b}{\cos A} = \frac{d}{\operatorname{vers} A} = \frac{e}{\operatorname{exsec} A} = \sqrt{a^2 + b^2}$$

$$d = c \operatorname{vers} A \qquad e = c \operatorname{exsec} A$$

OBLIQUE TRIANGLES

Given	Sought	Formulas
A, B, a	b, c	$b = \frac{a}{\sin A} \cdot \sin B$ $\quad c = \frac{a}{\sin A} \cdot \sin (A + B)$
A, a, b	B, c	$\sin B = \frac{\sin A}{a} \cdot b$ $\quad c = \frac{a}{\sin A} \cdot \sin C$
C, a, b	$\frac{1}{2}(A + B)$	$\frac{1}{2}(A + B) = 90° - \frac{1}{2}C$
	$\frac{1}{2}(A - B)$	$\tan \frac{1}{2}(A - B) = \frac{a - b}{a + b} \cdot \tan \frac{1}{2}(A + B)$
a, b, c	A	If $s = \frac{1}{2}(a + b + c)$, $\sin \frac{1}{2}A = \sqrt{\frac{(s - b)(s - c)}{bc}}$
		$\cos \frac{1}{2}A = \sqrt{\frac{s(s - a)}{bc}}$, $\tan \frac{1}{2}A = \sqrt{\frac{(s - b)(s - c)}{s(s - a)}}$
		$\sin A = 2 \frac{\sqrt{s(s - a)(s - b)(s - c)}}{bc}$
		$\operatorname{vers} A = \frac{2(s - b)(s - c)}{bc}$
	area	*area* $= \sqrt{s(s - a)(s - b)(s - c)}$
C, a, b	*area*	*area* $= \frac{1}{2}ab \sin C$
A, B, C, a	*area*	*area* $= \frac{a^2 \sin B \cdot \sin C}{2 \sin A}$

INDEX